Biochemical Microcalorimetry

EDITED BY

HARRY DARROW BROWN

Cancer Research Center and
the University of Missouri
Columbia, Missouri

ACADEMIC PRESS New York London 1969

ACADEMIC PRESS, INC.
111 Fifth Avenue, New York, New York 10003

United Kingdom Edition published by
ACADEMIC PRESS, INC. (LONDON) LTD.
Berkeley Square House, London W1X6BA

LIBRARY OF CONGRESS CATALOG CARD NUMBER: 69-12284

PRINTED IN THE UNITED STATES OF AMERICA

List of Contributors

Numbers in parentheses indicate the pages on which the authors' contributions begin.

T. Ackermann (121, 235), University of Munster, Munster, Germany.

Robert L. Berger (221, 275, 297), Laboratory of Technical Development, National Heart Institute, Bethesda, Maryland.

Harry Darrow Brown (149, 207, 291), Cancer Research Center and the University of Missouri, Columbia, Missouri.

William J. Evans (117, 257), Oilseed Crops Laboratory, U.S. Department of Agriculture, New Orleans, Louisiana.

W. W. Forrest (165), Division of Nutritional Biochemistry, Commonwealth Scientific and Industrial Research Organization, Adelaide, South Australia.

Sôzaburo Ono (99), College of Agriculture, University of Osaka Prefecture, Sakai, Japan.

Sam N. Pennington (207), Cancer Research Center and the University of Missouri, Columbia, Missouri.

Julius Praglin* (199), Keithley Instruments, Inc., Cleveland, Ohio.

Henri Prat** (181), Department of Botany, University of Aix, Marseille, France.

H. A. Skinner† (1), Department of Chemistry, University of Colorado, Boulder, Colorado.

Katsutada Takahashi (99), College of Agriculture, University of Osaka Prefecture, Sakai, Japan.

Ingemar Wadsö (83), Thermochemistry Laboratory, University of Lund, Lund, Sweden.

R. C. Wilhoit (33), Thermodynamics Research Center, Texas A & M University, College Station, Texas.

* Present address: 22 Cherry Oca Lane, Framingham, Massachusetts.

** Present address: Laboratoire de Botanique, Faculte des Sciences, Place Victor Hugo, Marseille, France.

† Present address: Chemistry Department, University of Manchester, Manchester, United Kingdom.

Preface

We have undertaken the compilation of an introduction to microcalorimetry for the biochemist who is uninitiated in either the basic characteristics of the instrumentation or in its application, or both, but who is sufficiently knowledgeable and enamored of the usefulness of the calorimeter in providing thermodynamic values or in its uniqueness as an analytic tool. I have described this hypothetical biochemist to the contributors, expert calorimetrists who, though not entirely sure that such an individual indeed exists, have nonetheless undertaken to write for him. Since our definition of the goal is sufficiently diffuse, it will possibly not be necessary to apologize for failing to reach it. In a sense, however, the task of constructing an *introduction* was, comparatively, not a difficult one. The reader, our "biochemist," brings much more with him than he might be prone to claim, and the field as we see it to exist in the scientific literature is comparatively small when restricted to studies applied rather directly to problems of biology and biological chemistry.

Each contributor was asked to review the literature, which in most areas is still quite modest, and to exercise the necessary and rightful freedom of the enthusiast to present his personal view of the potential application of microcalorimetry to biology. Consequently, there has been some overlap of material from chapter to chapter. However, despite the physical bounds within which the book has been conceived, there has been no attempt to subtract from these expressions. It has been my experience that, virtually to a man, calorimetrists, perhaps despite rather than because of their preoccupation with the technique, see in biological application a potentially huge and important field. If a well-defined description of this vision is included within the goal of the book, then we shall surely have achieved one aim.

The selection of topics is relatively restricted, and the scope of the work has been taken to exclude consideration of the simply beautiful work that has been done in physiological calorimetry or in whole-organism calorimetry. Even within the framework that we have chosen there are omissions: the absence of a lipid section was basically brought about by the very small existing literature. Other omissions or abridgments, particularly in the presentation of the technologies that might contribute to the success of the calorimetrist, were necessitated by the physical dimensions which we have accepted in order that the book be, with reasonable economy, available to the individual scientist. My most profound regret is for the omission of the chapter that was to be written by the late Professor M. Edouard Calvet. Professor Calvet remains a giant in the development of biochemical microcalorimetry. His enthusiasm in the acceptance of my invitation to contribute an essay to this book was in fact the encouragement that sustained the initial work. Professor Calvet's death not only prevented the inclusion of his insightful comments but, much more lamentable, interfered with a life that was devoted to the development of this area of knowledge. We have made no attempt to offer a substitute for his planned essay.

The authors are of diverse backgrounds—biochemical calorimetry is still at that pioneering stage wherein physicists, biologists, physical chemists, and biochemists work together not only harmoniously but with warmth.

H. D. Brown

Columbia, Missouri
May, 1969

Contents

Biochemical Microcalorimetry

CHAPTER I

Theory, Scope, and Accuracy of Calorimetric Measurements

H. A. SKINNER

I. Introduction

At the beginning of the eighteenth century, the nature of heat was not understood and remained a major problem in physical science. The necessary instrument for the measurement of quantities of heat had yet to be developed, but before the end of the century the calorimeter had evolved. Many prominent scientists of the era interested themselves in heat and its measurement—notably Fahrenheit, Lomonosov, Black, Watt, Lavoisier, Cavendish, and Dalton. Heat capacities and latent heats were measured by Black around 1760, and an ice calorimeter was designed by Lavoisier in 1784. An effective rotating reaction calorimeter was used successfully by Hess in 1839, and 1885 saw the invention of the bomb calorimeter by Bertholet.

Developments during the present century have aimed to place the calorimeter in the category of precision instruments and in large measure have succeeded. The range of applicability of calorimetry has been extended virtually down to the absolute zero, and to temperatures well in excess of 1000°C. Significant from the viewpoint of the biochemist has been the recent development of highly sensitive microcalorimeters, capable of measuring heat outputs of 1 mcal $hour^{-1}$ or less. Instrument manufacturers are showing interest in calorimeters of all types, and automatic controls, adiabatic shields, and pen recorders are now commonplace accessories to the modern calorimeter. The experimentalist in thermodynamics and thermochemistry is better served technically than would have seemed possible a few decades ago—but, of course, the problems he must tackle are now more complex and difficult. In fact, the limitations of the modern calorimeter are more frequently determined by the difficulties of definition and control of the process under investigation than by the inadequacies of the instrument serving to investigate it.

II. Calorimeters: Classification

The role of the calorimeter as the essential experimental instrument in thermodynamics calls for a wide range of calorimeter design and function. Effectively, however, calorimeters belong (or tend to belong) to one of two extreme types—either the adiabatic, or the conduction calorimeter. The ideal adiabatic calorimeter is perfectly insulated, thermally, from its surroundings, so that heat developed inside the calorimeter is totally retained therein. With this calorimeter one measures the temperature change of the calorimeter brought about by the process under investigation. By contrast, the ideal conduction calorimeter is perfectly connected, thermally, with its surrounding heat sink, so that heat developed in the calorimeter is totally transferred from it. With this type of calorimeter one measures the heat flow from the calorimeter to the heat sink: ideally, the heat sink is of infinite capacity, and there is no final change in temperature of the calorimeter resulting from the process under investigation.

Between the two extremes lies the well-known and commonly used isothermally jacketed calorimeter. In this arrangement the calorimeter is imperfectly insulated from its surroundings, but the surrounding heat sink is maintained throughout at constant temperature. Heat developed in the calorimeter brings about a temporary increase in temperature of the calorimeter, which is followed by a gradual return toward the temperature of the environment. One measures the calorimeter temperature as a function of time before, during, and after the process. The maintenance of a constant temperature environment is essential to proper calculation of heat transfer between calorimeter and surrounding jacket.

The isothermal-jacket calorimeter has occupied a predominant position in calorimetry, but it is now challenged by both the adiabatic and conduction types of calorimeter, primarily because of technical developments that have markedly improved the performance of adiabatic shields and thermostats and the construction of multiple thermocouples. As pointed out later, each type of calorimeter has its own areas of specific application in which it has superiority.

III. Heat Exchange between Calorimeter and Environment

Heat exchange between a calorimeter and its surrounding jacket is due to (1) conduction via solid connections between them (e.g., support pegs and electrical lead wires), (2) convection and conduction by gas molecules (normally air) occupying the interspace, (3) radiation, and (4) evaporation losses from an unsealed calorimeter containing liquid. The first three of these are dependent upon the existence of a temperature difference between calorimeter and jacket. The objective in adiabatic calorimetry is automatically to reduce the thermal head effectively to zero throughout the entire experiment and thus eliminate heat transfer by these mechanisms. Evaporation losses are preferably to be eliminated by using a sealed calorimeter. If it is inconvenient to do this, it is advisable to reduce evaporation losses by holding the jacket temperature at a level above that of the calorimeter.

Provided the thermal head between the calorimeter and its jacket is no more than a few degrees, the heat exchange by radiation and by gas conduction obeys Newton's law, so that

$$\frac{d\theta}{dt} = k(\theta_j - \theta) \tag{1}$$

where θ = calorimeter temperature at time t, θ_j = jacket temperature (maintained at a constant value), and k = leakage modulus of the system. Heat exchange by convection does not follow Eq. (1), and since this complicates accurate calculation of the heat transfer, the best solution is to render convection transfer small by suitable design of the calorimeter and jacket assembly.

A well-designed isothermal-jacket (or adiabatic) calorimeter aims at a constant and small value for the leakage modulus k. To achieve this the following design features should be met:

1. The outer surface of the calorimeter and the inner surface of the surrounding jacket should be highly polished to minimize radiative heat transfer between them.
2. The air gap between the calorimeter and its jacket should preferably be evacuated to minimize conduction and convection transfer.

3. If it is impracticable to evacuate the air gap, the jacket should be so designed that the jacket-calorimeter separation is no more than 12 mm in thickness at any point: This reduces convection transfer sufficiently to allow Eq. (1) to remain a good approximation.
4. If the calorimeter contains a liquid, the stirring must be sufficient to ensure uniformity of the calorimeter temperature, and the rate of stirring must be constant throughout the experiment to give a constant heat of stirring. The latter should not exceed 5×10^{-4} deg minute^{-1}.

Calculation of Heat Exchange

The adiabatic calorimeter starts with the advantage that calculation of heat exchange is not necessary. However, this presumes that the jacket can be made to behave as a perfect adiabatic shield from start to finish, and this is not always so. Adiabatic control is nowadays sought by using judicially disposed thermocouples affixed to the outer wall of the calorimeter and inner wall of the jacket as sensors of the thermal head and employing the thermal emf (via dc amplification and proportional control) to activate the jacket heaters. The continuous matching of jacket and calorimeter temperatures is relatively easy to achieve in case the rate of change of the calorimeter temperature is slow—but for rapid heating (e.g., in bomb-combustion calorimeters), perfect matching is hardly

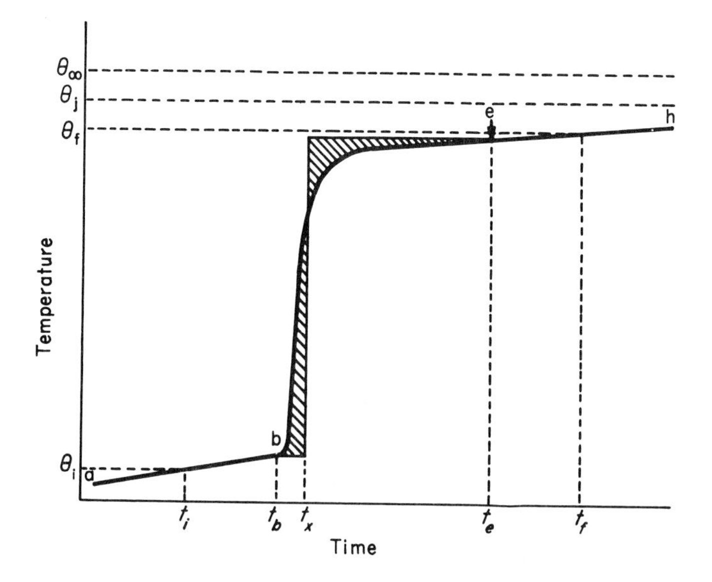

FIG. 1. Time-temperature curve for a calorimetric experiment. (Reprinted from Coops, Jessup, and van Nes, 1956.)

possible. The adiabatic method has distinct advantage for the study of slow processes, whereas the isothermal-jacket calorimeter is at its best for rapid processes.

The measurements made when using an isothermal-jacket calorimeter are presented graphically as a time-temperature curve (Fig. 1), which typically divides into three periods: (1) a fore period, during which the temperature change of the calorimeter is entirely due to heat transfer between jacket and calorimeter, and to stirring heat; (2) a main period, coinciding with the onset and continuance of the process under investigation in the calorimeter; and (3) an after period, in which the calorimeter temperature change is again entirely due to thermal leakage and stirring heat.

θ_b and θ_e denote the calorimeter temperatures at the beginning and end of the main period, and $(\theta_e - \theta_b)$ is the "observed" temperature change. This includes contributions resulting from stirring heat and thermal leakage and requires adjustment in order to isolate the temperature change solely due to the process under investigation. We accept that the total rate of temperature rise due to stirring and leakage is given by

$$\frac{d\theta}{dt} = u + k(\theta_j - \theta) \tag{2}$$

which differs from Eq. (1) only by addition of the constant stirring heat u.

From Fig. 1 we note that the time-temperature curve levels off ($d\theta/dt = 0$) at a particular temperature denoted by θ_∞. From Eq. (2),

$$\theta_j = \theta_\infty - u/k \tag{3}$$

Substitution of this expression for θ_j in Eq. (2) gives

$$\frac{d\theta}{dt} = k(\theta_\infty - \theta) \tag{4}$$

Let g_i and g_f be the measured values of $d\theta/dt$ at the mean temperatures θ_i and θ_f of the fore and after periods. Then, from Eqs. (2) and (4), we obtain

$$k = \frac{g_i - g_f}{\theta_f - \theta_i} \tag{5}$$

$$u = g_f + k(\theta_f - \theta_j) \tag{6}$$

$$\theta_\infty = \frac{g_f}{k} + \theta_f = \frac{g_i\theta_f - g_f\theta_i}{g_i - g_f} \tag{7}$$

The combination of Eqs. (6) and (2) provides

$$\frac{d\theta}{dt} = g_f + k(\theta_f - \theta) \tag{8}$$

The "correction" $\Delta\theta$ to be added to the observed temperature change $(\theta_e - \theta_b)$ is obtained by integration of Eq. (8) over the range of the main period,

$$\Delta\theta = -g_f(t_e - t_b) - k\int_{t_b}^{t_e} (\theta_f - \theta)\,dt \tag{9a}$$

$$= -[g_f + k(\theta_f - \theta_m)](t_e - t_b) \tag{9b}$$

where θ_m is the average temperature of the calorimeter wall during the main period.

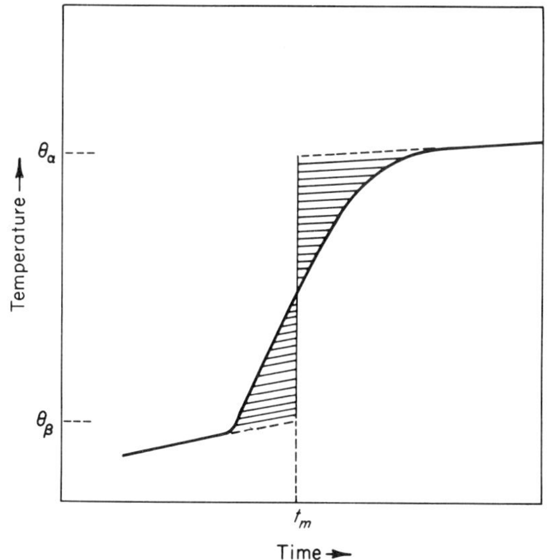

FIG. 2. Corrected temperature rise may be approximated for rapid processes in simple reaction calorimeters by linear extrapolation of the fore- and after-rating curves to the time t_m chosen to make the shaded areas above and below the rating curves equal in magnitude.

The value of the average temperature θ_m has to be evaluated by graphical or numerical integration. The Regnault–Pfaundler method is recommended: This requires the measurement of n temperatures θ_r at equal time intervals Δt over the main period. The average temperature θ_m is given by

$$\theta_m = \left(\sum_{r=2}^{n-1} \theta_r + \frac{\theta_b + \theta_e}{2}\right)\frac{\Delta t}{t_e - t_b} = \left(\sum_{r=2}^{n-1} \theta_r + \frac{\theta_b + \theta_e}{2}\right)\frac{1}{n-1} \tag{10}$$

This method of evaluation of the heat exchange is based on a treatment by Coops, Jessup, and van Nes (1956), which was designed specifically for application to bomb-combustion calorimetry where high accuracy is essential. Simpler procedures may suffice for less exacting studies. For rapid processes in simple reaction calorimeters, satisfactory results are obtained by linear extrapolation of the fore- and after-rating curves to the time t_m, chosen to make the shaded areas (see Fig. 2) above and below the rating curves equal in magnitude. The "corrected" temperature rise is given by $(\theta_\alpha - \theta_\beta)$. It is important, however, to employ the same method of evaluation in calibration as in all other experimental measurements.

IV. Conduction Calorimeters

Phase-change calorimeters, including the well-known Bunsen ice calorimeter, constitute a special type of conduction calorimeter. In these the heat developed in the active part of the calorimeter is totally transferred to the surrounding medium, where it is absorbed isothermally and brings about a proportional amount of phase change in the medium, measured by the accompanying volume change. There are advantages—including reasonably high sensitivity—in the use of phase-change calorimeters, and the modified ice calorimeter described by Ginnings and Corruccini (1947) and the diphenylether calorimeters of Jessup (1955) and of Giguère *et al.* (1955) are excellent examples of this particular art.

Current interest, however, centers on the development of mechanical-conduction calorimeters in which a thermoelectric pile serves to transfer heat from the calorimeter to its surrounding heat sink. The heat transfer via the

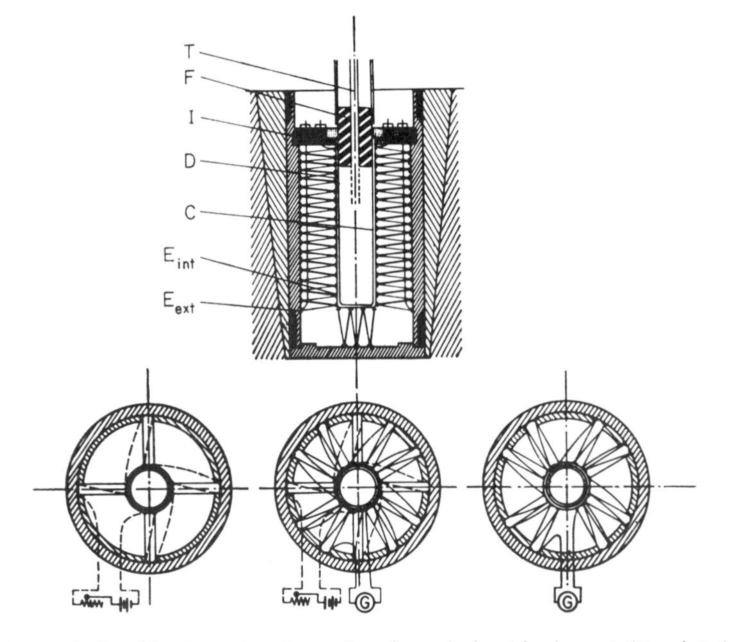

FIG. 3. Vertical and horizontal sections of a microcalorimetric element. (Reprinted from Calvet and Prat, 1963.)

thermocouple wires represents only a part of the total heat transfer from the calorimeter: Hence it is essential for meaningful results that heat transfer via the pile constitute a constant fraction of the total, irrespective of the process taking place in the calorimetric cell. The Tian–Calvet calorimeter (now commercially available from the Setaram Co. at Lyon, France) attempts to achieve this end by surrounding the outside wall of the calorimetric cell with identical thermoelectric junctions, each separated from one another by equal intervals. The couples are affixed normal to the calorimeter wall, and a sufficient number is used so that each junction covers only a minute portion of the total wall area. Figure 3 gives a diagrammatic representation of a typical Tian–Calvet calorimetric element. The process under investigation takes place in the cylindrical cell C, which push fits in a thin silver cavity D, covered by a very thin insulating layer of mica. The thermocouple junctions are attached to D (which constitutes the internal boundary E_{int} of the pile) and to the inside cylindrical wall of a hollow truncated cone (constituting the external boundary E_{ext} of the pile), which is wedged into a cavity of the same shape in the surrounding thermostated block (heat sink).

Let each junction occupy area S, and the uncovered area between junctions be S'. Let θ_i be the temperature of a given junction at the internal boundary. Since the combined area $(S + S')$ is small, it is permissible to accept that θ_i measures the temperature of the junction and its adjacent uncovered area—despite the part that there may be considerable temperature variation over the entire expanse of the internal boundary surface.

The heat flux from an area $(S + S')$ is made up of that conducted by the thermocouple wire,

$$\phi_i = c(\theta_i - \theta_e) \tag{11a}$$

(where c = thermal conductivity of the wire and θ_e = the constant temperature of the external heat sink), and of heat transfer from the surface by radiation and convection,

$$\phi_i' = \gamma(\theta_i - \theta_e) \tag{11b}$$

giving a total flux of

$$(\phi_i + \phi_i') = (c + \gamma)(\theta_i - \theta_e) \tag{12}$$

For n identical thermocouples, the total heat flux is

$$\Phi = (c + \gamma) \sum_{i=1}^{n} (\theta_i - \theta_e) \tag{13}$$

However, the emf generated in a thermocouple is given by

$$e_i = \epsilon_0(\theta_i - \theta_e) \tag{14}$$

where ϵ_0 is the thermoelectric power of the thermocouple; and since all the elements are connected in series, the total emf recorded by n thermocouples is

$$E = \epsilon_0 \sum_{i=1}^{n} (\theta_i - \theta_e) \quad (15)$$

The combination of Eqs. (13) and (15) gives

$$E = \frac{\epsilon_0}{(c + \gamma)} \cdot \Phi \quad (16)$$

so that the emf generated by the thermoelectric pile is proportional to the total heat flux from the internal boundary, irrespective of temperature gradients within the calorimetric cell.

Equation (16) is valid on the condition that the external boundary temperature θ_e remains constant throughout and that the calorimetric cell is totally surrounded by thermoelements. Neither requirement is fully met in the Calvet calorimeter. The upper part of the calorimetric element is closed by an insulating stopper, allowing entrance into the active cell of control rods for mixing of the reactants. Heat losses through the uncovered part are reduced by use of cells in the form of long, thin cylinders, so that the stopper takes up only about 4% of the external surface area. Nevertheless, this feature of the Calvet calorimeter remains as a potential source of inaccuracy.

The requirement of constant θ_e is nearly met, provided the calorimeter is used for microcalorimetric studies (very small heat outputs) and is surrounded by a well-thermostated metallic block (heat sink) of relatively high heat capacity. To improve reproducibility, Calvet introduced two additional features to the early Tian model by constructing twin microcalorimetric elements with compensation; the twin system is advantageous in microcalorimetry in general, whether adiabatic or conduction calorimeters are used.

The "heatburst" microcalorimeter, introduced by Benzinger (1965; Benzinger and Kitzinger, 1954, 1963), and made available commercially by Beckman Instruments (Spinco Division, Palo Alto), is similar in principle to the Calvet microcalorimeter but is significantly different in detail. It aims to transfer heat from the active zone to the surrounding heat sink rapidly and is so constructed that the entire calorimeter can be rotated (to mix reagents in the active cell). It is a twin system, but it differs from Calvet in that the thermoelectric pile surrounding the cell consists of a very large number (10^4) of junctions whereas the Calvet instrument operates on a limited number. The essential parts of the instrument are shown in equatorial section in Fig. 4 and in longitudinal section in Fig. 5.

The freedom to rotate the calorimeter, and thus mix reagents rapidly, and the employment of numerous thermocouples render the Benzinger microcalorimeter more rapid in response than the Calvet instrument. This is advantageous

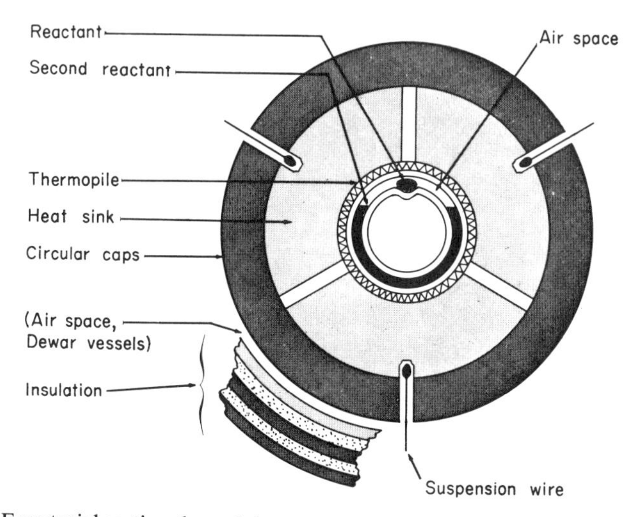

FIG. 4. Equatorial section through instrument. (Reprinted from Benzinger, 1965.)

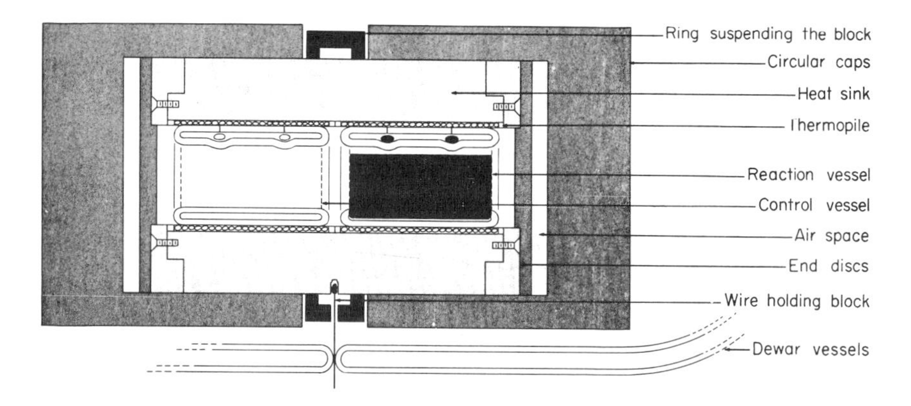

FIG. 5. Longitudinal section through instrument. (Reprinted from Benzinger, 1965).

in itself, and the Benzinger instrument is both convenient and relatively simple to use. On the other hand, it is doubtful if it is capablc of thc accuracy of the Calvet instrument for slow thermal processes.

V. Twin Calorimeters

The twin-calorimeter principle was first used by Joule in 1845 and is almost essential when high precision is sought for the thermal study of slow processes. The apparatus consists of two physically identical calorimeters symmetrically disposed within the same jacket, which may be either of the constant-temperature or of the adiabatic type. The process under investigation is carried out in one of the calorimeters (the "active" element), the other serving as the tare or "reference" element. Twin calorimeters may be operated with or without "compensation" of the thermal effect generated in the active cell.

The Benzinger microcalorimeter, designed primarily to measure heats of reaction, dilution, or mixing of liquid reagents and in particular heats of reaction in aqueous solution, is normally used without compensation. The active cell is charged with reactants, each in their respective compartments (Fig. 6), and fits pistonlike into one half of the cylindrical cavity of the surrounding aluminium heat sink. The reference cell is similarly charged with nonreactive reagents, and fits into the other half of the cylindrical cavity. On rotation of the heat-sink assembly, the reagents in the active cell are made to mix together and the heat developed is transferred via the surrounding thermopile to the heat sink; simultaneously the nonreactive reagents admix in the reference cell, and small "spurious" heat effects generated by mechanical rotations, wetting of the cell surface, etc. are transferred via the reference thermopile surrounding the twin cell. The active and reference thermopiles being connected in opposition, the resultant emf recorded by the twin thermopiles is attributed solely to the process under investigation in the active cell.

The Calvet microcalorimeter is designed for use with or without internal compensation of the heat effect produced in the active element. When needed, compensation is effected by Joule heating or Peltier cooling in the active element, according to the process—endothermic or exothermic—taking place. In constructing the apparatus, two thermoelectric piles are built around the active cell, one of which is used for Peltier compensation by passing a constant current of known intensity through the pile for a measured time. Total compensation is not sought, but Peltier cooling is intended to balance the major part of the heat produced.

The prime purpose of the twin system in the Calvet instrument is to achieve a stable zero reading and to eliminate fluctuations of the zero arising from imperfect thermostating of the jacket. Suppose the temperature of the external

FIG. 6. Reaction cells (*a–e*) for use in the Beckman microcalorimeter. (Reprinted from Benzinger, 1965.)

boundary θ_e is the same at a given time for both active and reference elements when the temperatures of the internal boundaries are θ_{i1} and θ_{i2} respectively. The emf's developed in the two piles are then

$$e_1 = n\epsilon_0(\theta_{i1} - \theta_e) \tag{17}$$

and

$$e_2 = n\epsilon_0(\theta_{i2} - \theta_e) \tag{18}$$

If the block temperature begins to vary by an amount $\Delta\theta_e$, the internal boundaries are affected identically by an amount $\tau\Delta\theta_e$ (the time constants τ of the elements being the same) and the modified emf's become

$$e_1 = n\epsilon_0[(\theta_{i1} + \tau\Delta\theta_e) - (\theta_e + \Delta\theta_e)] \tag{19}$$

$$e_2 = n\epsilon_0[(\theta_{i2} + \tau\Delta\theta_e) - (\theta_e + \Delta\theta_e)] \tag{20}$$

so that the resultant emf recorded by the piles connected in opposition is

$$e = e_1 - e_2 = n\epsilon_0(\theta_{i1} - \theta_{i2}) \tag{21}$$

which is independent of the variation $\Delta\theta_e$ in the jacket temperature. Thus the twin-differential system does not require absolute constancy of the jacket temperature but only that the *external boundaries of the twin cells should have identical temperatures at any given instant.* Compensation of the heat effect in the active cell is helpful in this respect, especially in case of rapid heat evolution in the active cell. For slow heat evolution, compensation is less necessary and is not usually attempted.

The construction of the heat sink and thermostat in the Calvet apparatus is shown in Fig. 7. The twin calorimetric elements fit into conical sockets (A)

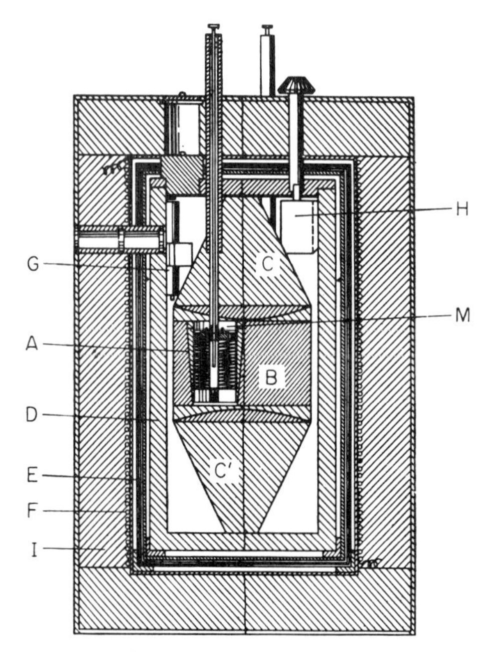

FIG. 7. Vertical cross section of the Calvet microcalorimeter. A, conical socket cut into the block B; B, the metal block; C and C′, metallic cones; D, thick metal cylinder; E, thermostat consisting of several metal canisters; F, electrical heater; G, galvanometer; H, switch; I, thermal insulation; and M, microcalorimetric element. (Reprinted from Calvet and Prat, 1963.)

cut in a cylindrical block of silver or aluminium (B). The block is positioned between the bases of two truncated cones (C and C′) placed centrally within a thick metal cylinder (D), surrounding the block. D is in turn enclosed by a series of metal canisters (E), and the whole assembly is lodged within an electrically controlled insulated thermostat jacket (F and I).

This arrangement is designed to transform lateral thermal perturbations arising from imperfect thermostatting into vertical perturbations equally divided between the twin elements. A lateral perturbation is, in the first place, damped by the multiple canisters (E) and is diffused in the thick cylinder D, thence to be conducted toward the twin elements by the metal cones (C and C′).

Thus the cones act as collimators, transforming the heat flux from the cone tops into a parallel vertical flux through the twin elements. Thermal lenses of poor conducting steel (attached below the cones) help to achieve a perfect equipartition of the thermostat perturbations between the twin calorimeters.

There are disadvantages in the Calvet arrangement in that the attainment of a steady state is time-consuming. It seems likely that a less heavy system based on electrically controlled shields could be made equally effective and more rapid in

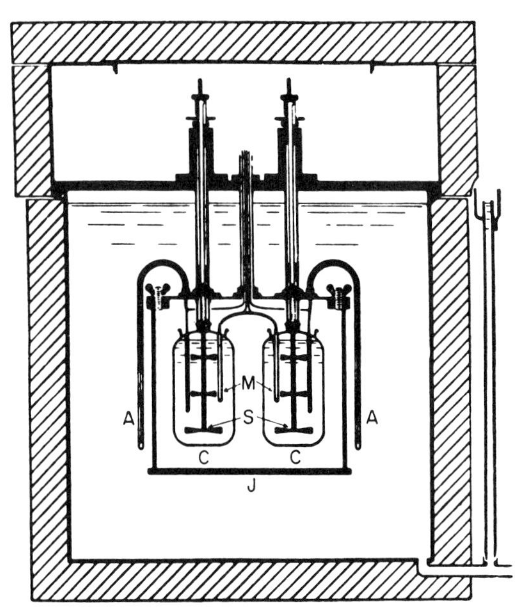

FIG. 8. Adiabatic twin calorimeter. (Reprinted from Sturtevant, 1949.)

response. Nevertheless, the objective of the Calvet design is sound and has not been attempted in other instruments including that of Benzinger.

The question of the sensitivity of the Calvet calorimeter has been discussed at length by Calvet and Prat (1956) in terms of the number of thermocouples used, the cross sections of the wires, and the thermoelectric power of the couples. On the basis of this analysis, the thermoelectric elements are designed to operate near maximum sensitivity. The Benzinger calorimeter employs many more couples than does the Calvet, which may limit its accuracy by developing excessive Johnson noise in the detector.

Twin calorimeters have been very successfully applied in conjunction with the adiabatic method. The adiabatic twin-calorimeter system described by Gucker, Ayres, and Rubin (1936) and used to measure the relative heat capacities of solutions provides a good example. The arrangement is shown schematically in Fig. 8. The twin calorimeters (C), made of gold-platinum alloy, are supported within the submarine J, which is totally immersed in an adiabatic water-jacket controlled by two thermals (A). The active calorimeter is charged with the solution under investigation, and the reference calorimeter contains pure solvent. The calorimeters are heated through about 1°C with their heaters in series, and the heat capacity of the solution is then evaluated from the known resistance ratio of the heaters, and the small residual temperature differential (measured by the main thermal M) remaining at the end of the heating period. The precision attained reached $\pm 0.01\,\%$. The differential adiabatic calorimeter of Gucker, Pickard, and Planck (1939) used to measure heats of dilution of solutions is essentially similar.

Buzzell and Sturtevant (1948, 1951) have described an adiabatic twin-calorimeter, with automatic compensation, suited for reaction calorimetric studies on processes with half-life periods of up to 1 hour. The significant feature here is the usage of a voltage integrator enabling variable electric power to be fed to the reference cell and thus to achieve continuous total compensation throughout an experiment. This instrument has served to measure the heats of several reactions and processes of biochemical interest, including denaturation studies on some proteins.

Stoesser and Gill (1967) have recently described a precision flow microcalorimeter that has several novel features and seems very well suited to measurements of the heats of mixing of liquids and of dilution of solutions (Gill *et al.*, 1967). The mixing of solutions in a flow arrangement has been used in titration calorimeters before, but not previously in a microcalorimetric system. The system demands differential twin operation in order to cancel spurious heat effects associated with the flow of liquids through the calorimeter.

Mixing of thermally equilibrated solutions in one of the twin cells is accomplished without a vapor space and without mechanical stirring by injecting the solutions from fixed-speed motor-driven syringes through Teflon solution tubes. The flow path of the liquids to be mixed from the syringes is as follows: (1) through separate Teflon tubes which wrap around an outer thermal shield for rough thermal equilibration, (2) to an inner shield for fine thermal equilibration, (3) to the mixing junction point, and subsequent equilibration path within the "active" twin, (4) back to the inner shield to reestablish fine thermal equilibration with the shield, and (5) finally to the "reference" twin and exit from the system. See Figs. 9 and 10.

A 200-junction thermopile detector between the twin cells is used to sense the differential between the cells, and automatic compensation is provided by pulsed electrical power input.

With this arrangement the two solutions may be injected at any desired ratio by selecting the appropriate speed for driving the syringes. Heat measurements can be made either by "steady-state" or by "pulse" techniques. In the former, a steady thermal state is approached after starting

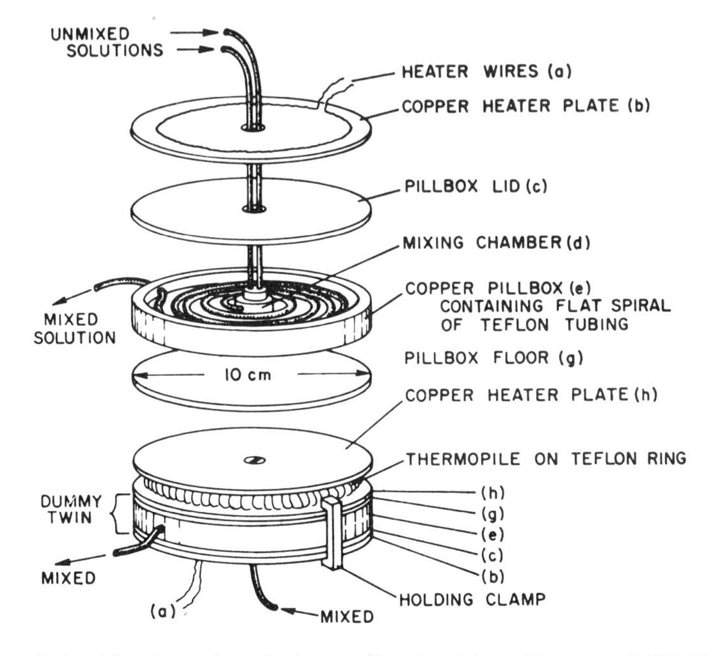

FIG. 9. Precision flow microcalorimeter. (Reprinted from Stoesser and Gill, 1967.)

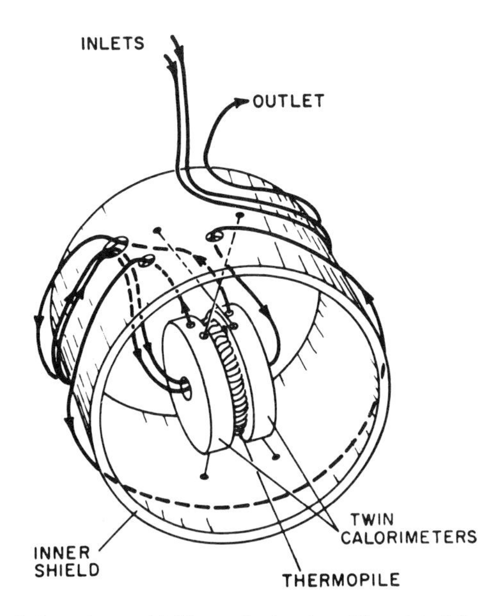

FIG. 10. Flow path from inner shield to calorimeters. (Reprinted from Stoesser and Gill, 1967.)

the flow and the heat evolution recorded over a fixed time interval. In the latter, the heat effect occurring on starting and stopping the flow for a given time period is measured.

VI. Isothermal-Jacket Calorimeters

Isothermal-jacket and adiabatic calorimeters serve to investigate a process that occurs over a range of temperature. The heat produced in a process $A(T_1^\circ) \rightarrow B(T_2^\circ)$ is totally retained within an adiabatic calorimeter. [The same process in an isothermal-jacket calorimeter results in an identical "corrected" temperature rise, $\Delta T = (T_2 - T_1)$.] Thermodynamics requires knowing the quantity of heat transferred from the calorimeter to the surroundings as a result of the isothermal process $A(T_1^\circ) \rightarrow B(T_1^\circ)$. To obtain this, a calibration experiment is performed on the calorimetric system, charged with the final products B. One measures the energy input into the calorimeter required to effect the change $B(T_1^\circ) \rightarrow B(T_2^\circ)$. The two steps are shown in the diagram below, from which it is apparent that the calibration experiment provides the required isothermal heat at temperature T_1°.

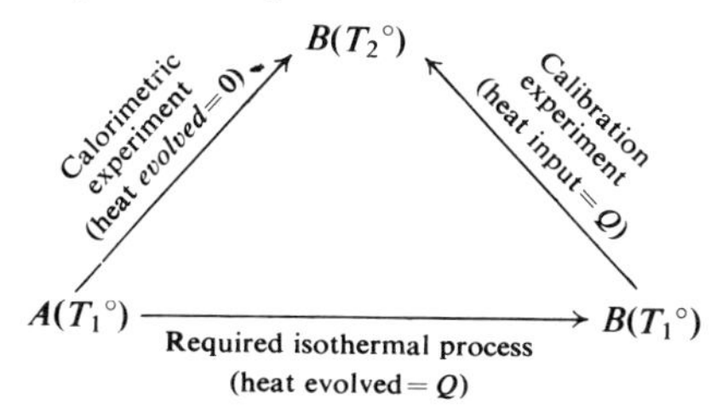

Alternatively, a calibration experiment may be carried on the calorimetric system charged with the initial reactants A. One measures the input of energy required to effect the change $A(T_1^\circ) \rightarrow A(T_2^\circ)$. The two steps are now as shown

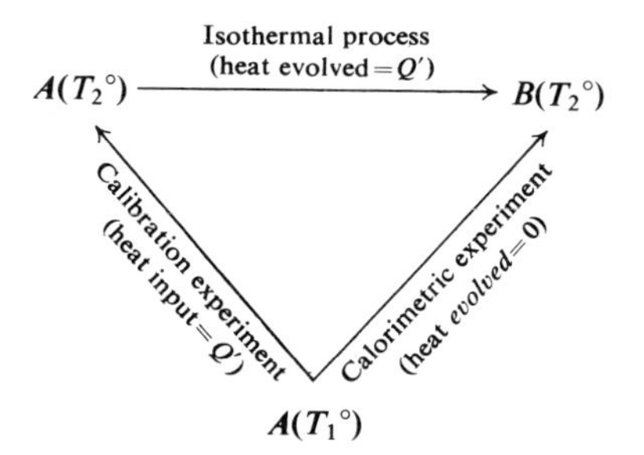

in the diagram above. The calibration experiment provides the isothermal heat of the process at T_2°. In general the measured heats Q and Q' will differ from one another and, provided the calorimeter is sufficiently precise, the value of $\Delta C_p [=\Delta(\Delta H)/\Delta T]$ for the process can be measured directly from calibration experiments on reactants and products over a limited temperature range (Wadsö and Åkesson, 1968).

The isothermal-jacket type of calorimeter is widely used for the measurement of heats of chemical reactions in solution. It consists essentially of a reaction vessel equipped with a temperature sensor, a heater for electrical calibration, a stirrer, and a mixing device—the whole system being enclosed within a carefully controlled thermostated jacket. Because of the range of chemical reactions, there are correspondingly many variations of design of reaction calorimeters. The reaction calorimetric system recently developed by Sunner and Wadsö (1966), and now commercially available from LKB Instruments (Rockville, Maryland) and from LKB-Produkter AB (Stockholm-Bromma 1) is both versatile in application and capable of high precision: It provides a ready-made solution to the technical problems that arise in most solution-reaction calorimetric studies.

The instrument is designed for the study of reactions in solution—or for other solution processes—which take place quickly (30 minutes maximum) in the temperature range 0–50°C at atmospheric pressure. The reaction vessel (capacity 100 ml) is made of thin-walled glass and fitted with a thermistor element (~2000 Ω resistance) as thermometer, a calibration heater (50 Ω), and a gold stirrer. The latter also acts as a holder for a cylindrical glass ampoule containing one of the reactants. To start the reaction, the stirrer rod is depressed, forcing the ampoule against a sapphire-tipped rod sealed to the bottom of the vessel. The reaction vessel fits into a surrounding jacket (chromium-plated brass), which can be evacuated to reduce heat exchange, and the whole assembly is submerged in a thermostated water bath (maintained at constant temperature within limits of 10^{-3}°C). The arrangement is shown schematically in Fig. 11. The thermistor forms one arm of a manually balanced dc Wheatstone bridge, and an electronic galvanometer is used for detection of the bridge off-balance.

To make time–temperature readings, the bridge is first set slightly off balance, and the chronometer (electrically connected to the bridge) is started. As the calorimeter temperature changes and the thermistor resistance reaches the balance point, one of the two chronometer hands halts and permits the exact balance-point time to be recorded. A resistance (usually 0.05 Ω) is switched in, again to offset the bridge and the "stopped" hand of the chronometer hand is restarted. When the calorimeter temperature once more reaches balance point, the chronometer hand stops again and the new balance time recorded: This procedure is continued throughout the fore-, main-, and after-rating periods of the experiment. When maximum precision is not necessary, the recording can be made automatic by connecting a potentiometer recorder to the outlet of the electronic galvonometer.

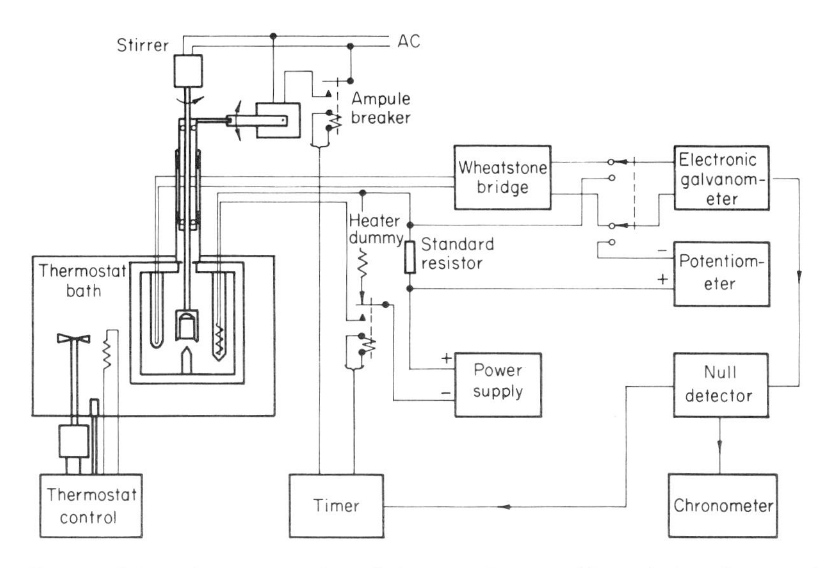

FIG. 11. Schematic representation of the LKB 8700 precision calorimetric apparatus. (Reprinted from Sunner and Wadsö, 1966.)

A. CALIBRATION OF REACTION CALORIMETERS

Reaction calorimeters normally are calibrated electrically by the "substitution method." This requires accurate measurement of the quantity of electrical energy needed to reproduce time–temperature recordings characteristic of the reaction under study. Precise matching of the reaction and electrical calibration time–temperature recordings is rarely possible unless the reactions occur slowly, but the calibration experiment can (and should) be set up to duplicate as far as possible the conditions of the reaction experiment—e.g., start and finish at the same temperatures, identical stirring rates in both experiments, and identical mechanical operations in both (such as ampoule crushing). Even with these precautions differences between the calibration and reaction experiments, residing in different thermal equilibration patterns in the two experiments, remain as a potential source of inaccuracy. It therefore is advisable to test the calorimeter against a standard process, well established thermally.

Test substances for combustion calorimetric studies—both by conventional and rotating-bomb techniques—are in common use, and in the case of certificated benzoic acid, the test substance has superceded electrical power as the normal means of calibration. For reaction solution calorimetry, suitable test reactions are being sought, but few can yet be said to have become "established." Irving and Wadsö (1964) proposed that the reaction between tris-hydroxymethylaminomethane (THAM) and excess 0.1 *M* aqueous HCl solution be used as a standard test for reaction calorimeters. The "best value" for this test reaction under recommended conditions is $\Delta H = -7.112$ kcal mole^{-1} at 25°C, and the reaction is convenient and precise provided that the calorimeter is intended for the study of rapid, moderately exothermic processes. Table I summarizes results obtained using the LKB reaction calorimeter on the THAM reaction.

TABLE I

HEAT OF SOLUTION OF THAM IN 100 ML 0.1 M HCl AT 25.00°C

Millimoles THAM in ampules	$-\Delta H$(25°C) (kcal mole^{-1})
3.7662	7.114
4.1888	7.109
4.3006	7.111
4.4093	7.114
4.5487	7.112
4.8816	7.107
	Mean 7.111 ± .001

Calibration electrically was performed before each experiment, the final temperature in test and calibration experiments being 25°C. Sample weights were corrected to true mass. The results indicate reproducibility to ±1 part in 7000, which is quite sufficient for most solution calorimetric purposes.

The THAM reaction is under study at the National Bureau of Standards (Washington, D.C.) and in due course may be officially recommended as a standard test reaction for reaction calorimeters. In this event "certified" THAM should become available to thermochemists as is certificated benzoic acid for combustion calorimetric studies.

B. Titration Calorimetry

The titration calorimeter is designed to measure the heat evolved during a titration process. Whereas in the normal reaction calorimeter the ampoule technique is used to isolate the reactants prior to mixing them together, the titration calorimeter allows controlled and gradual addition of one reactant to another. The availability of high-precision thermostats (constancy ±0.001°C) and the use of piston-driven pipettes has simplified the technical problem of mixing reactants in controlled amounts inside the calorimetric vessel from stock solutions held outside it. The LKB titration calorimeter uses a 10-ml piston burette equipped with a two-way stopcock to control the flow of liquid.

A measured quantity (accuracy ±0.002 ml) flows through a 2-mm capillary tube into a Teflon spiral, immersed in the water of the thermostat and thence via a capillary tube into the glass calorimetric reaction vessel. A multijunction thermocouple and sensitive galvonometer act as sensor of the temperature differential between the liquid being added and the liquid in the reaction vessel. The precision titration calorimeter described by Christensen *et al.* (1965) uses a similar system: The burette operates by driving a stainless-steel piston into a

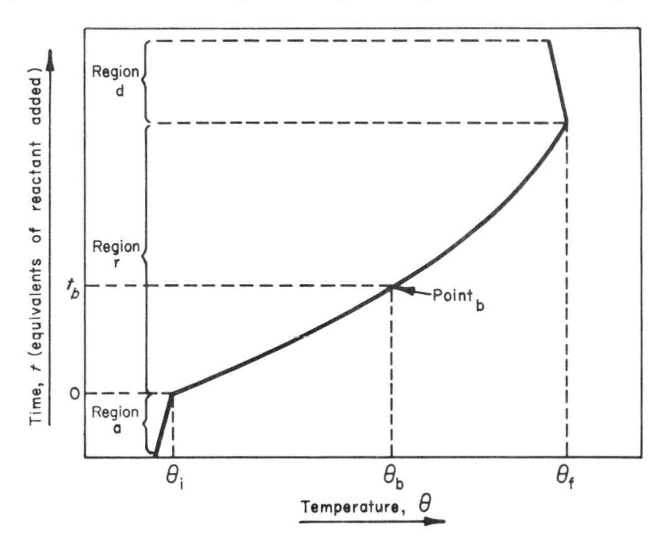

FIG. 12. Calorimetric time-temperature curve.

mercury chamber, displacing mercury, which in turn displaces titrant from the burette to the calorimeter. The burette is contained inside an airtight compartment within a water thermostat.

The titration calorimeter has been used to measure heats of formation of complexes in solution, enthalpies of stepwise dissociation of protons from acids, heats of dilution of solutions, and differential and integral heats of mixing of liquids and solutions. In special cases an enthalpy titration can provide information on ΔH and the equilibrium constant of the process under investigation from a single set of measurements (Christensen *et al.*, 1966). As a simple example, consider the reaction between a solution of A and titrant B to form AB, where the reaction conditions and equilibrium constant allow a measurable but not quantitative amount of reaction to take place (log K in the range 0.5–3.0). A typical time–temperature curve for this process carried out in an isothermal–jacket calorimeter is shown in Fig. 12. At a point b on the curve (temperature θ_b and t_b), the calorimeter contains unreacted A of concentration $[A]_b$, unreacted B of concentration $[B]_b$, and product AB of concentration

$[AB]_b$, determined by Eq. (22), where $K =$ the equilibrium constant for the reaction and the factors γ are activity coefficients of the indicated species (concentrations expressed as moles/liter).

$$K = [AB]_b \gamma_{AB}/[A]_b \gamma_A [B]_b \gamma_B \tag{22}$$

The region a is the fore-rating period, r is the main period during which titrant B is added, and d is the after-rating period. Let Q_b be the heat produced in the calorimeter by the reaction from time $0 \rightarrow t_b$ of the main period r, and let us anticipate that proper analysis of Fig. 12 enables us to isolate Q_b from spurious heat quantities (e.g., dilution heats of unreacted species A and B). We may write Eq. (23), where Q_{total} is the heat produced for complete reaction

$$Q_b = Q_{\text{total}}[AB]_b V_b \tag{23}$$

between A and B to form 1 mole AB and V_b is the volume (in liters) of solution in the calorimeter at point b. Furthermore, the material balance gives Eq. (24) and (25), $[A_{\text{total}}]$ and $[B_{\text{total}}]$ being known from the conditions of the experiment.

$$[A]_b + [AB]_b = [A_{\text{total}}]_b \tag{24}$$

$$[B]_b + [AB]_b = [B_{\text{total}}]_b \tag{25}$$

The combination of Eqs. (22), (23), (24), and (25) and setting $\Gamma_b = \gamma_A \gamma_B / \gamma_{AB}$ gives Eq. (26).

$$\frac{Q_{\text{total}}}{K} = \{V_b[B_{\text{total}}]_b [A_{\text{total}}]_b \cdot \Gamma_b | Q_b\} Q^2_{\text{total}} - \{[B_{\text{total}}]_b + [A_{\text{total}}]_b\} \Gamma_b Q_{\text{total}} + \Gamma_b Q_b | V_b \tag{26}$$

Writing

$$D_b = \{V_b[B_{\text{total}}]_b [A_{\text{total}}]_b \Gamma_b | Q_b\} \tag{27}$$

$$E_b = \{[B_{\text{total}}]_b + [A_{\text{total}}]_b\} \Gamma_b \tag{28}$$

and

$$F_b = \Gamma_b Q_b | V_b \tag{29}$$

Eq. (26) may be expressed in the form

$$\frac{Q_{\text{total}}}{K} = D_b Q^2_{\text{total}} + E_b Q_{\text{total}} + F_b \tag{30}$$

Let us now consider a different point at time t_c on the time–temperature curve. In similar manner we may obtain

$$\frac{Q_{\text{total}}}{K} = D_c Q^2_{\text{total}} + E_c Q_{\text{total}} + F_c \tag{31}$$

where D_c, E_c, and F_c are evaluated from Eqs. (27), (28), and (29) with appropriate changes in subscript. Hence it follows that

$$(D_c - D_b)\, Q_{\text{total}}^2 + (E_c - E_b)\, Q_{\text{total}} + (F_c - F_b) = 0 \tag{32}$$

and since the parameters D, E, and F are measured quantities (presuming Γ_b and Γ_c are known), Q_{total} may be calculated.

For ionic reactions in aqueous solution, Q_{total} is dependent on ionic strength (μ), and a series of measurements at different ionic strengths should be made to obtain Q_{total}° (by extrapolation) at $\mu = 0$. The latter, substituted back into Eq. (30), gives the true equilibrium constant K, and thence ΔG° of reaction, from which ΔS° may be derived.

The details of analysis of the time–temperature curves to obtain Q_b and Q_c are given elsewhere by Christensen *et al.* (1966) and are omitted here. The analysis is complicated since both the volume of solution in the calorimeter and the calorimeter heat capacity vary throughout the main period of the experiment. Moreover, the mixing and dilution heats of reactants A and B (which remain in part unreacted in the calorimeter) contribute to the overall heat produced in the calorimeter and must be duly allowed for.

C. Combustion Calorimetry

The most general method of measuring the heats of formation of organic substances is by measurement of their standard heats of combustion using bomb-combustion techniques. The experimental methods have been described in detail elsewhere in two volumes (Rossini, 1956; Skinner, 1962), to which the reader is referred. From the biochemical viewpoint, the bomb calorimeter may seem somewhat remote and of little direct interest. Nevertheless, it has and will continue to provide the most reliable and accurate ΔH_f° data on pure materials, including the "units" (amino acids, peptides, etc.) from which complex structures of biochemical importance are built up. The development of rotating-bomb calorimetry has extended the scope of the combustion method well beyond its early limitation to organic substances of carbon, nitrogen, hydrogen, and oxygen. Organic substances of sulphur, the halogens, boron, silicon, and certain metals now can be satisfactorily studied by rotating-bomb calorimetry, and a method suited to organic compounds of phosphorus is actively being sought.

Bomb-combustion calorimeters are capable of a precision approaching 0.01% in favorable cases. The limitations arise not from the calorimeter itself but from the process under investigation. Evidently, to achieve an overall precision of 1 part in 10,000 requires that the combustion process be thoroughly understood and that there be no undetected side reactions (e.g., incomplete combustion) or irrelevant heat production arising from impurities in the combustion sample. The need for an adequate supply of starting material and for thorough analysis of the combustion products imposes a limitation that will apply to many compounds of biochemical interest. There is, however, a growing interest in microbomb calorimetry, requiring only very small quantities of combustion material, yet capable of good precision.

Mackle and O'Hare (1963) have described an aneroid microbomb calorimeter of low heat capacity (100 cal deg^{-1}), which they used to measure combustion heats on hydrocarbon samples weighing ~10 mg. A precision of 0.2% was claimed—but the authors point out that there are difficulties arising from incomplete combustion of the sample, caused by insufficient heating of the sample container (crucible) during the combustion process. There is a lower limit to the mass of sample, below which combustion is incomplete. Calvet *et al.* (1960) have developed a microbomb calorimetric element for use in the Calvet instrument and have similarly reported unsatisfactory results when the sample weight was reduced below a critical limit. Steel microbomb elements were first used, but more recently Tachoire (1962) has made use of glass microbombs and obtained encouraging results using oxygen pressures slightly in excess of an atmosphere. Measurements of the heats of combustion of a number of sterols have been reported by Tachoire (1965); currently differential microbomb techniques are under investigation, using twin microbombs to measure differences in the heat of combustion of isomers.

The measured energy of combustion for a reaction carried out in a bomb calorimeter will normally differ significantly from the thermodynamically desired quantity—i.e., ΔU° or the energy change for the reaction under standard conditions. The necessary corrections to standard states were detailed by Washburn (1933) for the case of combustion in oxygen of compounds containing carbon, hydrogen, and oxygen and have come to be known generally as the "Washburn corrections." The standard states are defined as follows:

1. For O_2, N_2, and CO_2, the pure substances in the "ideal" gas state at 1 atm pressure.
2. For water, the pure liquid at 1 atm pressure.
3. For the substance under investigation, its normal state (solid or liquid) under 1 atm pressure. The "standard temperature" is nowadays taken at 25°C.

In the case of combustion of a liquid compound $C_aH_bO_cN_d$, the "standard" reaction is considered to be Eq. (33), occurring isothermally at 25°C. In

$$C_aH_bO_cN_d\text{ (liquid under 1 atm pressure)} + \tfrac{1}{4}(4a+b-2c)\,O_2\text{(gas, 1 atm)} \rightarrow a\,CO_2\text{(gas, 1 atm)} + \tfrac{1}{2}bH_2O\text{(liquid under 1 atm)} + \tfrac{1}{2}dN_2\text{(gas, 1 atm)} \quad (33)$$

practice, the actual bomb process used moist O_2 at high pressure (usually 30 atm): The products formed are a mixture of moist gases under pressure, and part of the CO_2 dissolves in the water present in the bomb. The corrections needed to reduce the actual process to standard conditions are the Washburn corrections: For the particular case of combustion of a compound $C_aH_bO_cN_d$ Cox *et al.* (1954) have derived a single equation, which is convenient to apply and incorporates all the Washburn corrections.

The reduction to standard states of experimental results obtained from rotating-bomb techniques is somewhat tedious. A detailed analysis of the standard state corrections needed for combustion of an organic sulphur compound, $C_aH_bO_cS_d$, has been given by Hubbard *et al.* (1956) and Bjellerup (1962) has dealt similarly with the case of bromine-containing organic compounds. These treatments serve as models from which the extension to cover compounds containing other elements should follow.

VII. High Temperature Calorimetry

Calorimetry at high temperatures (of the order 1000°C) presents difficult technical problems, and high precision is hardly to be expected. The biochemist, however, is unlikely to meet problems calling for study at very elevated temperatures—but he may have some interest in heat-capacity measurements extending to temperatures well above normal, and he also may find an interest in recent developments in the techniques of differential thermal analysis. Heat capacity measurements at normal (and low) temperatures are made using an adiabatic calorimeter, fitted with a separate heater and thermometer: The adiabatic shield can be controlled automatically. West and Ginnings (1958) have described a heat-capacity calorimeter of this type, capable of a limiting precision of 0.01%, which was devised for operation at temperatures up to 500°C. The calorimeter (Fig. 13) is contained within a massive thermal block (G) and makes use of multiple radiation shields (S_2 and L_2), which constitute the adiabatic shield, and of a second set of shields (S_1 and L_1), which surround the calorimeter (C). At higher temperatures, when radiation becomes a major heat transfer mechanism, the adiabatic calorimeter loses precision and the "drop" method is preferred. In this method the sample under investigation is heated in a furnace to a known high temperature and then is quickly "dropped" into a receiving calorimeter at (or near to) room temperature. The receiving calorimeter may be a metal block or it may be an ice calorimeter, and a series of measurements are made, varying the temperature of the furnace at regular intervals over a particular range. There are some limitations to this method. If the sample has a solid–solid transition, it is possible that the transition will not complete in the time of drop, and part of the transition energy may be "frozen in" and thus give erroneous results. If the heat capacity changes significantly with temperature the accuracy of measurement of the heat capacity (which is the derivative of enthalpy) becomes much less than the accuracy achieved in measuring the overall heats (relative enthalpies): Thus it is important to make the calorimetric measurements with high precision.

The drop calorimeter described by Ginnings and Corruccini (1947) is representative of a precision drop calorimeter suited to studies up to about

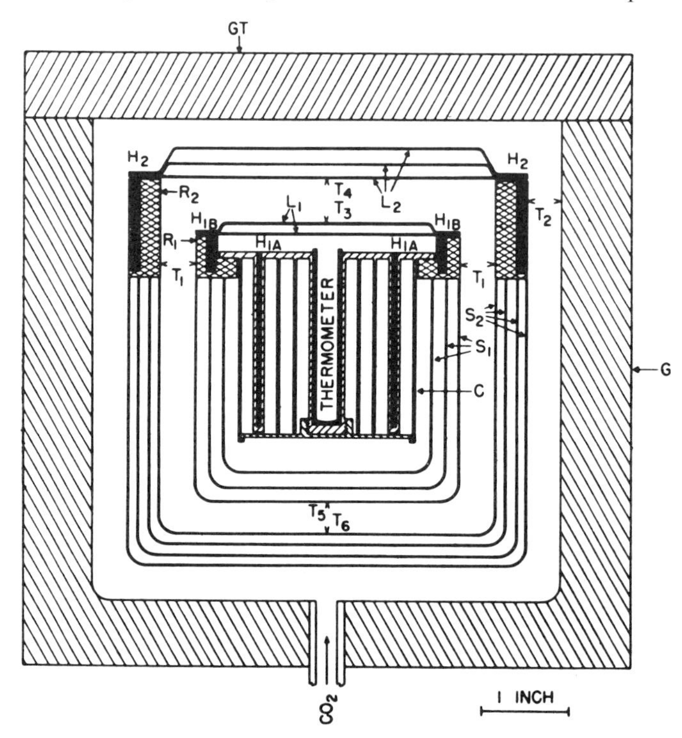

FIG. 13. Adiabatic calorimeter for heat capacity measurements at temperatures up to 500°C. (Reprinted from West and Ginnings, 1958.)

900°C. The high temperature drop calorimeter described by Hoch and Johnston (1961), designed to operate at temperatures up to 2700°C, employed radio-frequency heating of the sample in high vacuum and used a Bunsen ice calorimeter as the receiver.

Differential thermal analysis (DTA) is becoming widespread in its application, and DTA equipment is available commercially from several manufacturers. In brief, DTA requires the measurement of the temperature differential between a sample of material under investigation and a contiguous reference material as both are heated within the same enclosure simultaneously.

A change in state of the sample (phase transition, melting, or decomposition), which occurs fairly rapidly, is marked by the appearance of a peak in the recording of the temperature differential. The main application is in qualitative or semiquantiative analysis.

Quantitative treatment of DTA curves by other than empirical methods is hardly practicable, despite the theoretical background that has accumulated in recent years (see, e.g., Smothers and Chiang, 1966). The relationship between DTA and calorimetry is thus hard to define precisely, although in prescribed circumstances the correlation between them is good.

Of the several instruments available, the Differential Scanning Calorimeter (Perkin Elmer Corporation, Norwalk, Connecticut) is perhaps of most direct interest to the calorimetricist. In this apparatus the sample and reference material are subjected to a controlled change in temperature, but a temperature differential between the sample and reference is not allowed to develop: This is prevented by supplying additional heat, as required, to either sample. The quantity of heat needed to maintain zero differential is recorded as a function of time, or of temperature.

Small samples (of the order of 10 mg or less) are placed in isolated metal pans, which act as sample holders. The instrument has two "control loops"—one for control of the average temperature, the other for differential temperature control. The programmer signal, which reaches the average temperature amplifier, is compared with signals received from platinum resistance thermometers built into the sample and reference holders. In the differential temperature loop, signals representing the sample and reference temperatures are fed to the differential temperature amplifier via a comparator circuit, so that the differential power increment can be directed to sample or reference heaters to nullify any difference of temperature between them. A signal proportional to the differential power is transmitted to the pen recorder, and the area under a peak is proportional to the heat energy absorbed or liberated by the process responsible for the peak. The instrument can be used at temperatures up to 500°C.

The majority of DTA studies have been made on inorganic materials, in particular on ceramics, soils, glasses, and minerals. The applications in polymer chemistry—including measures of degree of crystallinity, glass transistion temperatues, thermal stability, onset of cross-linking and other characteristics of polymer structures—relate more closely to the potential interests of the biochemist in DTA methods. The thermal decomposition of certain proteins has been examined by DTA, and the application of DTA to protein denaturation in solution by Steim (1965) points to further applications to systems of biochemical interest.

Mention should be made of modifications to the ordinary Calvet microcalorimeter that allow it to be applied to studies at elevated temperatures up to

1000°C. The high temperature microcalorimeter uses Pt-to-Pt/Rh thermocouples, platinum cells, and a steel block encased in a refractory muffle. The apparatus is well suited to DTA studies and should find increasing application in the near future: The calorimeter is now commercially available.

VIII. Low Temperature Calorimetry

Modern low temperature calorimetry is a precision technique, developed to meet the demands—both practical and theoretical—for accurate heat-capacity data at temperatues extending downward virtually to absolute zero. The prototype low temperatue calorimeter, due to Eucken and Nernst, was of the isothermal-jacket type and used a liquid-nitrogen filled Dewar vessel to provide the low temperature environment: It has been largely superceded by the adiabatic shield calorimeter, in which the sample container is surrounded by a metal shield, adjustable in temperature and controlled throughout the experiment to match the temperature of the container. The cryostat uses an exterior bath of liquid nitrogen to allow liquid hydrogen or helium to be kept in the inner liquid bath. Typical examples of this type of calorimeter include those described by Scott *et al.* (1945), Westrum *et al.* (1953), and Furukawa *et al.* (1956). These are suited to studies in the range of 10°–300°K. Cole *et al.* (1960) have described an isothermal-type low temperature calorimeter of comparable precision to the adiabatic shield instruments. Hill (1959) has given a thorough and critical review of low temperature calorimetry, and Westrum (1961) has dealt similarly with cryostats. A recent book (Scott and McCullough, 1968) describes the experimental techniques for accurate heat-capacity measurements on solids, liquids, solutions, and gases over the range of temperatures currently accessible to experiment.

Heat-capacity measurements down to temperatures approaching the absolute zero require removal of all metallic connections between calorimeter and jacket and make use of the cooling mechanism provided by the adiabatic demagnetization of paramagnetic salts. Examples of calorimeters of this type are described by Edwards *et al.* (1962) and by Heltemes and Swenson (1961).

IX. Accuracy of Calorimetric Measurements

In considering the probable error limits to be attached to a series of calorimetric measurements, it is important to distinguish between the precision of the measurements and their accuracy. The term "precision" is a measure of the reproducibility of the experimental results, whereas "accuracy" defines how close the experimental results are to reality. Evidently, the measure of precision is based on the random errors in a series of experiments and will not reveal systematic errors that may be present. Precision is thus an upper limit to accuracy, which will normally be less than the precision suggests. The usual measure of precision in modern calorimetry is the standard deviation of the mean s, defined by Eq. (34), where x_i = value of the ith measurement, $\bar{x}$ = average value from n measurements, and n = total number of measurements.

$$s = \sqrt{\sum_{i}^{n} (x_i - \bar{x})^2 / n(n-1)} \tag{34}$$

A calorimetric study necessarily involves calibration of the calorimeter—either directly by electrical energy or indirectly by using a standard calibration process. If electrical calibration experiments yield a mean value $\bar{E}$ (subject to a standard deviation $= s_E$) for the energy equivalent of the calorimeter and separate calorimetric measurements yield a mean value $\bar{Q}$ (subject to a standard deviation $= s_Q$) for the heat of the process under investigation, the overall standard deviation to be assigned to $\bar{Q}$ is Eq. (35).

$$s = \bar{Q}\sqrt{(s_E/\bar{E})^2 + (s_Q/\bar{Q})^2} \tag{35}$$

The uncertainty interval, equal to $2s$, is normally accepted as the overall uncertainty in the precision of measurement of the heat quantity $\bar{Q}$.

The overall accuracy of a series of measurements cannot be evaluated systematically in like manner to overall precision, and to some extent the "overall accuracy" reported by an investigator is subjective. However, many of the items contributing to overall accuracy can be independently assessed. These include the following:

1. Accuracy of weights and volume measures.
2. Accuracy of measurement of time intervals.
3. Accuracy of temperature measurement.
4. Accuracy of electrical equipment used in the electrical calibration.
5. Accuracy of the calorimeter in specific test processes.
6. Accuracy of "corrections" applied (e.g., Washburn corrections).
7. Purity of chemical reagents used.
8. Analysis of the process presumed to occur in the calorimeter.

Of these items, 1, 2, 3, and 4 can be satisfactorily evaluated in most cases, and high accuracy can be attained. Item 1 may be a limiting factor in microcalorimetric studies involving minute volumes of dilute solutions. Item 3 meets special difficulties at very low temperatures and to some extent at moderately high temperatures (>1000°C). Over "normal" temperature ranges, the platinum resistance thermometer both is reliable and is an "absolute" thermometer. Thermistor resistance elements are subject to drift with usage and maltreatment and should be checked from time to time against a platinum-resistance thermometer. Item 4 can be troublesome, particularly in the case of microcalorimeters, where heat conduction losses via the inlet electrical leads may be a serious source of error. Ginnings and West (1964) have examined the heater lead problem in calorimetry in some detail and have pointed out that careful consideration must be given to heat lead design to ensure accuracies of better than 0.1 %.

Evidently, the most satisfactory check on calorimeter performance is provided by measuring the heat of a well-defined standard process, and this is now recommended procedure for specific types of calorimeter, including bomb calorimeters of both conventional and rotating designs, low-temperature and high-temperature calorimeters for the measurement of heat capacities, and solution-reaction calorimeters. Efforts to define suitable standard liquid mixtures for the heat of mixing measurements are well advanced (McGlashan, 1962). However, an investigator will often find that there is no "standard" process comparable with the one he wishes to investigate, and the search for reliable standard test procedures to cover all eventualities is likely to continue for some time.

The most frequent causes of systematic error, however, arise from items 7 and 8, due to impurities in the chemical reagents used and to insufficient (or inaccurate) analysis of the process taking place in the calorimeter. Impurities—even in small traces—are particularly ruinous in bomb-combustion calorimetry, and the preparation of a pure sample and the estimation of its purity frequently are the most arduous and time-consuming parts of a combustion-calorimetric investigation. In reaction calorimetric studies, where an impurity may not interfere seriously with the process under investigation, purity limits tend to be substantially reduced, but they never must be ignored. The biochemist may face this difficulty often and may well find that the purity of samples is the important limiting factor in the overall assessment of accuracy. There is little point in designing or using a calorimeter capable of a precision of a fraction of a percent unless the chemicals used, and the process under investigation, can be defined to at least a similar order of precision in thermodynamic terms. An investigator might profit by designing his calorimeter only after he has determined the true accuracy limits of the chemistry of the process he intends to investigate.

REFERENCES

Benzinger, T. H. (1965). *Fractions* **2**, 2.

Benzinger, T. H., and Kitzinger, C. (1954). *Federation Proc.* **13**, 11.

Benzinger, T. H., and Kitzinger, C. (1963). "Temperature—Its Measurement and Control", Vol. 3, Chapt. 5. Reinhold, New York.

Bjellerup, L. (1962). *In* "Experimental Thermochemistry" (H. A. Skinner, ed.), Vol. 2, Chapt. 3. Wiley, New York.
Buzzell, A., and Sturtevant, J. M. (1948). *Rev. Sci. Instr.* **19**, 688.
Buzzell, A., and Sturtevant, J. M. (1951). *J. Am. Chem. Soc.* **73**, 2454.
Calvet, E., and Prat, H. (1956). "Microcalorimétrie." Masson, Paris.
Calvet, E., Chovin, P., Moureau, P., and Tachoire, H. (1960). *J. Chim. Phys.* **57**, 593.
Calvet, E., and Prat, H. (1963). *In* "Recent Progress in Microcalorimetry" (H. A. Skinner, ed.). Macmillan (Pergamon), New York.
Christensen, J. J., Izatt, R. M., and Hansen, L. D. (1965). *Rev. Sci. Instr.* **36**, 779.
Christensen, J. J., Izatt, R. M., Hansen, L. D., and Partridge, J. A. (1966). *J. Phys. Chem.* **70**, 2003.
Cole, A. G., Hutchens, J. O., Robie, R. A., and Stout, J. W. (1960). *J. Am. Chem. Soc.* **82**, 4807.
Coops, J., Jessup, R. S., and van Nes, K. (1956). *In* "Experimental Thermochemistry" (F. D. Rossini, ed.), Vol. 1, Chapt. 3. Wiley (Interscience), New York.
Cox, J. D., Challoner, A. R., and Meetham, A. R. (1954). *J. Chem. Soc.* p. 265.
Edwards, D. O., McWilliam, A. S., and Daunt, J. G. (1962). *Phys. Letters* **1**, 101, 218.
Furukawa, G. T., Douglas, T. B., McCoskey, R. E., and Ginnings, D. C. (1956). *J. Res. Natl. Bur. Std.* **57**, 67.
Giguère, P. A., Morisette, B. G., and Olmos, A. W. (1955). *Can. J. Chem.* **33**, 657.
Gill, S. J., Downing, M., and Sheats, G. F. (1967). *Biochemistry* **6**, 272.
Ginnings, D. C., and Corruccini, R. J. (1947). *J. Res. Natl. Bur. Std.* **38**, 583.
Ginnings, D. C., and West, E. D. (1964). *Rev. Sci. Instr.* **35**, 965.
Gucker, F. T., Ayres, F. D., and Rubin, T. R. (1936). *J. Am. Chem. Soc.* **58**, 2118.
Gucker, F. T., Pickard, H. B., and Planck, R. W. (1939). *J. Am. Chem. Soc.* **61**, 459.
Heltemes, E. C., and Swenson, C. A. (1961). *J. Chem. Phys.* **35**, 1264.
Hill, R. W. (1959). *Progr. Cryog.* **1**, 179.
Hoch, M., and Johnston, H. L. (1961). *J. Phys. Chem.* **65**, 855.
Hubbard, W. N., Scott, D. W., and Waddington, G. (1956). *In* "Experimental Thermochemistry" (F. D. Rossini, ed.), Vol. 1, Chapt. 5. Wiley (Interscience), New York.
Irving, R. J., and Wadsö, I. (1964). *Acta Chem. Scand.* **18**, 195.
Jessup, R. S. (1955). *J. Res. Natl. Bur. Std.* **55**, 317.
McGlashan, M. L. (1962). *In* "Experimental Thermochemistry" (H. A. Skinner, ed.), Vol. 2, Chapt. 15. Wiley, New York.
Mackle, H., and O'Hare, P. A. G. (1963). *Trans. Faraday Soc.* **59**, 2693.
Rossini, F. D., ed. (1956). "Experimental Thermochemistry," Vol. 1. Wiley (Interscience), New York.
Scott, D. W., and McCullough, J. P. (1968). "Experimental Thermodynamics," Vol. 1. Butterworth, London and Washington, D.C.
Scott, R. B., Meyers, C. H., Rands, R. D., Brickwedde, F. G., and Bekkedahl, N. (1945). *J. Res. Natl. Bur. Std.* **35**, 39.
Skinner, H. A., ed. (1962). "Experimental Thermochemistry," Vol. 2. Wiley, New York.
Smothers, W. J., and Chiang, Y. (1966). "Handbook of Differential Thermal Analysis," Chem. Publish Co., New York.
Steim, J. M. (1965). *In* "Thermal Analysis" (J. P. Redfearn, ed.), p. 84. Macmillan, New York.
Stoesser, P. R., and Gill, S. J. (1967). *Rev. Sci. Instr.* **38**, 422.
Sturtevant, J. M. (1949). *In* "Calorimetry. Physical Methods of Organic Chemistry" (A. Weissberger, ed.), 2nd ed., Vol. 1 Wiley (Interscience), New York.
Sunner, S., and Wadsö, I. (1966). *J. LKB Inst.* **13**, 1.
Tachoire, H. (1962). *Compt. Rend.* **255**, 2950.
Tachoire, H. (1965). Ph.D. Thesis, Univ. of Aix-Marseille.

Wadsö, I., and Åkesson, G. (1968). To be published.
Washburn, E. W. (1933). *J. Res. Natl. Bur. Std.* **10**, 525.
West, E. D., and Ginnings, D. C. (1958). *J. Res. Natl. Bur. Std.* **60**, 309.
Westrum, E. F., Jr. (1961). *Advan. Cryog. Eng.* **7**, 1.
Westrum, E. F., Jr., Hatcher, J. B., and Osborne, D. W. (1953). *J. Chem. Phys.* **21**, 419.

CHAPTER II

Thermodynamic Properties of Biochemical Substances

R. C. WILHOIT

I. Introduction

Nearly all physical and chemical changes absorb or evolve heat. During the past century the measurement and interpretation of these effects has played an important role in many branches of science and technology. The knowledge of heat effects that accompany transitions in well-defined systems is required for thermodynamic calculations. It also gives some clue about the nature of the transitions and, in favourable cases, some help in unraveling the molecular mechanisms that bring about the transformations. For example, heats of chemical reactions in the gas phase may be interpreted in terms of the formation and breaking of chemical bonds. Although no simple interpretation can be given about processes that occur in liquid solutions, especially if ions are involved, heats of solution and dilution furnish some information about solute–solute and solute–solvent interactions. Even when heats of chemical reactions cannot directly relate to a molecular interpretation, they furnish important subsidiary evidence for or against a proposed mechanism. Heat effects that accompany transformations in macromolecules are related to changes in intramolecular ordering.

Although the central importance of energy changes in the study of living organisms has long been recognized, it is only recently that enough biochemical knowledge has accumulated to make a detailed quantitative study of such energy changes possible. Calorimeters are used in biochemical studies to obtain thermodynamic data and also to detect the presence and extent of exothermic or endothermic processes. Because of the demand for quantitative information about biochemical systems and the advances in instrumentation which have been made, it is likely that calorimeters will soon become a major tool of the biochemist.

However, the principal emphasis in this chapter will be on the use of calorimetric data to study energy transformations that occur in living organisms. These will be described in terms of the appropriate thermodynamic formulas. These energy transformations constitute one of the basic distinctions between living and nonliving matter. In fact, it may be argued that the principal difference between living and nonliving systems lies in the nature of such transformations and that the principal function of the complex physical and chemical organization that is found in living organisms is to provide a framework for bringing about these energy transformations. Although it is easy to list characteristics common to most living organisms and to recognize whether an organism is alive or not, the distinction in terms of simple physical or chemical phenomena is a subtle one. All evidence indicates that living things are subject to the same physical laws, including those of thermodynamics, as are nonliving systems. When studied in isolation, biochemical reactions do not demonstrate any exceptional energetic phenomena compared to reactions that occur in the nonliving world. It is only in the complete living organisms that something special can be found. A living cell is characterized by constant activity. Thousands of physical and chemical processes, each accompanied by an energy transformation, are in continuous progress. These processes interact with each other, not in a random fashion, but in a definite, controlled pattern that results in a systematic sequence of energy transformations. The principle that the whole organism is more than the simple sum of its parts is a familiar one to biologists. Although the complex structure and organization of component parts seems to be a requirement for life, it is the pattern of energy transformations, rather than the material composition, which constitutes the most fundamental difference between a living and a nonliving system.

The overall requirements for energy by living organisms are obvious. They extract nutrients from their environment, move substances from place to place within the organism, and excrete waste products into their surroundings. When these movements occur against a concentration gradient, the process is known as active transport. Living organisms maintain an internal state that is constant, within certain limits, in the face of much larger changes that may occur in the environment. Living organisms also grow in size by synthesizing substances of complex molecular structures from simpler materials taken in as food. Living organisms also may produce mechanical movement, generate electrical potentials, and exhibit purposeful behavior and learning. The energy required to carry out these processes is furnished by sunlight for plants and by the metabolism of nutrients extracted from the surroundings for animals.

Two kinds of energy—heat and work—are recognized in thermodynamics. The amount of heat and work evolved or absorbed by a system when it changes from an initial to a final state depends on the nature of the system, on the initial and final states, and—within the limits set by the first and second laws of thermodynamics—on the process that produces the change. It is a common observation that certain changes may occur spontaneously in nature. Examples are expansion of a gas from a region of high pressure to one of low pressure, diffusion of a solute from a region of high concentration to one of low concentration, conduction of heat from a high temperature to a low temperature, and the occurrence of chemical reactions. Although such changes may occur by processes in which no work is done, it is always theoretically possible to obtain work energy from spontaneous processes by an appropriate mechanism. The reverse of any spontaneous change never occurs unless work energy is supplied from the outside. The maximum work that can be obtained from any spontaneous change may be calculated from the thermodynamic properties of the initial and final states; however, this amount of work can be recovered only by a hypothetical reversible process. Therefore, any actual process that occurs in nature always results in the conversion of some work energy to heat or in the loss of some potential for doing work. These observations are included in a more explicit and quantitative manner in the second law of thermodynamics. It is possible to convert some heat energy to work, but only within the limitations set by the second law. A machine for converting heat to work is called a heat engine.

The previously listed activities of living organisms are all nonspontaneous changes that require work energy to be supplied. However, living organisms are not heat engines; they cannot convert heat energy to work. The work energy needed to maintain life, therefore, must be extracted from the environment by harnessing spontaneous changes that occur there. Any heat produced in this process is lost to the organism as far as the generation of work is concerned. Many organisms, particularly the higher plants and animals, obtain work energy from the oxidation of nutrients such as carbohydrates and fats with molecular oxygen to produce, principally, carbon dioxide and water. This same result can be obtained by directly burning these materials in air. Some of the heat produced in this manner could be converted to work by use of a heat engine, such as a steam engine, but a living organism cannot operate in this manner. The oxidation of nutrients is brought about through a long sequence of intermediate reactions, and work energy is extracted at several places in this sequence.

Work is recovered from spontaneous chemical reactions in living organisms by coupling them to nonspontaneous processes. This coupling is the result of certain common intermediates in the mechanisms of these reactions. The major portion of work energy in all organisms is funneled through the production of adenosine triphosphate (ATP) from the reaction of adenosine diphosphate (ADP) and phosphate ions. The reversal of this reaction then produces the work needed to carry out the functions characteristic of life. The elucidation of the sequence of transformations that result in the production of adenosine triphosphate has been accomplished during the past several decades and represents one of the triumphs of modern biochemistry. The entire sequence of reactions and their associated enzymes is discussed in most modern textbooks on biochemistry. [Fruton and Simmonds (1960) is a typical example.]

The production of adenosine triphosphate is coupled to several of the spontaneous reactions of these sequences. However, only a portion of the theoretically available work energy is recovered, since many of the other steps proceed without coupling. Of course, some irreversibility is necessary if the reactions are to proceed at an appreciable rate. The rates of most of the reactions in these sequences are controlled by appropriate enzymes. In this way control of the rate of energy production is attained at many of the noncoupled reaction steps.

Certain quantitative aspects of adenosine triphosphate production are illustrated in Table I. Shown in the first six columns are the number of moles of various products formed from 1 mole of glucose at several stages in the sequence of reactions of intermediary metabolism. The next two columns list the changes in Gibbs energy and enthalpy for the transformation of 1 mole of glucose, present in the initial solution described at the bottom of the table, to the final states indicated. These quantities have been calculated from published data. $-\Delta G$ is equal to maximum work that could be obtained from the reaction, provided some means of carrying it out reversibly could be devised. ΔH is the heat absorbed when the reaction is carried out irreversibly. The next two columns show the number of moles of either adenosine triphosphate or guanosine triphosphate formed from phosphoric acid and adenosine diphosphate or guanosine diphosphate, respectively. The formation of these two compounds is coupled at various stages to other reactions and serves to recover some of the work energy made available. Under the conditions listed in Table I, the formation of 1 mole of adenosine triphosphate from adenosine diphosphate requires 11.2 kcal of work. The total work recovered in this way is shown in the last column. It was assumed that the work required to form guanosine triphosphate is the same as that required for adenosine triphosphate. The fact that ΔG for the formation of glycogen and the glucose phosphates are positive with respect to glucose means that work energy can be obtained from the formation of glucose from these compounds. Also, an additional mole of ATP is produced in the sequence of reactions when starting from glycogen—as compared to starting from glucose.

TABLE I

THERMODYNAMICS OF CARBOHYDRATE METABOLISM[a]

	Number of moles of products formed at various stages					Kcal $mole^{-1}$ of glucose		Number of moles of triphosphates		Work energy recovered
	$H_2O(1)$	CO_2 0.02 *m*	$NADH_2$ 10^{-3} *m*	$NADPH_2$ 10^{-1} *m*	FH_2	ΔG	ΔH	ATP	GTP	
Glycogen (*s*)	1	0	0	0	0	+7.3		−1	0	−11.2
1 Glucose-1-phosphate (2 × 10^{-4} *m*)	1	0	0	0	0	+6.7		−1	0	−11.2
1 Glucose-6-phosphate (2 × 10^{-4} *m*)	1	0	0	0	0	+4.9		−1	0	−11.2
1 Glucose (10^{-2} *m*)	0	0	0	0	0	0	0	0	0	0
2 Pyruvic acid (2 × 10^{-4} *m*)	0	0	2	0	0	−42.9		2	0	22.3
2 Lactic acid (2 × 10^{-4} *m*)	0	0	0	0	0	−48.9	−26.2	2	0	22.3
2 Acetaldehyde (2 × 10^{-4} *m*)	0	2	2	0	0	−49.7		2	0	22.3
2 Ethanol (2 × 10^{-4} *m*)	0	2	0	0	0	−60.4	−30.1	2	0	22.3
2 Acetyl-coenzyme A (2 × 10^{-4} *m*)	0	2	4	0	0	−51.6		2	0	22.3
	−6	6	8	2	2	−70.5		2	2	44.7
	6	6	0	0	0	−686.7	−692.8	36	2	424.5

[a] All dissolved species at equilibrium in buffer solution at pH = 7 containing 0.02 *m* Mg^{+2} ions. Thermodynamic quantities are relative to a reference state consisting of a solution of 0.01 *m* glucose, 10^{-3} *m* NAD, 10^{-3} *m* NADP, 10^{-3} *m* flavin, 10^{-3} *m* Coenzyme A, 10^{-3} *m* ADP, 10^{-3} *m* GDP and 10^{-3} *m* H_3PO_4, and oxygen (*g*, 0.2 atm).

The reaction can take place spontaneously only if there is a decrease in the Gibbs energy. Therefore, the sum of the seventh and the last columns in Table I should decrease for each successive step in a sequence of reactions. This is seen to be true. The concentrations chosen to calculate the data in Table I are somewhat representative of the concentrations that might be present in living cells. However, the actual state of these compounds at the place where the reactions take place is not known and, in fact, undoubtedly changes as the activity of the cell changes. In most cases the value of ΔG for the reactions as they occur in the cell probably do not differ by more than a few kilocalories from those shown in Table I.

A similar analysis of the metabolism of trimyristin, a typical fat, is shown in Table II. An increase in Gibbs energy is shown on going from the second to the third state, even considering the three moles of adenosine triphosphate that are consumed. Since the reaction does take place in the living organisms, this indicates that the choice of states does not very well represent the actual states present. Because of the large number of moles of acetyl coenzyme A formed at this step, the value of ΔG for the reaction is quite sensitive to the pH of the medium and to the concentration of the products.

Tables I and II show the thermodynamic quantities per mole of starting materials. It is interesting to compare the amount of work energy recovered by the organism in the form of ATP for the complete oxidation of 1 gm of starting material to carbon dioxide and water. These values are 2.36 kcal gm^{-1} of glucose, 2.69 kcal gm^{-1} of glycogen, and 5.11 kcal gm^{-1} of trimyristin. In each case this is about 60% of the total work energy that could be recovered in a completely reversible oxidation.

Tables I and II illustrate one way of using thermodynamic data to obtain quantitative information about energy transfer in a living organism. However, this kind of calculation does not tell the whole story. A living organism is a dynamic system. A more complete picture would be given by the knowledge of the rate of energy production and consumption that occurs in the organism at each step in the sequence of metabolic reactions. This, in turn, requires the application of both thermodynamic and kinetic methods of analysis. In principle, each irreversible process, such as chemical reactions, diffusion, electrical current flow, and fluid flow, can be described by a differential equation. All these equations, along with relations that represent control mechanisms, interactions among different process, and boundary conditions, would form a system of simultaneous differential and algebraic equations. The mathematical solution of this system of equations would furnish a complete detailed prediction of the behavior of the organism, including energy utilization, under any stated external conditions. It is unlikely that such a complete mathematical description of a living cell, together with all the complex structural details, will be possible. However, it may be possible to construct

TABLE II

THERMODYNAMICS OF FAT METABOLISM[a]

	H_2O(1)	CO_2 0.02 *m*	Glycerol 10^{-4} *m*	$NADH_2$ 10^{-3} *m*	Kcal mole^{-1} of trimyristin		Number of moles of triphosphate		Work energy recovered
					ΔG	ΔH	ATP	GTP	
1 Trimyristin (*s*)	18	0	0	0	0	0	0	0	0
3 Myristic Acid (10^{-5} *m*)	15	0	1	0	−54.1	−34.0	−0	0	0
21 Acetyl Coenzyme A (10^{-4} *m*)	0	0	1	36	+198.3		−3	0	−33.5
	57	42	1	0	−6149.5	−6242.8	354	21	4193

[a] All dissolved species are at equilibrium in buffer solution at a pH of 7 and contain 0.02 *m* Mg^{+2} ions. Thermodynamic quantities are relative to a reference state consisting of solid trimyristin, a solution of 10^{-3} *m* NAD, 10^{-3} Coenzyme A, 10^{-3} *m* ADP, 10^{-3} *m* GDP and 10^{-3} *m* H_3PO_4, and oxygen (*g*, 0.2 atm).

simplified mathematical models that will elucidate certain aspects of the behavior of living organisms. Recent progress in these endeavors has been reviewed in the symposium edited by Chance, Estabrook, and Williamson (1965).

II. Review of Thermodynamics

Classical thermodynamics consists of logical deductions, largely in mathematical form, from the three laws of thermodynamics. The principal use of thermodynamics in biochemistry, as in other applications, is the derivation of useful mathematical relations among the properties of the system being studied. These equations in themselves do not produce numerical values of properties, yet they can be used to generate a large amount of information from a comparatively small amount of data obtained by experimental measurement or by the use of nonthermodynamic theories. The equations of classical thermodynamics are almost entirely restricted to the description of systems at equilibrium or to the changes from an initial to a final state of equilibrium. However, many of the variables that have been found useful for the description of systems at equilibrium are also useful for the description of nonequilibrium states. Also, many thermodynamic equations can be modified or generalized to apply to nonequilibrium states, at least within certain limitations. An extension of classic thermodynamics called the "thermodynamics of irreversible processes" has been developed to deal explicitly with nonequilibrium states. The evaluation of changes in enthalpy and Gibbs energy, and related quantities such as equilibrium constants and electrical cell potentials, are essential to the quantitative study of the energetics of living organisms. The two main aspects of this calculation are the reduction of experimental measurements made under controlled conditions to standard thermodynamic properties that can be conveniently tabulated and the conversion of tabulated data to properties of real systems of interest. Living organisms consist of complex multi-component reacting systems. The calculation of the thermodynamic properties of such systems presents a formidable challenge.

This section is a condensed review of thermodynamic calculations of the type required for biochemical applications. It consists of a collection of formulas and computational procedures arranged in a logical manner for ready reference, for use in interpreting calorimetric and related measurements. It is not intended as a complete discussion of the principles of thermodynamics and does not include complete derivations of many of the equations. For additional background information and detailed derivations, the reader should consult one of the many standard textbooks on chemical thermodynamics.

In vitro studies of the thermodynamic properties of biochemically important compounds and reactions are usually in terms of equilibrium states, and therefore classical thermodynamic formulas are applicable. Comparison of the thermodynamic properties of substances ingested by living organisms with those excreted give much valuable insight into the energetic transformations that take place. Nevertheless, since living organisms are not themselves at equilibrium, the limitations of classical thermodynamics must be kept in mind when applying these principles to the description of processes that occur in living organisms. Although the properties of equilibrium states may be taken as a starting point, the complete description of those processes characteristic of the living state requires a combination of thermodynamic and kinetic theories.

The properties of systems in equilibrium are described in terms of variables of state, such as temperature, pressure, volume, concentrations, and various other derived properties. When a system changes from an initial to a final state, some or all of these variables undergo a change that is characteristic of the initial and final states but is independent of the intermediate states or of the process which produces the change. The word "change" will be used to designate some particular initial and final state, irrespective of how the change occurs. The word "process" implies a particular procedure for getting from the initial to the final state. Although changes in variables of state depend only on the change that occurs in the system, the exchange of heat and work between the system and its surroundings depends on the particular process that takes place. Changes in variables of state are indicated by placing a Δ in front of the symbol, while differentials are indicated by the letter d. Differentials of variables of state are exact, in the mathematical sense. The heat and work transferred to a system are designated by q and w, respectively. These are considered as positive when energy is added to the system and negative when energy is given off to the surroundings. Heat and work transferred in infinitesimal change are designated by δq and δw. These differentials are inexact in the mathematical sense. Thermodynamic equations can be written in the form of finite differences or in the form of differentials.

A. The First Law

The first law of thermodynamics relates the change in internal energy when a system changes from an initial to a final state to the amount of heat and work transferred to the system [Eq. (1)]. The heat q is transferred by con-

$$\Delta E = E_f - E_i = q + w \tag{1}$$

duction from a high to a low temperature, while the work w is transferred by one or more forces that displace some quantity. In the applications of interest here, work is transferred either by expansion or compression of the system [Eq. (2)] or by transfer of electrical charges to or from the system [Eq. (3)].

$$w = -\int P dV \tag{2}$$

$$w = -\int \epsilon dQ \tag{3}$$

Work may be transferred by either a reversible or an irreversible process. In either case the pressure P, or electrical potential, is that which acts on the surroundings. If the process is reversible, P or ϵ are the values for the system at equilibrium. Equation (1) states that the internal energy is a variable of state, so that ΔE depends only on the initial and final states of the system.

The enthalpy is another variable of state defined by Eq. (4). If the system

$$H = E + PV \tag{4}$$

changes by a process in which work is zero, then $\Delta E = q$. However, if $w = \Delta(PV)$, then $\Delta H = q$. The most common example of this process is one in which only reversible expansion work is done at constant pressure.

The heat capacity of the system is defined as the heat absorbed divided by the temperature change when it undergoes an infinitesimal change under some specified conditions. The two most important types of heat capacity are those at constant volume and at constant pressure [Eqs. (5) and (6)].

$$C_v = \frac{\delta q_v}{dT} = \left(\frac{\partial E}{\partial T}\right)_V \tag{5}$$

$$C_p = \frac{\delta q_p}{dT} = \left(\frac{\partial H}{\partial T}\right)_P \tag{6}$$

B. Calorimetric Experiments

A thermodynamic description of a calorimetric experiment may be given by considering the calorimeter and its contents to be the system and the jacket and other parts of the apparatus to be the surroundings. When the calorimeter and its contents change from an initial to a final state, we may write Eq. (7) for a

$$q + w' = \Delta E_v \tag{7}$$

constant volume calorimeter and Eq. (8) for a constant pressure calorimeter.

$$q + w' = \Delta H_p \tag{8}$$

w' is work, other than compression work, transferred to the calorimeter during experiment. Usually w' represents electrical energy used to operate an electrical heater, but it also can represent mechanical energy transferred by stirring or other means. q represents the heat transferred by conduction, convection, and radiation from the jacket to the calorimeter.

One type of calorimetric measurement is made to determine the change of enthalpy or energy for a chemical or physical change at constant temperature. Typical changes studied are chemical reactions, changes of phase, or mixing of components to form a solution. Let T_i and T_f be the initial and final temperatures of the calorimeter in such an experiment, and let C_o, C_i, and C_f be the average heat capacities of the calorimeter and the contents before and after the change takes place, respectively. Then, for a constant pressure calorimeter,

$$q + w' = C_o(T_f - T_i) + C_i(T_r - T_i) + C_f(T_f - T_r) + \Delta H_r \qquad (9)$$

where ΔH_r is the change in enthalpy of the contents of the calorimeter at the reference temperature T_r. Another type of experiment may be done to determine the enthalpy of the contents of the calorimeter as a function of temperature. If H_i and H_f represent the enthalpy at the initial and final temperature at the same pressure, then

$$q + w' = C_o(T_f - T_i) + H_f - H_i \qquad (10)$$

If the change in temperature is sufficiently small, the result may be expressed as a heat capacity [Eq. (11)]. Analogous equations, in terms of internal energy

$$C_p = \frac{H_f - H_i}{T_f - T_i} \qquad (11)$$

and constant volume heat capacity, may be written for constant volume calorimeters. The heat capacity of the empty calorimeter C_o is calculated from a separate calibration experiment that is conducted with a standard substance in the calorimeter in as nearly the same manner as possible as the measurement experiment.

It is usually possible to measure the work energy transferred to the calorimeter with high accuracy, but the determination of heat to transfer is much more difficult. Most calorimeters are designed either to reduce the heat exchange as much as possible or to keep it the same for the calibration and the measurement experiments.

C. Thermochemical Calculations

It is sometimes desirable to convert a change in energy measured at constant volume to a change in enthalpy measured at constant pressure, or vice versa. Strictly speaking, these two quantities do not refer to the same change since the initial and final states for the constant pressure change cannot be the same as those for the constant volume change. However, if a gas is present in either or both the initial or final states, the relation (12) is a good approximation.

$$\Delta H_p = \Delta E_v + \Delta n RT \tag{12}$$

Δn is the change in number of moles of gas resulting from the transformation.

The change in ΔH for a chemical reaction produced by a change in temperature may be calculated from the constant pressure heat capacities by Eq. (13),

$$\Delta H(T_2) = \Delta H(T_1) + \int_{T_1}^{T_2} \Delta C_p \, dT \tag{13}$$

where ΔC_p is the total heat capacity of the products of the reaction minus the total heat capacity of the reactants. Since heats of reaction are not ordinarily strongly dependent on temperature, it is usually sufficient to consider that ΔC_p is constant when calculating the change ΔH resulting from a temperature change of around 50° or less.

Since E and H are both variables of state, the ΔE's or ΔH's for a series of intermediate steps may be added together to obtain the ΔE or ΔH for the net overall change. This is very important in thermochemical calculations since the changes in enthalpy for a large number of reactions may be calculated from a relatively small number of measured values. For example, the heat of a chemical reaction can be calculated from the heat of formation of the reactants and

$$\nu_1 A + \nu_2 B \rightarrow \nu_3 C + \nu_4 D$$

products [Eq. (14)].

$$\Delta H(\text{reaction}) = \nu_3 \Delta Hf(C) + \nu_4 \Delta Hf(D) - \nu_1 \Delta Hf(A) - \nu_2 \Delta Hf(B) \tag{14}$$

The heat of a chemical reaction is dependent upon the physical states (gas, liquid, solid, and solution) of the reactants and products, as well as upon conditions such as temperature, pressure, and concentrations. The superscript *o* written after the symbol of a thermodynamic property indicates that it represents the substance in the standard state. The choice of standard states is purely arbitrary, but certain conventions have become commonly accepted. The standard states for pure solids and liquids are taken at a pressure of 1 atm. The standard state of a gas is the hypothetical ideal gas state at 1 atm. Standard states for solutions will be considered later.

Heats of formation of organic compounds are usually calculated from measured heats of combustion, rather than by direct reaction of the elements. The idealized combustion of a compound containing carbon, hydrogen, oxygen, and nitrogen is

$$C_aH_bO_cN_d(s, \text{l, or } g) + (a + \tfrac{1}{4}b - \tfrac{1}{2}c)\,O_2(g) \rightarrow aCO_2 + \tfrac{1}{2}bH_2O(\text{l}) + \tfrac{1}{2}dN_2(g)$$

Such reactions are carried out in a constant-volume oxygen-bomb calorimeter. The heat evolved is equal to $-\Delta Ec/n$, where n is the number of moles burned.

This may be converted to ΔHc by use of Eq. (12). However, the conversion of the measured heat of combustion, for the reaction that actually occurs in the bomb, to the standard state value requires the application of numerous corrections. Many laboratories now report heats of combustion accurate to one part in ten thousand. The heat of formation may be calculated from the heat of combustion by

$$\Delta Hf(C_aH_bO_cN_d) = a\Delta Hf(CO_2, g) + \tfrac{1}{2}b\Delta Hf(H_2O, 1) - \Delta Hc(C_aH_bO_cN_d) \quad (15)$$

D. The Second and Third Laws of Thermodynamics

The second law evolved about a century ago from attempts to determine the maximum possible efficiency of engines for converting heat energy to work. Its principal value in thermodynamics is in distinguishing processes that may occur spontaneously in nature from those that are impossible. The second law may be stated in several logically equivalent ways, but the following three statements, which include a definition of entropy and temperature, are convenient for practical calculation:

1. The differential of entropy S is defined as Eq (16), where δq_r is the heat

$$dS = \frac{\delta q_r}{T} \quad (16)$$

   absorbed by the system when it undergoes an infinitesimal change by a reversible process.
2. The entropy is a variable of state, and thus dS is an exact differential. This implies that $1/T$ is an integrating factor for the inexact differential δq_r.
3. When a change takes place in a completely isolated system, $dS = 0$ if it occurs by a reversible process and $dS > 0$ if it occurs by an irreversible (or spontaneous) process. This implies that the entropy of an isolated system can never decrease and that it is a maximum in a state of equilibrium.

In general the total change in entropy of any system may be represented as Eq. (17), where dS_e is the change in entropy that results from a reversible

$$dS_t = dS_i + dS_e \quad (17)$$

transfer of heat or matter to or from the system (this may be positive or negative) and dS_i is the increase in entropy (always positive) resulting from the irreversible processes that occur in the system. The change in the entropy of a system may be calculated by applying Eq. (16) to a reversible path from the initial to the final state. The change in entropy is the same for any type of process, however. For a reversible process at constant pressure such as a phase change, Eq. (16) gives Eq. (18), where T is the equilibrium temperature.

$$\Delta S = \frac{\Delta H}{T} \tag{18}$$

For a nonisothermal change at constant pressure,

$$S(T_2) - S(T_1) = \int_{T_1}^{T_2} \frac{\delta q_r}{T} = \int_{T_1}^{T_2} \frac{C_p}{T}\, dT \tag{19}$$

The third law of thermodynamics states that the entropy of any perfectly crystalline solid is zero at absolute zero. Thus the absolute entropy of a solid, liquid, or gas at any given temperature may be calculated by substituting values of heat capacity measured down to near absolute zero in Eq. (19) and observed heats of phase changes in Eq. (18). Since experimental measurements cannot be made all the way to absolute zero, the heat capacity must be extrapolated from the lowest experimental temperature to zero degrees. In recent years absolute entropies of organic compounds are routinely calculated from experimental measurements down to 15–20°K (boiling point of liquid hydrogen).

The requirement that dS_i in Eq. (17) be greater than zero implies that the heat absorbed by a system when it undergoes a certain isothermal change by a reversible process be algebraically greater than the heat absorbed when it undergoes the same change by any irreversible process. Consequently, the work done on the system during the reversible process is algebraically less than the work done on it during any irreversible process. Therefore, the work done by the system on the surroundings when the change is brought about reversibly is always greater than it is when the change is brought about irreversibly. Thus an irreversible process results in the loss of potentially available work. Let $-w_r$ be the reversible work done by the system, $-w_i$ the irreversible work done, and ΔS_i the increase in entropy due to irreversible process. Then

$$(-w_r) - (-w_i) = T\Delta S_i \tag{20}$$

Equation (20) may be generalized to apply to a system in a nonequilibrium state by writing Eq. (21), where dS_i/dt is the rate of increase of entropy that results

$$\Phi = T\frac{dS_i}{dt} \tag{21}$$

from irreversible processes within the system. The dissipation function, Φ, plays a key role in the thermodynamics of irreversible processes. Equation (20) implies that the dissipation function is the rate at which a system loses the ability to generate work. Thus it is an important consideration in the thermodynamics of living organisms.

Two additional thermodynamic functions that also are variables of state are useful mainly because of their relation to reversible work. The Helmholtz energy is defined as

$$A = E - TS \tag{22}$$

The change in Helmholtz energy when a system changes reversibly at constant temperature is

$$\Delta A = \Delta E - T\Delta S = q_r + w_r - T\Delta S = w_r \tag{23}$$

ΔA may be used to predict the possibility of occurrence of a change in state by a process in which no work is done. If ΔA is negative, the work done in this process (zero) is greater than the reversible work, and the process is an irreversible or spontaneous one. If ΔA is positive, the change cannot occur unless work is done on the system. Therefore, the process in which no work is done is impossible, or forbidden, by the second law. If ΔA is zero, the change can occur by a reversible process, and the initial and final states are at equilibrium under these conditions.

The Gibbs energy is defined as in Eq. (24). The change in Gibbs energy when

$$G = H - TS \tag{24}$$

a system undergoes a change at constant temperature and pressure is given in Eq. (25), where w_r' is the net reversible work, or work other than compression–

$$\begin{aligned} \Delta G &= \Delta H - T\Delta S = q_r + w_r + P\Delta V - T\Delta S \\ &= w_r + P\Delta V = w_r' \end{aligned} \tag{25}$$

expansion work. Since most laboratory experiments are done at constant temperature and pressure and the net reversible work that can be done by the system is of more interest than the total work, the Gibbs energy is of more general use in calculations than is the Helmholtz energy. For a change at constant temperature and pressure by a process in which only compression–expansion work is done, it follows that the process is spontaneous if ΔG is negative, reversible if ΔG is zero, and forbidden if ΔG is positive.

The four differential equations, (26)–(29), which relate the basic thermodynamic variables of state, can be used to generate many useful relations.

$$dE = TdS - PdV \tag{26}$$

$$dH = TdS + VdP \tag{27}$$

$$dA = -SdT - PdV \tag{28}$$

$$dG = -SdT + VdP \tag{29}$$

A great deal of information about a chemical reaction may be obtained from the value of ΔG on going from reactants to products. It is the negative of the maximum net work that could be obtained if the reaction is carried out reversibly. If ΔG is positive, then work must be done on the reaction to make it proceed at constant temperature and pressure. It is related to the change in enthalpy and entropy by Eq. (30). According to the previous discussion,

$$\Delta G = \Delta H - T\Delta S \tag{30}$$

ΔH is the heat absorbed by the reaction when it is conducted irreversibly at constant pressure and $T\Delta S$ is the heat absorbed when it is conducted reversibly at constant pressure. Both ΔH and ΔS may be derived from appropriate calorimetric measurements. ΔG also may be obtained by a direct measurement of the net reversible work. This is most often done by causing the reaction, to occur in an electrochemical cell in which oxidation occurs at the anode and reduction at the cathode. If ϵ is the reversible potential of the cell measured at zero current, n is the number of equivalents of electricity transferred from cathode to anode by the reaction, and $\mathscr{F}$ the Faraday constant, then

$$\Delta G = -n\mathscr{F}\epsilon \tag{31}$$

The change in Gibbs energy and entropy for a chemical reaction may be calculated from the sum of the ΔG's and ΔS's of the individual steps leading to the final reactions in a manner similar to that previously described for ΔH. The effect of temperature on the heat of reaction may be represented by Eq. (13). Corresponding equations for the change in entropy and Gibbs energy are given below. They can be readily derived from Eqs. (19)–(29).

$$\left(\frac{\partial \Delta S}{\partial T}\right)_P = \frac{\Delta C_p}{T} \tag{32}$$

$$\Delta S(T_2) - \Delta S(T_1) = \int_{T_2}^{T_1} \frac{\Delta C_p}{T}\, dT \tag{33}$$

$$\left(\frac{\partial \Delta G}{\partial T}\right)_P = -\Delta S \tag{34}$$

$$\left[\frac{\partial\left(\frac{\Delta G}{T}\right)}{\partial T}\right]_P = -\frac{\Delta H}{T^2} \tag{35}$$

$$\frac{\Delta G(T_2)}{T_2} - \frac{\Delta G(T_1)}{T_1} = -\int_{T_1}^{T_2} \frac{\Delta H}{T^2}\, dT \tag{36}$$

If ΔC_p is considered to be constant for the reaction, Eq. (36) can be integrated with the help of Eq. (13) to give Eq. (37). Analogous equations for calculating

$$\Delta G(T_2) = \Delta G(T_1)(T_2/T_1) + \Delta H(T_1)(1 - T_2/T_1) + \Delta C_p\{T_2[1 - \ln(T_2/T_1)] - T_1\} \tag{37}$$

the effect of temperature on a reversible electrochemical cell potential ϵ may be obtained by substituting ΔG from Eq. (31) into Eqs. (34)–(37).

Gibbs energies of formation ΔGf^o, absolute entropies S^o, and heat capacities $C_p{}^o$ are often tabulated for pure compounds in the same standard state that have been already described for heats of formation. Data of this type at 25°C will be found listed in Table I, Appendix. These may be used to calculate changes in Gibbs energy, entropy, or heat capacity for chemical reactions by equations analogous to (14). Values of ΔH^o, ΔG^o, and ΔS^o at other temperatures in the vicinity of 25°C may be calculated through the use of Eqs. (13), (33), (34), (36), and (37). Values of $(G^o - H_0{}^o)/T$ and $(H^o - H_0{}^o)/T$ are frequently tabulated for individual compounds over a wide range of temperature. They may be combined to generate heats and Gibbs energy changes for chemical reactions at any temperature within this range.

$$\Delta H^o = T\Delta\left(\frac{H^o - H_0{}^o}{T}\right) + \Delta H_0{}^o \tag{38}$$

$$\Delta G^o = T\Delta\left(\frac{G^o - H_0{}^o}{T}\right) + \Delta H_0{}^o \tag{39}$$

$\Delta H_0{}^o$ is the change in enthalpy for the reaction at absolute zero temperature.

Irreversible processes may be described in terms of appropriate flows of energy or matter. Each such process is governed by a corresponding force. The appropriate force for chemical reactions at constant temperature and pressure is $-\Delta G$. For this reason $-\Delta G$ is sometimes called the "affinity" and given the symbol A. The dissipation function defined by Eq. (20) is equal to the product of the flow and its conjugate force. If R is taken as the rate of a chemical reaction that occurs in a homogeneous system and is given in terms of the moles of product formed per unit time, then the dissipation function for the reaction is Eq. (40). The second law requires that this be positive, or zero.

$$\Phi = R(-\Delta G) \tag{40}$$

The total dissipation function for a system that contains two or more reactions in progress is equal to the sum of corresponding products for each such reaction [Eq. (41)]. If each of these reactions are physically independent of the

$$\Phi = \sum_i R_i(-\Delta G_i) \tag{41}$$

others, each term in Eq. (41) also must be positive. Nevertheless, if some of the reactions are coupled, it is possible that some of the terms will be negative, since the second law requires only that the total dissipation function be positive. Since a positive value of $R_i(-\Delta G_i)$ means that the potential for performing work is being lost, a negative value means that some work energy is being recovered. In systems containing coupled processes, those that have positive values of $R_i(-\Delta G_i)$ are said to be the "driving" processes and those with negative values, the "driven" processes. Coupling may occur not only among chemical reactions but also among processes such as diffusion, electrical current flow, and heat flow. The relations between flows and forces and the phenomena that results from coupling of rate processes are described by the equations of the thermodynamics of irreversible processes. An introduction to this subject is given in the treatises by Prigogine (1955), Fitts (1962), and Katchalsky and Curran (1965). These equations provide the framework for keeping track of the transfer of work energy in living organisms. They do not, however, give direct information about the molecular mechanisms responsible for the coupling phenomena.

E. Thermodynamics of Systems of Variable Composition

1. *Partial Molar Quantities*

The equations given up to this point apply to systems of constant composition. Additional variables are needed to represent the properties of systems of variable composition. If J represents any of the extensive variables of state of a system composed of k independent components, then the total differential of J may be represented as in Eq. (42). The derivatives with respect

$$dJ = \left(\frac{\partial J}{\partial T}\right)_{P,n_i} dT + \left(\frac{\partial J}{\partial P}\right)_{T,n_i} dP + \sum_{i=1}^{k} \left(\frac{\partial J}{\partial n_i}\right)_{T,P,n_j} dn_i \tag{42}$$

to temperature and pressure are taken with the number of moles of all components constant. The derivatives with respect to each of the components n_i are taken at constant temperature and pressure and with a constant number of moles of all components except the ith component. The number of terms in Eq. (42) is greater by one than the number of degrees of freedom, in the sense used in the Gibbs phase rule. The partial molal quantity for component i corresponding to J is then defined as in Eq. (43), where, again, the symbol n_j

$$\bar{J}_i = \left(\frac{\partial J}{\partial n_i}\right)_{T,P,n_j} \tag{43}$$

indicates that all the $k - 1$ components other than the ith one are held constant. Thus the differential of the total J for the system, at constant temperature and pressure, in terms of the moles of the k components is Eq. (44). Since the $\bar{J}_i$'s

$$dJ = \sum_{i=1}^{k} \bar{J}_i \, dn_i \tag{44}$$

are intensive quantities that are functions only of the temperature, pressure, and composition, Eq. (45) may be integrated at constant composition to give Eq. (45), which gives the value of the total J for the system in terms of the partial molal J and the number of moles of the independent components in the system.

$$J = \sum_{i=1}^{k} \bar{J}_i n_i \tag{45}$$

It should be emphasized again that the number of terms in Eqs. (44) and (45) is equal to the number of independent components. This is the number of components whose concentration can be varied independently of the concentration of the other components. Furthermore, partial molal quantities can be defined only for components that can be isolated in a pure state and added to the system as such, either directly or indirectly. In general the partial molal quantities have no simple relationship to the properties of the molecular species actually present in the solution. To illustrate the significance of these remarks, consider a solution composed of a solvent and three components that participate in a chemical reaction represented by A $\rightarrow$ B + C. Assume that the concentrations of A, B, and C correspond to equilibrium values, with respect to this reaction. If the reaction can now be stopped, partial molal quantities can be defined and measured for the solvent and for each of the three solutes, A, B, and C. This system now has four independent components. However, if the reaction is continued so that equilibrium is maintained, the system has only three independent components. Any three components that can be obtained in a pure state and can be mixed to produce the same state at equilibrium can be chosen as the independent ones. This means, e.g., that the partial molal quantity of component A, when measured with the reaction stopped, may not be the same as the partial molal quantity of A with the reaction present. If the additional restriction, that the number of moles of B is equal to the number of moles of C, is placed on the system, then it will have only two independent components. This is the case if B and C are formed only by the dissociation of A. Thus the physical significance of partial molal quantities in systems that may include chemical reactions or other possible restrictions must be interpreted with care since different values and different meanings may be associated with the same symbol, depending upon whether the restriction is kept in force when the component is added to the system. Another important conclusion that follows is that partial molal quantities cannot be defined or evaluated, without further conventions or assumptions, for individual ionic species. This is because such species cannot be isolated in a pure state and added to the system while holding other components constant.

Taking the total differential of (45) and comparing the result with (44) shows that

$$\sum_{i=1}^{k} n_i \, d\bar{J}_i = 0 \tag{46}$$

which is a general form of the Gibbs–Duhem equation. For a system composed of two components, Eq. (47) can be used to derive a relationship between the two corresponding partial molal quantities. In Eq. (47) com-

$$d\bar{J}_2 = -\frac{n_1}{n_2}\,d\bar{J}_1 = -\frac{X_1}{X_2}\,d\bar{J}_1 = -\frac{1000}{M_1\,m}\,d\bar{J}_1 \tag{47}$$

ponent 1 is the solvent, component 2 is the solute, X_1 and X_2 are the mole fractions, m is the molality, and M_1 is the molecular weight of the solvent. Integration of Eq. (47) between the limits of the final solution and the pure solvent gives Eq. (48), where $\bar{J}_2{}^\circ$ refers to the solute at infinite dilution. Since

$$\bar{J}_2 - \bar{J}_2{}^o = -\int \frac{n_1}{n_2}\,d\bar{J}_1 = -\int \frac{X_1}{X_2}\,d\bar{J}_1 = -\frac{1000}{M_1}\int \frac{d\bar{J}_1}{m} \tag{48}$$

the functions being integrated in Eq. (48) approach infinity at the lower limit, special procedures are used to evaluate the integral. Partial molal quantities for the solvent may be evaluated from a knowledge of the partial molal quantities for the solute in an analogous manner.

Apparent molal quantities are more directly related to most experimental measurements than are partial molal quantities. They are defined by Eqs. (49) and (50), where J is the total value of the property for a solution composed on n_1

$$\phi J_1 = \frac{J - n_2 J_2}{n_1} \tag{49}$$

$$\phi J_2 = \frac{J - n_1 J_1}{n_2} \tag{50}$$

moles of solvent and n_2 moles of solute, J_2 is the value of the property for 1 mole of pure solute, and J_1 is the value for 1 mole of pure solvent. Partial molal quantities may be calculated from apparent molal quantities by Eqs. (51) and (52). J in Eqs. (42)–(52) may be directly replaced by V, C_p, or S. However,

$$\bar{J}_1 = -m^2 \frac{d(\phi J_1/m)}{dm} \tag{51}$$

$$\bar{J}_2 = \frac{d(m\phi J_2)}{dm} \tag{52}$$

equations for E, H, A, and G only have significance relative to some standard state.

2. *Heats of Solution and Heats of Reactions in Solution*

Heat effects in solution are represented by relative enthalpies. The total relative enthalpy of a solution is given in Eq. (53), and the relative partial molal

$$L = H - H^o \tag{53}$$

enthalpies are given in Eqs. (54) and (55), where the superscript *o* represents

$$\bar{L}_1 = \bar{H}_1 - \bar{H}_1{}^o \tag{54}$$

$$\bar{L}_2 = \bar{H}_2 - \bar{H}_2{}^o \tag{55}$$

the appropriate standard state. The standard state for the solvent in aqueous solutions is commonly chosen as pure solvent, and the enthalpy of the solute in the standard state is equal to the enthalpy at infinite dilution. Equations (56)–(58) may be easily derived. These properties may be related to various types of mixing phenomena.

$$L = n_1 \bar{L}_1 + n_2 \bar{L}_2 \tag{56}$$

$$\phi L_1 = \frac{L - n_2 L_2}{n_1} = \phi H_1 - \phi H_1{}^o = \frac{n_2}{n_1}(\bar{L}_2 - L_2) + \bar{L}_1 \tag{57}$$

$$\phi L_2 = \frac{L - n_1 L_1}{n_2} = L/n_2 = \phi H_2 - \phi H_2{}^o = \bar{L}_2 + \frac{n_1}{n_2}\bar{L}_1 \tag{58}$$

a. Differential Heat Solution. An infinitesimal amount of solute is added to a solution composed of n_1 moles of solvent and n_2 moles of solute [Eq. (59)].

$$dn_2 \text{ solute} + (n_1 + n_2) \text{ solution} \rightarrow (n_1 + n_2 + dn_2) \text{ solution}$$

$$q_2 = \frac{\Delta H}{dn_2} = \bar{L}_2 - L_2 \tag{59}$$

b. Integral Heat of Solution. n_2 moles of pure solute are mixed with n_1 moles of pure solvent to give the final solution.

$$n_2 \text{ solute} + n_1 \text{ solvent} \rightarrow (n_1 + n_2) \text{ solution}$$

$$Q_2 = \frac{\Delta H}{n_2} = \phi L_2 - L_2 = \frac{H^E}{X_2} = \bar{L}_2 - L_2 + \frac{n_1}{n_2}\bar{L}_1 \tag{60}$$

$$Q_1 = \frac{\Delta H}{n_1} = \phi L_1 = \frac{H^E}{X_1} = \frac{n_2}{n_1}(\bar{L}_2 - L_2) + \bar{L}_1 \tag{61}$$

H^E is the excess enthalpy of mixing. If $n_1 = \infty$,

$$Q_2{}^o = \frac{\Delta H}{n_2} = -L_2 \tag{62}$$

c. Differential Heat of Dilution. An infinitesimal amount of solvent is added to the solution.

$$dn_1 \text{ solvent} + (n_1 + n_2) \text{ solution} \rightarrow (n_1 + dn_1 + n_2) \text{ solution}$$

$$q_1 = \frac{\Delta H}{dn_1} = \bar{L}_1 \tag{63}$$

d. Integral Heat of Dilution. n_1' moles of solvent are added to a solution composed of n_1 moles of solvent and n_2 moles of solute.

$$n_1' \text{ solvent} + (n_1 + n_2) \text{ solution} \rightarrow (n_1 + n_1' + n_2) \text{ solution}$$

$$\Delta Q_2 = \frac{\Delta H}{n_2} = \phi L_2' - \phi L_2 \tag{64}$$

$$\Delta Q_1 = \frac{\Delta H}{n_1} = \phi L_1' - \phi L_1 \tag{65}$$

If $n_1' = \infty$,

$$\Delta Q_2{}^o = -\phi L_2 \tag{66}$$

Quantities such as Q_2, Q_1, ΔQ_2, and ΔQ_1 can be obtained by direct calorimetric measurements. The other quantities must be calculated from these measured quantities. The differential heat of solution can be calculated from the integral heat of solution as in Eq. (67). According to (62) L_2 is approximately equal to the heat of solution of 1 mole of the solute in a large excess of water.

$$q_2 = -L_2 + \left(\frac{\partial(n_2 \phi L_2)}{\partial n_2}\right)_{n_1} = \frac{d(mQ_2)}{dm} \tag{67}$$

A somewhat more precise way of calculating these quantities from experimental measurements is to express the apparent relative molal enthalpy as some function of molality, such as a power series with integral or non-integral powers of m [Eq. (68)]. The series must approach zero at zero molality.

$$\phi L_2 = \sum_i a_i m^{p_i} \tag{68}$$

The coefficients in this series may be determined by a graphical or least square fit to a series of experimental heats of dilution from an initial concentration m to a final concentration m' [Eq. (69)]. The other quantities may then be calculated as in Eqs. (70) and (71). L_2 then may be calculated from one

$$\Delta Q_2 = \sum a_i (m'^{p_i} - m^{p_i}) \tag{69}$$

$$\bar{L}_2 = \phi L_2 + \sum p_i a_i m^{p_i} \tag{70}$$

$$\bar{L}_1 = -\frac{M_1 m}{1000} \sum p_i a_i m^{p_i} \tag{71}$$

observed integral heat of solution through the use of (60) and (68). The apparent molal enthalpy of the solvent is then obtained as

$$\phi L_1 = \frac{mM_1}{1000}(\phi L_2 - L_2) \tag{72}$$

Since,

$$\left(\frac{\partial \bar{H}_2}{\partial T}\right)_P = \bar{C}_{p2} \tag{73}$$

and

$$\left(\frac{\partial \bar{H}_1}{\partial T}\right)_P = \bar{C}_{p1} \tag{74}$$

the temperature derivatives of the relative partial molal enthalpies and the heats of solution and dilution are

$$\left(\frac{\partial \bar{L}_2}{\partial T}\right)_P = \bar{C}_{p2} - \bar{C}^o_{p2} \tag{75}$$

$$\left(\frac{\partial \bar{L}_1}{\partial T}\right)_P = \bar{C}_{p1} - C_{p1} \tag{76}$$

$$\left(\frac{\partial q_2}{\partial T}\right)_P = \bar{C}_{p2} - C_{p2} \tag{77}$$

$$\left(\frac{\partial Q_2}{\partial T}\right)_P = \bar{C}_{p2} - C_{p2} + \frac{1000}{M_1 m}(\bar{C}_{p1} - C_{p1}) \tag{78}$$

$$\left(\frac{\partial \Delta Q_2{}^\circ}{\partial T}\right)_P = \bar{C}^o_{p2} - \bar{C}_{p2} + \frac{1000}{M_1 m}(C_{p1} - \bar{C}_{p1}) \tag{79}$$

The relation of heat of solution to the temperature coefficient of solubility will be taken up in a later section.

The heat of a chemical reaction in solution at infinite dilution is related to the heats of formation of the reactants and products in solution in the standard state in the manner illustrated by Eq. (14). The heat of formation of a solute in the standard state solution is related to its heat of formation in the pure solid, liquid, or gaseous state by

$$\Delta Hf^\circ(\text{aq}) = \Delta Hf^\circ(s, l, \text{or } g) + Q_2{}^\circ = \Delta Hf^\circ(s, l, \text{or } g) - L_2 \tag{80}$$

The change in enthalpy when a reaction takes place under conditions other than the standard state can be represented by

$$\Delta H(\text{aq}) = \Delta H^\circ(\text{aq}) + \Delta L_i \tag{81}$$

where ΔL_i is the sum of terms such as $n_i L_i$ for the final solution, including the solvent, minus the sum of similar terms for the initial solution, or solutions if more than one solution is mixed to initiate the reaction.

In line with the remarks previously made about the partial molal properties of components of a solution, it follows that the heat of formation of individual ions of an electrolyte are not subject to direct measurement. However, since the enthalpy of the solute in the standard state is equal to its value at infinite dilution, it is reasonable to assume that, for an electrolyte, $M_{\nu+}X_{\nu-}$, which dissociates into ν_+ cations and ν_- anions, that

$$\Delta Hf^o(M_{\nu+}X_{\nu-}, aq) = \nu_+ \Delta Hf^o(M^{+x}, aq) + \nu_- \Delta Hf^o(X^{-y}, aq) \tag{82}$$

Standard state heats of formation of individual ions in aqueous solution are frequently tabulated. Usually these are subject to the convention that the standard heat of formation of hydrogen ions is zero. Thus the heat of formation of the anion X^{-y} is equal to the heat of formation of the acid $H_y X$. The heat of formation of the cation M^{+x} can then be calculated from that of the salt $M_{v+} X_{v-}$ by the use of Eq. (82). These conventional heats of formation should not be confused with the "true" heats of formation of aqueous ions, which are much more difficult to determine. The principal advantage in tabulating these data is that the standard state heats of formation of a large number of electrolytes may be recovered from a relatively small number of data on individual ions. The additivity of heats of formation of individual ions, as expressed in Eq. (82), is strictly valid only at infinite dilution; however, this relation is frequently adopted as an approximation for real solutions.

The heat of ionization of a weak acid may be obtained from the measurement of the heat of neutralization with a strong base. To illustrate this procedure consider that ΔHn is the change in enthalpy that results from the addition of 1 mole of a strong base to a solution containing 1 mole of a weak acid HA. The initial solution contains n_2 moles of undissociated acid and $(1 - n_2)$ moles of A^- anions and hydrogen ions. The final solution contains n_2' moles of undissociated acid. The quantities can be calculated from the dissociation constants of the acid. If it is assumed that the enthalpies of the various components of these solutions are the same as those at infinite dilution, then the heat of ionization of 1 mole of the acid is given by

$$\Delta Hi = \frac{\Delta Hn + (1 - n_2') \Delta Hw^o}{n_2 - n_2'} \tag{83}$$

ΔHw^o is the heat of ionization of water at infinite dilution.

Many reactions of importance in biochemistry involve weak acids in either or both the reactants and products. These reactions are frequently studied in buffer solution. Such reactions, when written in terms of the various ionic species present, may be very complex, and therefore the interpretation of the observed heat of reaction may be correspondingly difficult. To illustrate this situation consider the reaction HA(aq) → HC(aq), where HA and HC are weak acids. Let ΔHr represent the change in enthalpy for the conversion of 1 mole of undissociated HA to 1 mole of undissociated HC. If this reaction takes place in a solution containing a large excess of a buffer $HB + B^-$, such that the concentration of hydrogen ions remains constant, then the actual reaction is

$$n_2\text{HA} + (1 - n_2)\,\text{A}^- + n_3\,\text{HB} + n_4\,\text{B}^- \rightarrow n_5\,\text{HC} + (1 - n_5)\,\text{C}^- + (n_2 + n_3 - n_5)\,\text{HB} + (n_4 + n_5 - n_2)\,\text{B}^-$$

Designate the heat of this reaction as ΔH_{obs}. It depends on the properties of the buffer present as well as on the reactants and products of the reaction in question. Let ΔH_a be the heat absorbed when a solution containing 1 mole of a strong acid is added to a large excess of buffer. If it is again assumed that the enthalpies of all species correspond to infinite dilution, then

$$\Delta Hr = \Delta H_{\text{obs}} + (n_5 - n_2)\,\Delta H_a + (1 - n_2)\,\Delta Hi(\text{HA}) - (1 - n_5)\,\Delta Hi(\text{HC}) \qquad (84)$$

3. *Gibbs Energies of Solution*

Equations (26) through (29) relate the differentials of energy, enthalpy, pressure, volume, temperature, entropy, Helmholtz energy, and Gibbs energy for a system of constant composition. The corresponding equations for a system of variable composition are

$$dE = TdS - PdV + \sum \mu_i\, dn_i \qquad (85)$$

$$dH = TdS + VdP + \sum \mu_i\, dn_i \qquad (86)$$

$$dA = -SdT - PdV + \sum \mu_i\, dn_i \qquad (87)$$

$$dG = -SdT + VdP + \sum \mu_i\, dn_i \qquad (88)$$

where μ_i is the chemical potential of the ith component and the summations in the above equations include all the independent components. Since these are all exact differentials, it follows that

$$\mu_i = \left(\frac{\partial E}{\partial n_i}\right)_{S,V,n_j} = \left(\frac{\partial H}{\partial n_i}\right)_{S,P,n_j} = \left(\frac{\partial A}{\partial n_i}\right)_{T,V,n_j} = \left(\frac{\partial G}{\partial n_i}\right)_{T,P,n_j} = \bar{G}_i \qquad (89)$$

and thus the chemical potential is the same as the partial molal Gibbs energy. Equations for the Gibbs energy, analogous to those for enthalpy given in the previous section, could be written. However, it is usual to represent Gibbs energy changes in solution in terms of fugacity or activity, defined as

$$\mu_i - \mu_i{}^o = RT\ln\frac{f_i}{f_i{}^o} = RT\ln a_i \qquad (90)$$

where, as before, the superscript o indicates the standard state of the ith component. By definition, the fugacity of the ith component in an *ideal* solution is

$$f_i = X_i f_i^{\square} \tag{91}$$

in which $f_i^{\square}$ is the fugacity of the pure compound. Therefore, the activity of a component in an ideal solution is equal to the mole fraction if the pure component is taken as the standard state. Before describing choices of standard states for real solutions, a few general relations involving activities will be given. The temperature derivative of the activity [Eq. (92)] follows from Eqs. (35),

$$\left(\frac{\partial \ln a_i}{\partial T}\right)_P = \frac{1}{R}\frac{\partial}{\partial T}\left(\frac{\mu_i}{T} - \frac{\mu_i{}^o}{T}\right) = -\frac{\bar{L}_i}{RT^2} \tag{92}$$

(89), and (90). The Gibbs energy change for a chemical reaction in any designated state can be related to the Gibbs energy change with reactants and products in their standard states by use of the corresponding activities [Eq. (93)].

$$\Delta G = \Delta G^o + RT\Delta(\ln a_i) = \Delta G^o + RT \ln \prod_i a_i^{\nu_i} \tag{93}$$

The product in the last term is taken over all the activities of reactants and products, each raised to the power of the exponent in the balanced equation. The exponents are considered to be positive for products of the reaction and negative for reactants. At equilibrium, $\Delta G = 0$, and (93) becomes Eq. (94).

$$\Delta G^\circ = -RT \ln \prod_i a_i(\mathrm{aq})^{\nu_i} = -RT \ln K \tag{94}$$

The activities in (94) are for the components in a state of chemical equilibrium, and K is the equilibrium constant. Since ΔG^o for a given reaction depends only on the temperature and the choice of standard states, the equilibrium constant also is a function of these conditions. The combination of Eqs. (35) and (94) gives

$$\left(\frac{\partial \ln K}{\partial T}\right)_P = \frac{\Delta H^o}{RT^2} \tag{95}$$

4. *Standard States for Real Solutions—Nonelectrolytes*

The activity, since it is a ratio of fugacities, is a dimensionless quantity. However, its value does depend on the choice of standard state. The properties of most solutions, especially those containing a highly polar solvent such as water, depart strongly from those of the ideal solution. As previously described for the relative partial molal enthalpy, the usual choice of standard state for the solvent is the pure solvent. However, since the chemical potential of a solute approaches infinity at infinite dilution, this cannot be used as the standard state for Gibbs energy, or related quantities, as it was for enthalpy. According to Henry's law the fugacity of a solute becomes proportional to its concentration at high dilution. The usual choice of the standard state for reporting thermodynamic data for solutes in solution is a hypothetical solution of unit concentration. The fugacity is obtained by a linear extrapolation of values observed at low concentrations and the enthalpy equals the value in the real solution at infinite dilution. This standard state depends on the method of expressing concentration. Since, for precision measurements, the molality is usually used for aqueous solutions, this basis will be used in the following equations. Other standard states can be defined by replacing molality with other concentration units in these equations.

According to Henry's law,

$$\lim_{m\to 0} \frac{f_2}{m} = k_2 \tag{96}$$

where k_2 is the Henry's law constant for component 2. k_2 can be derived from experimental measurements of vapor pressure of component 2 in dilute solutions. Thus,

$$\lim_{m\to 0} \frac{a_2}{m} = \lim_{m\to 0} \frac{f_2}{f_2{}^o m} = \frac{k_2}{f_2{}^o} = 1 \tag{97}$$

and therefore the fugacity in the standard state is equal to the Henry's law constant. The activity coefficient of component 2 is defined as

$$\gamma_2 = \frac{a_2}{m} \tag{98}$$

If x_2 and k_2' are the activity coefficient and Henry's law constant on the mole fraction basis and y_2 and k_2'' are the activity coefficient and Henry's law constant on the molarity basis, then

$$\frac{x_2}{\gamma_2} = \frac{m}{X_2}\frac{k_2}{k_2'} = 1 + \frac{mM_1}{1000} \tag{99}$$

$$\frac{y_2}{\gamma_2} = \frac{m}{M}\frac{k_2}{k_2''} = 1 + \frac{mM_2}{1000} \tag{100}$$

5. *Standard States for Real Solutions—Strong Electrolytes*

Let A designate an electrolyte which dissociates in solution into cations B and anions C as

$$A \rightarrow \nu_+ B + \nu_- C$$

If A is a strong electrolyte, the concentration of undissociated A in solution is extremely small. According to the remarks made on page 51 the partial molal properties, including derived quantities such as fugacity and activity, of the ions B and C cannot be measured since their concentrations are not individually independent. Two kinds of partial molal quantities may be defined for undissociated A, depending on whether the undissociated species is considered to be in equilibrium with the ions. Let the subscript e be used to designate the partial molal property when equilibrium is maintained as the concentration of A is changed. The relation between the two kinds of relative partial molal enthalpies and chemical potentials for A may be obtained by considering the change in Gibbs energy and enthalpy for the following transformation: Dissolve 1 mole of pure A in a large amount of solution, with B and C held constant:

	ΔG	ΔH
$A(s) \rightarrow A(aq)$	$\mu_A - G_A{}^{\square}$	$\bar{L}_A - L_A{}^{\square}$

Transform α moles of A(aq) into α moles of B and α moles of C at equilibrium:

$\alpha A(aq) \rightarrow \alpha B(aq) + \alpha C(aq)$	0	$\alpha \Delta H_i$

The net change is

$A(s) \rightarrow (1-\alpha) A(aq) + \alpha B(aq) + \alpha C(aq)$	$\mu_{Ae} - G_A{}^{\square}$	$\bar{L}_{Ae} - L_A{}^{\square}$

Thus, $\mu_A = \mu_{Ae}$ and, therefore, $f_A = f_{Ae}$ and $a_A = a_{Ae}$. But for the enthalpy,

$$\bar{L}_{Ae} = \bar{L}_A + \alpha \Delta H_i \tag{101}$$

Since the undissociated species is nearly always in equilibrium with its ions in experimental measurements, it is the quantities such as μ_{Ae}, f_{Ae}, a_{Ae}, and L_{Ae} that are measured. With this understanding, the subscript e will be omitted for properties of electrolytes in the rest of the equations.

For electrolytes, m will designate the stoichiometric molality, which is the molality of A if there were no dissociation. Experiment shows that for electrolytes the fugacity becomes proportional to m^{ν} at low concentrations, where $\nu = \nu_+ + \nu_-$. Henry's law for either strong or weak electrolyte may be written as

$$\lim_{m \to 0} \frac{f_A}{m^{\nu}} = \nu_+^{\nu_+} \nu_-^{\nu_-} k_A \tag{102}$$

The standard state for a strong electrolyte is now defined as the hypothetical solution at unit stoichiometric molality m, in which the fugacity has a value such that the activity of A approaches

$$(\nu_+^{\nu_+} \nu_-^{\nu_-}) m^{\nu} = m_+^{\nu_+} m_-^{\nu_-}$$

at infinite dilution. Therefore,

$$\lim_{m \to 0} \frac{a_A}{\nu_+^{\nu_+} \nu_-^{\nu_-} m^{\nu}} = \lim_{m \to 0} \frac{f_A}{\nu_+^{\nu_+} \nu_-^{\nu_-} f_A{}^{o} m} = \frac{k_A}{f_A{}^{o}} = 1 \tag{103}$$

Even though the concentration of undissociated species is very small in a strong electrolyte, it is still considered to be in equilibrium with the ions, so that

$$f_A = f_+^{\nu_+} f_-^{\nu_-} \tag{104}$$

The following convention is now made for both strong and weak electrolytes:

$$\mu_A{}^o = \nu_+ \mu_+{}^o + \nu_- \mu_-{}^o, \text{ which implies } f_A{}^o = (f_+{}^o)^{\nu_+}(f_-{}^o)^{\nu_-} \tag{105}$$

Equation (105) is a convention that is part of the definition of the standard state of strong electrolytes, since it is not subject to experimental verification. Equations (104) and (105) imply that

$$a_A = \frac{f_A}{f_A{}^o} = \left(\frac{f_+}{f_+{}^o}\right)^{\nu_+} \left(\frac{f_-}{f_-{}^o}\right)^{\nu_-} = a_+^{\nu_+} a_-^{\nu_-} \tag{106}$$

The mean ionic activity of an electrolyte is defined as

$$a_\pm = a_A^{1/\nu} \tag{107}$$

and the corresponding activity coefficients are defined as

$$\gamma_A = \frac{a_A}{m} \tag{108}$$

$$\gamma_+ = \frac{a_+}{m_+} \quad \text{and} \quad \gamma_- = \frac{a_-}{m_-} \tag{109}$$

$$\gamma_\pm = \frac{a_\pm}{m_\pm} = \frac{a_\pm}{(\nu_+^{\nu_+} \nu_-^{\nu_-})^{1/\nu} m} = (\gamma_+^{\nu_+} \gamma_-^{\gamma_-})^{1/\nu} \tag{110}$$

$$= \frac{\gamma_A^{1/\nu}}{(\nu_+^{\nu_+} \nu_-^{\nu_-})^{1/\nu} m^{1-1/\gamma}}$$

The activity coefficients γ_A and $\gamma_\pm$ can be measured experimentally, but the ionic activity coefficients γ_+ and γ_- cannot be measured individually. In the limit of infinite dilution $\gamma_\pm$ goes to unity but γ_A goes to zero.

6. *Standard States for Real Solutions—Weak Electrolytes*

An appreciable fraction of a weak electrolyte remains undissociated at non-zero concentrations. If α is the fraction that is undissociated, the concentrations of individual species are related to the stoichiometric molality by

$$m_A = (1-\alpha)m \qquad m_+ = \nu_+ \alpha m \qquad m_- = \nu_- \alpha m \tag{111}$$

The equilibrium constant for the dissociation is defined as

$$K = \lim_{m \to 0} \frac{m_+^{\nu_+} m_-^{\nu_-}}{m_A} = \lim_{m \to 0} \frac{\nu_+^{\nu_+} \nu_-^{\nu_-} m^{\nu-1} \alpha^\nu}{1-\alpha} \tag{112}$$

All rigorous methods of measuring the dissociation constant require an extrapolation to infinite dilution at some stage of the calculations. The standard state of a weak electrolyte is a hypothetical solution at unit stoichiometric molality in which the fugacity has a value such that the activity of A approaches the molality of the undissociated A at infinite dilution. Thus,

$$\lim_{m\to 0} \frac{a_A}{m_A} = \lim_{m\to 0} \frac{a_A K}{m_+^{\nu_+} m_-^{\nu_-}} = \lim_{m\to 0} \frac{a_A K}{\nu_+^{\nu_+} \nu_-^{\nu_+} m^\nu \alpha^\nu}$$
$$= \frac{K k_A}{f_A{}^o} = 1 \tag{113}$$

At any nonzero concentration, in view of (104) and (105),

$$a_A = \frac{f_A}{f_A{}^o} = \left(\frac{f_+}{f_+{}^o}\right)^{\nu_+} \left(\frac{f_-}{f_-{}^o}\right)^{\nu_-} \frac{1}{K} = \frac{a_+^{\nu_+} a_-^{\nu_-}}{K} \tag{114}$$

The activity coefficient of a weak electrolyte is defined as

$$\gamma_A = \frac{a_A}{m_A} = \frac{a_+^{\nu_+} a_-^{\nu_-}}{K m_A} = \frac{\gamma_\pm{}^\nu m_+^{\nu_+} m_-^{\nu_-}}{K m_A} \tag{115}$$

Then the dissociation constant may be written as

$$K = \frac{a_+^{\nu_+} a_-^{\nu_-}}{a_A} = \frac{m_+^{\nu_+} m_-^{\nu_-}}{m_A} \frac{\gamma_+^{\nu_+} \gamma_-^{\nu_-}}{\gamma_A}$$
$$= K' \frac{\gamma_+^{\nu_+} \gamma_-^{\nu_-}}{\gamma_A} \tag{116}$$

where K' is the apparent dissociation constant in a particular solution. These three types of standard states, for nonelectrolytes, strong electrolytes, and weak electrolytes, will be called thermodynamic standard states to distinguish them from other "practical" standard states, which will be defined later and used to represent properties of solutes in certain types of solutions of particular interest in biochemical studies. Additional comments concerning standard states in solutions consisting of complex chemical equilibria will be made later.

Equation (80) relates the heat of formation of a solute in the standard state in solution to the heat of formation of the pure solute. The corresponding Gibbs energy of formation is

$$\Delta Gf^o(\text{aq}) = \Delta Gf^o(s, l, \text{or } g) - RT \ln a_2(\text{sat'd}) \tag{117}$$

where a_2(sat'd) is the activity of the solute in the saturated solution. If the solute is completely miscible with the solvent, a_2(sat'd) is replaced by the activity of the pure solute $a_2{}^\square$, which can be related to the vapor pressure of the pure solute and the appropriate Henry's law constant [Eq. (118)]. The entropy of formation

$$a_2{}^\square = \frac{p^\square}{k_2} \tag{118}$$

in the standard state is then obtained as

$$\Delta Sf^{o}(\text{aq}) = [\Delta Hf^{o}(\text{aq}) - \Delta Gf^{o}(\text{aq})]/T \tag{119}$$

For an electrolyte, Eq. (105) means that

$$\Delta Gf^{o}(\text{A, aq}) = \nu_{+}\Delta Gf^{o}(\text{B, aq}) + \nu_{-}\Delta Gf^{o}(\text{C, aq}) \tag{120}$$

In order to permit the tabulation of the Gibbs energy of formation of individual ions, an additional convention, analogous to the one made for enthalpy, is used. That is, $\Delta Gf^{o}(\text{aq})$ is taken as zero for hydrogen ions. Thus the standard-state Gibbs energy of formation of an anion B is equal to the standard-state Gibbs energy of formation of the strong acid HB, or to $1/n$ times $\Delta Gf^{o}(\text{aq})$ for H_nX. A somewhat different convention is adopted for defining the entropy of ions. According to this convention $S^{o}(\text{H}^{+}, \text{aq})$, rather than $\Delta S^{o}(\text{H}^{+}, \text{aq})$ is set to zero. It should be realized that when calculating the change in entropy for formation of an ion X^{-n} from the absolute entropies of ions tabulated in this way it is related to the Gibbs energies and enthalpy of formation by

$$\Delta Sf^{o}(X^{-n}, \text{aq}) = [\Delta Hf^{o}(X^{-n}, \text{aq}) - \Delta Gf^{o}(X^{-n}, \text{aq})]/T + \tfrac{1}{2}nS^{o}(\text{H}_2, g) \tag{121}$$

7. *Additional Mathematical Relations Involving Activities*

The osmotic coefficient is often used in reporting results of measurements of vapor pressure and freezing points of solutions. It is related to the activity of the solvent by

$$\phi = -\frac{X_1}{X_2}\ln a_1 = -\frac{1000}{\nu m M_1}\ln a_1 \tag{122}$$

where M_1 is the molecular weight of the solvent. For a nonelectrolyte, $\nu = 1$, while for an electrolyte ν is the total number of ions formed by dissociation of 1 molecule and m is the stoichiometric molality. The osmotic coefficient is also convenient for calculating the activity of the solute from that of the solvent. By using Eqs. (47) and (122),

$$d\ln a_2 = -\frac{1000}{mM_1}d\ln a_1 = \left(d\phi + \frac{\phi}{m}dm\right)\nu \tag{123}$$

Integration of this equation gives

$$\ln\gamma = \phi - 1 + \int_0^m \frac{\phi - 1}{m}dm \tag{124}$$

For nonelectrolytes γ in Eq. (124) corresponds to the activity coefficient γ_2, as defined by Eq. (98), while for electrolytes it corresponds to $\gamma_{\pm}$, as defined by Eq. (110).

Activities of components in solution are obtained either through the measurements of phase equilibria, such as vapor pressure, freezing points, boiling points, or osmotic pressure, or from the reversible electrical potential of electrochemical cells without liquid junction. Details of these calculations can be found in books by Pitzer and Brewer (1961) or Harned and Owen (1958). The use of phase equilibria to measure activities depends on the fact that the fugacities of any component are the same in all phases that are in mutual equilibrium. Assuming that the gas phase in equilibrium with the solution is ideal, it follows that the activity of the ith component in solution is

$$a_i = \frac{P_i}{P_i^o} \tag{125}$$

where P_i is the vapor pressure of the component in the solution and P_i^o is the vapor pressure of the component in the standard state. This corresponds to the pure solvent for the solvent and to the Henry's law constant for the solute.

The fugacities of any component are equal in all phases that are in mutual equilibrium. If the same standard state is chosen for all phases, the activities are also equal. Thus, in general, for equilibrium of a solution with a pure solid component,

$$\left(\frac{\partial \ln a_i}{\partial T}\right)_{n_i} dT + \left(\frac{\partial \ln a_i}{\partial n_i}\right)_T dn_i = \frac{d \ln a_i^{\square}}{dT} dT \tag{126}$$

where a_i is the activity of component i in solution and $a_i^{\square}$ is the activity of the pure component. The combination of Eqs. (92) and (126) gives Eq. (127).

$$d \ln a_i = \frac{\bar{L}_i - L_i^{\square}}{RT^2} dT \tag{127}$$

The application of (127) and (123) to the freezing point of a solution (equilibrium between the solution and the solid solvent) gives

$$d \ln a_2 = -\frac{1000}{mM_1} d \ln a_i = \frac{dT}{m\lambda} \tag{128}$$

where dT is the freezing point depression and λ is the cryoscopic constant.

$$\lambda = \frac{RT^2 M_1}{1000 \Delta H_m} \tag{129}$$

where ΔH_m is the heat of fusion of the solvent.

Application of (127) to the equilibrium between a solution and the pure solute gives

$$\left(\frac{\partial \ln a_2}{\partial m}\right)_T \frac{dm}{dT} = \left[\left(\frac{\partial \ln \gamma_2}{\partial m}\right)_T + \frac{1}{m}\right] \frac{dm}{dT} = \frac{q_2}{RT^2} \tag{130}$$

where m, a_2, γ_2, and q_2 all refer to the saturated solution. In terms of the osmotic coefficient,

$$\left[\left(\frac{\partial \phi}{\partial m}\right)_T + \frac{\phi}{m}\right]\frac{dm}{dT} = \frac{q_2}{\nu RT^2} \tag{131}$$

Equations (130) and (131) are useful for calculating the differential heat of solution in the saturated solution from a knowledge of the temperature coefficient of the solubility and the activity coefficient. Additional applications of these, and related equations, are discussed by Williamson (1944).

An electrochemical cell without liquid junction consists of two electrodes in contact with the same homogeneous solution. Passage of an electric current through this cell is accompanied by a chemical reaction. The change of Gibbs energy for this reaction is related to the reversible potential between the two electrodes by Eq. (31). Combination of Eqs. (31) and (94) gives an equation that relates the reversible potential of cell without liquid junction when measured with any combination of activities of reactants and products to the reversible potential measured when all reactants and products are in their standard states [Eq. (132)]. Methods of determining standard cell potentials

$$\epsilon = \epsilon^o - \frac{RT}{n} \ln \prod_i a_i^{\nu_i} \tag{132}$$

and of using them to obtain activities of electrolyte solutions and dissociation constants of weak acids have been thoroughly reviewed by Harned and Owen (1958).

A cell in which the two electrodes are immersed in different solutions that are separated physically but connected electrically are known as cells with liquid junction. Such cells are often used in various kinds of analytical procedures. Although there is no rigorous thermodynamic relationship between the potential developed by these cells (except for concentration cells) and the thermodynamic properties of the solutions and electrodes, they are often used to obtain approximate thermodynamic data. These calculations are subject to various kinds of assumptions, such as those concerning the magnitude of the liquid junction potentials.

pH is an important property of aqueous solutions that may be formally related to the activity of hydrogen ions by Eq. (133). Although this equation

$$\mathrm{pH} = -\log a(\mathrm{H}^+) \tag{133}$$

may be used to relate the potentials of cells with liquid junction to pH, it has no rigorous thermodynamic significance since activities of individual ions cannot be measured. However, pH is an important and useful property of aqueous solutions, and it can be measured by a perfectly definite procedure using the hydrogen electrode or certain other electrodes. The significance of pH and experimental methods of measuring it have been reviewed by Bates (1964). There is no rigorous general relationship between pH and the thermodynamic properties of aqueous solutions. In dilute solutions approximate equations may be obtained by replacing the activity of hydrogen ions in (133) by the molality of hydrogen ions.

F. Standard States and Activities of Weak Acids in Buffered Solutions

Many biochemical reactions include weak acids as reactants, products, or both. Furthermore, polybasic acids that ionize in a series of steps are quite common, and in addition many of these ionic species form complexes with metal ions that are normally found in the solutions. As a consequence, biochemical reactions typically involve very complicated ionic equilibria. In the presence of polyelectrolytes such as proteins or nucleic acids this complexity becomes extreme. The interpretation of thermodynamic measurements on such solutions, and the reduction of the data to well-defined standard states, often presents great difficulties. Most enzymes are effective only within a narrow range of pH, usually in the vicinity of 7, and the solutions that exist in living organisms are buffered at a pH near the optimum for the enzymes present. These reactions also are usually studied in buffered solutions in the laboratory.

Some comments have been made on the measurement of heats of reactions in buffered solutions. To illustrate some of the relationships among the activities of weak acids in buffered solutions, consider the acid HA, which dissociates as $HA \rightarrow H^+ + A^-$. The dissociation constant is

$$K_a = \frac{a_{H^+} a_{A^-}}{a_{HA}} = \frac{m_{H^+} m_{A^-}}{m_{HA}} \frac{\gamma_\pm}{\gamma_{HA}} \tag{134}$$

and the stoichiometric molality is the sum of the molalities of the undissociated species and the anions [Eq. (135)]. As for weak electrolytes of any type,

$$m = m_{HA} + m_{A^-} \tag{135}$$

two kinds of standard states are used for the weak acid in aqueous solution. The activity based on the standard state defined by Eq. (113) is represented by a_{HA}, and the Gibbs energy of formation in this state, by $\Delta Gf°(HA, aq)$. The activity based on the standard state defined by Eq. (103) is represented by $a_2 = a_\pm^2$. The Gibbs energy of formation in this state is equal to the sum of the Gibbs energies of formation of the two ions, but since the usual conventions set the Gibbs energy of formation of hydrogen ions at zero, the Gibbs energy of formation in this standard state is $\Delta Gf°(A^-, aq)$. The Gibbs energy of formation in either of these states may be related to the Gibbs energy of formation of the pure acid by Eq. (117), with the corresponding activity of the saturated solution. The two standard state Gibbs energies of formation are related by Eq. (136).

$$\Delta Gf^{o}(\mathrm{A}^-, \mathrm{aq}) = \Delta Gf^{o}(\mathrm{HA, aq}) - RT \ln K_a \tag{136}$$

Now consider this same acid to be dissolved in a solution that also contains a buffer of constant concentration. This solution can be described as consisting of the buffer solution as the solvent and the weak acid as the solute. Properties of species in the buffer solvent will be designated by the use of the subscript b. If the stoichiometric molality of the acid is decreased while holding the buffer concentration constant, the ratio of $m_{b\mathrm{A}}$ to $m_{b\mathrm{HA}}$ will approach a constant [Eq. (137)], where $m_{b\mathrm{H}^+}$ is a characteristic of the buffer and $\gamma_{b\mathrm{HA}}$ and $\gamma_{b\pm}$ are the

$$K_b = \lim_{m \to 0} \left(\frac{m_{b\mathrm{A}}}{m_{b\mathrm{HA}}} \right) = \frac{K_a \gamma_{b\mathrm{HA}}}{m_{b\mathrm{H}^+} \gamma_{b\pm}^2} \tag{137}$$

activity coefficients approached at low molality of the acid in the buffer. Equation (137) defines K_b in terms of experimentally measurable quantities. In this solvent Henry's law for the acid takes the form of that for a nonelectrolyte, rather than for an electrolyte [Eq. (138)], where m_b is the stoichiometric

$$\lim_{m \to 0} \left(\frac{f_{b2}}{m_b} \right) = k_{b2} \tag{138}$$

molality and k_{b2} is the Henry's law constant in the buffer solvent. In view of the similarity of (138) to (96), a standard state may be defined by a method analogous to (97). The Gibbs energy of formation in this state will be represented by ΔGf^{o}(HA, eq buf) and the corresponding activity by a_{b2}. Henry's law may also be written as Eq. (139). The Gibbs energy of formation in the

$$\lim_{m \to 0} \frac{f_{b2}}{m_{b\mathrm{HA}}} = k_{b\mathrm{HA}} \tag{139}$$

standard state based on $k_{b\mathrm{HA}}$ will be designated by ΔGf^{o}(HA, buf), and the corresponding activity, by $a_{b\mathrm{HA}}$. The substitution of Eqs. (135) and (137) into Eq. (139) shows that

$$k_{b\mathrm{HA}} = (1 + K_b)\, k_{b2} \tag{140}$$

and

$$a_{b2} = (1 + K_b)\, a_{b\mathrm{HA}} \tag{141}$$

The activity of the anion in the buffer solvent may be related formally to the activity of the undissociated acid by

$$K_b = \frac{a_{b\mathrm{A}^-}}{a_{b\mathrm{HA}}} \tag{142}$$

Substitution of K_b in Eq. (141) gives

$$a_{b2} = a_{b\mathrm{HA}} + a_{b\mathrm{A}^-} \tag{143}$$

Since both $a_{b\mathrm{HA}}$ and K_b can be experimentally measured, it appears that the activity of the anion A^- can be determined, contrary to what has been previously said about the evaluation of the thermodynamic properties of individual ions. However, $a_{b\mathrm{A}^-}$ can be obtained only through equations such as Eqs. (142) or (143) and does not have any direct experimental significance. For an ideal solution, Eq. (143) is equivalent to Eq. (135).

ΔGf^o(HA, eq, buf) is the Gibbs energy of formation of the various species that are formed from 1 m of the acid HA at equilibrium in the buffered solution. It is sometimes convenient to consider the Gibbs energy of formation of species that would result from the addition of 1 mole of the anion A^- to the buffer solution. The Gibbs energy of formation of the anion on this basis is related to that of the acid by

$$\Delta Gf^o(\mathrm{A}^-, \mathrm{eq}, \mathrm{buf}) = \Delta Gf^o(\mathrm{HA}, \mathrm{eq}, \mathrm{buf}) - \Delta Gf^o(\mathrm{H}^+, \mathrm{buf}) \tag{144}$$

Six different standard states have now been defined for the acid HA. Symbols that have been used for the Gibbs energy of formation in these states,

TABLE III

GIBBS ENERGY SYMBOLS

Designation	Gibbs energy of formation	Corresponding activity	Defining equation
Undissociated acid in aqueous solution	ΔGf^o(HA, aq)	a_{HA}	(113)
Anion in aqueous solution	$\Delta Gf^o(\mathrm{A}^-$, aq)	a_2	(103)
Equilibrium acid in buffer solution	ΔGf^o(HA, eq, buf)	a_{b2}(HA)	(138)
Equilibrium anion in buffer solution	$\Delta Gf^o(\mathrm{A}^-$, eq, buf)	$a_{b2}(\mathrm{A}^-)$	(144)
Undissociated acid in buffer solution	ΔGf^o(HA, buf)	$a_{b\mathrm{HA}}$	(139)
Anion in buffer solution	$\Delta Gf^o(\mathrm{A}^-$, buf)	$a_{b\mathrm{A}^-}$	(142)

the corresponding activities, and the equations that define them are summarized in Table III. In general the Gibbs energies of formation in any two of these states may be related through the corresponding Henry's law constants.

$$\Delta Gf^o - \Delta Gf^{o\prime} = RT \ln \frac{k}{k'} \tag{145}$$

The activity coefficient $\gamma_{b\mathrm{HA}}$ that appears in (137) is related to the activity of HA in the buffer referred to undissociated acid in aqueous solution, $a'_{b\mathrm{HA}}$, by

$$\gamma_{b\mathrm{HA}} = \lim_{m\to 0}\left(\frac{a'_{b\mathrm{HA}}}{m_{b\mathrm{HA}}}\right) = \frac{1}{f^{o}_{\mathrm{HA}}}\lim_{m\to 0}\left(\frac{f_{b\mathrm{HA}}}{m_{b\mathrm{HA}}}\right) = \frac{k_{b\mathrm{HA}}}{k_{\mathrm{HA}}} \tag{146}$$

Thus, the combination of (145) and (146) gives

$$\Delta Gf^{o}(\mathrm{HA, buf}) = \Delta Gf^{o}(\mathrm{HA, aq}) + RT\ln\gamma_{b\mathrm{HA}} \tag{147}$$

Some other useful relations are

$$\Delta Gf^{o}(\mathrm{HA, eq, buf}) = \Delta Gf^{o}(\mathrm{HA, buf}) + RT\ln\frac{1}{1+K_b} \tag{148}$$

$$\Delta Gf^{o}(\mathrm{HA, eq, buf}) = \Delta Gf^{o}(\mathrm{HA, aq}) + RT\ln\frac{\gamma_{b\mathrm{HA}}}{1+K_b} \tag{149}$$

$$\Delta Gf^{o}(\mathrm{A^-, eq, buf}) = \Delta Gf^{o}(\mathrm{A^-, aq}) + RT\ln\frac{K_b}{1+K_b} \tag{150}$$

All of these equations can be put in a more general form to represent the properties of polyvalent acids. For example, the expanded form of Eq. (149) for the Gibbs energy of formation of all the species resulting from the dissociation of the acid H_nA is

$$\Delta Gf^{o}(\mathrm{H}_n\mathrm{A, eq, buf}) = \Delta Gf^{o}(\mathrm{H}_n\mathrm{A, aq}) + RT\ln\frac{\gamma_b\,\mathrm{H}_n\mathrm{A}}{1+K_{b1}+K_{b1}K_{b2}+K_{b1}K_{b2}K_{b3}+\cdots} \tag{151}$$

The constants K_{bi} are related to the acid dissociation constants K_{ai} for the formation of the ions $H_{n-i}A^{-i}$ by equations analogous to Eq. (137). For an ampholyte, such as an amino acid, which dissociates as

$$\begin{aligned} &\mathrm{H_2A^+} \to \mathrm{HA^{+-}} + \mathrm{H^+}, \text{ dissociation constant } K_{a1} \\ &\mathrm{HA^{+-}} \to \mathrm{A^-} + \mathrm{H^+}, \text{ dissociation constant } K_{a2} \end{aligned} \tag{152}$$

$$\Delta Gf^{o}(\mathrm{HA^{+-}, eq, buf}) = \Delta Gf^{o}(\mathrm{HA^{+-}, aq}) + RT\frac{K_{b1}}{1+K_{b1}+K_{b1}K_{b2}}$$

HA^{+-} represents the dipolar ion that exists in solution.

Thermodynamic properties of weak and strong acids are often tabulated for the undissociated acid and the anion in their hypothetical standard states in aqueous solution at unit molality. The Gibbs energies of formation in these states have been designated as $\Delta Gf^{o}(\mathrm{HA, aq})$ and $\Delta Gf^{o}(\mathrm{A^-, aq})$. The use of these states has become well established and should be continued. However, they are not always convenient for reporting thermodynamic properties of reactions of the type often found in biochemistry. Experimental measurements of heats of reaction, equilibrium constants, and cell potentials in buffered solutions are usually obtained in terms of the total stoichiometric concentrations of all ionic species derived from each component of the reaction. The thermodynamic properties of components in such solutions often differ considerably from those of individual species in aqueous solutions. Furthermore, the auxiliary data needed to convert measurements made in buffered solutions to properties of individual species in aqueous solutions are seldom available. Even approximate calculations, based on the assumption that all activity coefficients are unity, require a knowledge of all the equilibrium constants involved and may become very tedious. As a result of these difficulties, different procedures have been used by different investigators; and it is sometimes difficult to decide what standard state has been used, or approximated, in reporting a particular thermodynamic property.

Several authors have made suggestions for reporting thermodynamic data on biochemical reactions in hopes of establishing a more orderly procedure. Carpenter (1960) has recommended that thermodynamic data for reactions involving weak acids be reported in terms of the undissociated forms of the acids in aqueous solution. In a series of publications, George *et al.* (1963), George and Rutman (1963, 1964), Phillips *et al.* (1966), and Rutman and George (1961) have converted the changes in Gibbs energies for the hydrolyses of several nucleotides to a form that is equivalent to the use of standard states defined by Eq. (153), where α is the number of moles of hydrogen ion produced

$$\Delta Gf^{\circ}(\text{HA, eq, buf, corr}) = \Delta Gf^{\circ}(\text{HA, eq, buf}) + 2{,}303\alpha\, RT\,\text{pH} \quad (153)$$

by the dissociation of 1 mole of the acid HA when placed in the buffered solution. This convention partly, but not entirely, removes the effect of the pH of the buffer on the Gibbs energy of formation. This is the purely arbitrary convention, as are all standard state definitions, and, while it may have some advantage in bringing out certain features of reactions at equilibrium, its physical significance is obscure.

The hypothetical standard state of an acid at equilibrium in a buffered solution is a convenient one for tabulating data on biochemical reactions since it is closely related to experimental measurements. The Gibbs energy of formation in this state has been designated as ΔGf°(HA, eq, buf).* It is analogous to the standard state of the anion in aqueous solution; the difference is that pure water is taken as the solvent in the one case and the buffer solution is taken as the solvent in the other case. On this basis the activity of the acid in the

* In effect, this standard state has frequently been used in the biochemical literature for reporting Gibbs energy changes of reactions. It has been given the symbol, $\Delta G'$, corresponding to a designated pH.

buffer approaches the stoichiometric molality (the sum of the molalities of all species derived from the acid by dissociation) at low concentrations of the acid.

It is obvious that the numerical value of ΔGf°(HA, eq, buf) for any given acid depends on the nature of the buffer solution. Its most important property is the pH, but other properties such as the concentration and nature of the components of the buffer solution also will exert an effect. The Gibbs energy of formation in the buffer may be obtained through the use of Eq. (117), in which a_i(sat'd) refers to a_{b2}(sat'd) for the acid in the saturated solution containing the buffer. It also may be calculated by use of Eq. (151), in which $\gamma_{b\text{HA}}$ is obtained from Eq. (146) and K_{bi}'s are calculated from the limiting values of $m_{b\text{H}_{n-i}\text{A}}/m_{b\text{H}_n\text{A}}$ as shown in Eq. (137). The enthalpy of the acid in this state will be considered to correspond to the enthalpy at infinite dilution in the buffer, analogous to the enthalpy of a component in pure water. The heat of formation likewise depends on the particular nature of the buffer used.

Since very few measurements, appropriate for calculating Gibbs energies of formation in a buffer solution, have been made, a hypothetical buffer that has certain defined properties has been adopted for reporting values of ΔGf°(HA, eq, buf) and ΔHf°(HA, eq, buf) listed in the Appendix. Such values do not correspond exactly to the properties in any real buffer solution but may be used to calculate approximately the properties of compounds in dilute solutions in buffers at pH of 7 where there is no specific interaction or complex formation between the species derived from the acid and the components of the buffer. The hypothetical buffer is composed of the weak acid HB, which dissociates as

$$\text{HB} \rightarrow \text{H}^+ + \text{B}^-$$

and the corresponding salt MB, where M^+ is a cation of unit charge. Other pertinent properties of the buffer are as follows:

1. The dissociation constant of HB is 10^{-7} exactly.
2. The buffer solution consists of 0.1 m HB and 0.2 m MB.*
3. ΔH and ΔC_p for the dissociation of HB are zero.
4. The activity coefficients of all uncharged species in this solution are one exactly, and the mean activity coefficients of electrolytes are given by the modified Debye–Huckel equation,

$$\log \gamma_{\pm} = -\frac{S\Gamma^{\frac{1}{2}}}{1 + A\Gamma^{\frac{1}{2}}} \tag{154}$$

* The standard state of the acid consists of a hypothetical 1 m solution in a buffer of only 0.1 m concentration. Under these conditions the solution is no longer buffered since the pH is not independent of the acid concentration. However, this does not affect the use of the properties in this state in calculating properties of dilute solutions of the acid in the buffer.

where Γ is twice the ionic strength defined by

$$\Gamma = \sum_i m_i z_i^2 \tag{155}$$

and is equal to 0.1 for the hypothetical buffer. S is the Debye–Huckel coefficient, which at 25°C is given by Eq. (156). The constant A is related to the

$$S = \frac{0.3600}{\nu} \sum_i \nu_i z_i^2 \tag{156}$$

"ion size parameter" a by $A = 0.2324\, a$ at 25°C. Taking $a = 4$ Å as an average size for organic ions gives $A = 0.93$. Substitution of these values into (154) gives Eq. (157). The Gibbs energy of formation of an acid in the hypothetical buffer

$$\log \gamma_{\pm} = -0.1137 \frac{\sum_i \nu_i z_i^2}{\nu} \tag{157}$$

can now be related to the Gibbs energy of formation in pure water by Eqs. (151) or (152), in which

$$K_{bi} = \frac{0.59 \times 10^7 \gamma_z K_{ai}}{\gamma_{b\pm}^2} \tag{158}$$

where K_{ai} is the acid dissociation constant for the ith ionization step ($HA^z \rightarrow H^+ + A^{z-1}$), γ_z is the activity coefficient of the ion HA^z, and $\gamma_{b\pm}$ the mean ionic activity coefficient of the pair of ions H^+ and A^{z-1}, calculated by (157). Table IV is helpful in carrying out this calculation.

TABLE IV

z	$\frac{K_{bi}}{K_{ai}} 10^{-7}$
2	0.459
1	0.770
0	1.000
−1	1.30
−2	2.19
−3	3.70
−4	6.25

The enthalpy of formation in the hypothetical buffer is equal to an appropriately weighted average of the enthalpies of formation of the various species present at equilibrium. The weighting factors are equal to the number of moles of each species formed from 1 mole of the acid added.

These conventions permit the tabulation of enthalpies and Gibbs energies of formation that can be easily related to experimental measurements made in dilute solutions buffered at pH of 7. They can be readily modified to approximate the thermodynamic properties in other buffer solutions.

REFERENCES

1. Ackermann, T., and Schreiner, F. (1958). *Z. Elecktrochem.* **62**, 1143.
2. Adriaanse, N., and Coops, J. (1962). *Bull. Thermodyn. Thermochem.* No. 2, p. 21.
3. Ahlberg, J. E., Blanchard, E. R., and Lundberg, W. O. (1937). *J. Chem. Phys.* **5**, 539.
4. Alberty, R. A., Smith, R. M., and Bock, R. M. (1951). *J. Biol. Chem.* **193**, 425.
5. Amador, A. (1965). Ph.D. Thesis, New Mexico Highlands Univ., Las Vegas, New Mexico.
6. Anderson, G. L., Jr., Higbie, H., and Stegman, G. (1950). *J. Am. Chem. Soc.* **72**, 3798.
7. Anderson, A. G., and Stegeman, G. (1941). *J. Am. Chem. Soc.* **63**, 2119.
8. Armstrong, G. T., Domalski, E. S., Furukawa, G. T., and Krivanec, M. A. (1964a). *Natl. Bur. Std.* (*U.S.*), *Rept.* No. 8595.
9. Armstrong, G. T., Furkawa, G. T., and Hilsenrath, J. (1964b). *Natl. Bur. Std.* (*U.S.*), *Rept.* No. 8521.
10. Armstrong, G. T., Domalski, E. S., Inscoe, M. N., Halow, I., Furukawa, G. T., and Buresh, M. K. (1965). *Natl. Bur. Std.* (*U.S.*), *Rept.* No. 8906.
11. Ashby, J. N., Clarke, H. B., Crook, E. M., and Datta, S. P. (1955). *Biochem. J.* **59**, 203.
12. Ashton, H. W., and Partington, J. R. (1934). *Trans. Faraday Soc.* **30**, 598.
13. Aveyard, R., and Lawrence, A. S. C. (1964). *Trans. Faraday Soc.* **60**, 2265.
14. Barron, E. S. G., and Hastings, A. B. (1934). *J. Biol. Chem.* **107**, 567.
15. Barry, F. (1920). *J. Am. Chem. Soc.* **42**, 1911.
16. Bates, R. G. (1951). *J. Res. Natl. Bur. Std.* **47**, 127.
17. Bates, R. G. (1964). "Determination of pH." Wiley, New York.
18. Bates, R. G., and Acree, S. F. (1943). *J. Res. Natl. Bur. Std.* **30**, 129.
19. Bates, R. G., and Hetzer, H. B. (1961). *J. Phys. Chem.* **65**, 667.
20. Bates, R. G., and Pinching, G. P. (1949a). *J. Res. Natl. Bur. Std.* **42**, 419.
21. Bates, R. G., and Pinching, G. P. (1949b). *J. Res. Natl. Bur. Std.* **43**, 519.
22. Bates, R. G., and Pinching, G. P. (1949c). *J. Am. Chem. Soc.* **71**, 1274.
23. Bauer, C. R., and Gemmill, C. L. (1952). *Arch. Biochem. Biophys.* **35**, 110.
24. Beare, W. G., McVicar, G. A., and Ferguson, J. B. (1930). *J. Phys. Chem.* **34**, 1310.
25. Benesi, H. A., Mason, L. S., and Robinson, A. L. (1946). *J. Am. Chem. Soc.* **68**, 1755.
26. Benzinger, T. H., and Hems, R. (1956). *Proc. Natl. Acad. Sci. U.S.* **42**, 896.
27. Benzinger, T. H., Kitzinger, C., Hems, R., and Burton, K. (1959). *Biochem. J.* **71**, 400.
28. Bernhard, S. A. (1956). *J. Biol. Chem.* **218**, 961.
29. Bertrand, G. L., Millero, F. J., Wu, C.-H., and Hepler, L. G. (1966). *J. Phys. Chem.* **70**, 699.
30. Blaschko, H. (1925). *Biochem. Z.* **158**, 428.
31. Bock, R. M., Ling, N., Morell, S. A., and Lipton, S. H. (1956). *Arch. Biochem. Biophys.* **62**, 253.
32. Bonner, O. D., and Breazeale, W. H. (1965). *J. Chem. Eng. Data* **10**, 325.
33. Borsook, H., Ellis, E. L., and Huffman, H. M. (1937). *J. Biol. Chem.* **117**, 281.
34. Borsook, H., and Schott, H. F. (1931a). *J. Biol. Chem.* **92**, 535.
35. Borsook, H., and Schott, H. F. (1931b). *J. Biol. Chem.* **92**, 559.

36. Bower, V. E., and Robinson, R. A. (1963). *J. Phys. Chem.* **67**, 1524.
37. Bresler, H. W. (1904). *Z. Physik. Chem. (Leipzig)* **47**, 613.
38. Britton, H. T. S. (1925). *J. Chem. Soc.* **127**, 1896.
39. Brown, H. T., and Pickering, S. U. (1897). *J. Chem. Soc.* **71**, 756.
40. Bunton, C. A., and Chaimovich, H. (1966). *J. Am. Chem. Soc.* **88**, 4082.
41. Buresh, M. K., Reilly, M. L., Furukawa, G. T., and Armstrong, G. T. (1965). *Natl. Bur. Std. (U.S.), Rept.* No. 8992.
42. Burton, K. (1952). *Biochim. Biophys. Acta* **8**, 114.
43. Burton, K. (1955). *Biochem. J.* **59**, 44.
44. Burton, K. (1958). *Nature* **181**, 1594.
45. Burton, K. (1959). *Biochem. J.* **71**, 388.
46. Burton, K., and Krebs, H. A. (1953). *Biochem. J.* **54**, 94.
47. Burton, K., and Wilson, T. H. (1953). *Biochem. J.* **54**, 86.
48. Butler, J. A. V., Ramchandani, C. N., and Thomson, D. W. (1935). *J. Chem. Soc.* p. 280.
49. Butwill, M. E., and Rockenfeller, J. D. (1966). Personal communication.
50. Campbell, A. N., and Gieskes, J. M. T. M. (1965). *Can. J. Chem.* **43**, 1004.
51. Canady, W. J., Papee, H. M., and Laidler, K. J. (1958). *Trans. Faraday Soc.* **54**, 502.
52. Cannan, R. K., and Shore, A. (1928). *Biochem. J.* **22**, 924.
53. Cantelo, R. C., and Billinger, R. D. (1928). *J. Am. Chem. Soc.* **50**, 3212.
54. Cantelo, R. C., and Billinger, R. D. (1930). *J. Am. Chem. Soc.* **52**, 869.
55. Carpenter, T. H. (1960). *J. Am. Chem. Soc.* **82**, 1111.
56. Chance, B., Estabrook, R. W., and Williamson, J. R., eds. (1965). "Control of Energy Metabolism." Academic Press, New York.
57. Chappel, F. P., and Hoare, F. E. (1958). *Trans. Faraday Soc.* **54**, 367.
58. Charbonnet, G. H., and Singleton, W. S. (1947). *J. Am. Oil Chemists' Soc.* **24**, 140.
59. Christensen, J. J., and Izatt, R. M. (1962). *J. Phys. Chem.* **66**, 1030.
60. Christensen, J. J., Izatt, R. M., Hansen, L. D., and Partridge, J. A. (1966). *J. Phys. Chem.* **70**, 2003.
61. Christensen, J. J., Izatt, R. M., and Hansen, L. D. (1967). *J. Phys. Chem.* **89**, 213.
62. Clarke, H. B., Cusworth, D. C., and Datta, S. P. (1954). *Biochem. J.* **58**, 146.
63. Clarke, T. H., and Stegeman, G. (1939). *J. Am. Chem. Soc.* **61**, 1726.
64. Cohen, P. P. (1940). *J. Biol. Chem.* **136**, 585.
65. Cohn, E. J., McMeekin, T. L., Edsall, J. T., and Weare, J. H. (1934). *J. Am. Chem. Soc.* **56**, 2270.
66. Cohn, E. J., McMeekin, T. L., Terry, J. D., and Blanchard, M. H. (1939). *J. Phys. Chem.* **43**, 169.
67. Coleman, C. F., and DeVries, T. (1949). *J. Am. Chem. Soc.* **71**, 2839.
68. Cori, C. F., Colowick, S. P., and Cori, G. T. (1937). *J. Biol. Chem.* **121**, 465.
69. Cori, C. F., Cori, G. T., and Green, A. A. (1943). *J. Biol. Chem.* **151**, 39.
70. Cottrell, T. L., Drake, G. W., Levi, D. L., Tully, K. J., and Wolfenden, J. H. (1948). *J. Chem. Soc.* p. 1016.
71. Cottrell, T. L., and Wolfenden, J. H. (1948). *J. Chem. Soc.* p. 1019.
72. Dahlgren, G., and Long, F. A. (1960). *J. Am. Chem. Soc.* **82**, 1303.
73. Dalman, L. H. (1937). *J. Am. Chem. Soc.* **59**, 2547.
74. Dalton, J. B., and Schmidt, C. L. A. (1933). *J. Biol. Chem.* **103**, 549.
75. Dalton, J. B., and Schmidt, C. L. A. (1935). *J. Biol. Chem.* **109**, 241.
76. Darke, W. F., and Lewis, E. (1928). *Chem. Ind. (London)* **47**, 1073.
77. Darling, S. (1945). *Acta Physiol. Scand.* **10**, 150.
78. Darling, S. (1947). *Nature* **160**, 403.
79. Datta, S. P., and Grzybowski, A. K. (1958). *Biochem. J.* **69**, 218.
80. Datta, S. P., and Grzybowski, A. K. (1963). *J. Chem. Soc.* p. 6004.

81. Datta, S. P., Gryzbowski, A. K., and Weston, B. A. (1963). *J. Chem. Soc.* p. 792.
82. Davies, M., and Kybett, B. (1965). *Trans. Faraday Soc.* **61**, 2646.
83. Davies, M., and Thomas, D. K. (1956). *J. Phys. Chem.* **60**, 41.
84. DeForcrand, R. (1884). *Ann. Chim. Phys.* **3**, 187.
85. Degani, C., and Halmann, M. (1966). *J. Am. Chem. Soc.* **88**, 4075.
86. Dehn, W. M. (1917). *J. Am. Chem. Soc.* **39**, 1399.
87. Dimmling, W., and Lange, E. (1951). *Z. Elektrochem.* **55**, 322.
88. Dolliver, M. A., Gresham, T. L., Kistiakowsky, G. B., Smith, E. A., and Vaughan, W. E. (1938). *J. Am. Chem. Soc.* **60**, 440.
89. Douglas, T. B., Ball, A. F., and Torgesen, J. L. (1951). *J. Am. Chem. Soc.* **73**, 1360.
90. Duboux, M. (1921). *J. Chim. Phys.* **19**, 179.
91. Dulitskaya, K. A. (1945). *Zh. Obshch. Khim.* **15**, 9.
92. Edgar, G., and Shiver, H. E. (1925). *J. Am. Chem. Soc.* **47**, 1179.
93. Egan, E. P., Jr., and Luff, B. B. (1961). *J. Phys. Chem.* **65**, 523.
94. Egan, E. P., Jr., and Luff, B. B. (1966). *J. Chem. Eng. Data* **11**, 192.
95. Egan, E. P., Jr., Luff, B. B., and Wakefield, Z. T. (1958). *J. Phys. Chem.* **62**, 1091.
96. Egan, E. P., Jr., and Wakefield, Z. T. (1957). *J. Phys. Chem.* **61**, 1500.
97. Ellenbogen, E. (1956). *J. Am. Chem. Soc.* **78**, 369.
98. Ellerton, H. D., and Dunlop, P. J. (1966). *J. Phys. Chem.* **70**, 1831.
99. Ellerton, H. D., Reinfelds, G., Mulcahy, D. E., and Dunlop, P. J. (1964). *J. Phys. Chem.* **68**, 398.
100. Emery, A. G., and Benedict, F. G. (1911). *Am. J. Physiol.* **28**, 301.
101. Ernst, R. C., Watkins, C. H., and Ruwe, H. H. (1936). *J. Phys. Chem.* **40**, 627.
102. Euler, H. von, Adler, E., Gunther, G., and Hellström, H. (1937). *Z. Physiol. Chem. Hoppe-Seylers* **245**, 217.
103. Evans, D. M., Hoare, F. E., and Mellia, T. P. (1962). *Trans. Faraday Soc.* **58**, 1511.
104. Evans, F. W., and Skinner, H. A. (1959). *Trans. Faraday Soc.* **55**, 260.
105. Everett, D. H., Landsman, D. A., and Pinsent, B. R. W. (1952). *Proc. Roy. Soc.* (*London*) **A215**, 403.
106. Fitts, D. D. (1962). "Nonequilibrium Thermodynamics," McGraw-Hill, New York.
107. Flitcroft, T., Skinner, H. A., and Whiting, M. C. (1957). *Trans. Faraday Soc.* **53**, 784.
108. Frenzel, C., Burian, R., and Haas, O. (1935). *Z. Elektrochem.* **41**, 419.
109. Fricke, R. (1929). *Z. Elektrochem.* **35**, 631.
110. Fruton, J. S., and Simmonds, S. (1960). "General Biochemistry," 2nd Ed. Wiley, New York.
111. Furtsch, F. F., and Stegeman, G. (1936). *J. Am. Chem. Soc.* **58**, 881.
112. Furukawa, G. T., Buresh, M. K., Reilly, M. L., and Armstrong, G. T. (1966a). *Natl. Bur. Std.* (*U.S.*), *Rept.* No. 9043.
113. Furukawa, G. T., Buresh, M. K., Reilly, M. L., Armstrong, G. T., and Mitchell, G. D. (1966). *Natl. Bur. Std.* (*U.S.*), *Rept.* No. 9089.
114. Furukawa, G. T., Reilly, M. L., Armstrong, G. T., Mitchell, G. D., and Halow, I. (1966c). *Natl. Bur. Std.* (*U.S.*), *Rept.* No. 9374.
115. Furukawa, G. T., Reilly, M. L., Mitchell, G. D., Domalski, E. S., Halow, I., and Armstrong, G. T. (1966d). *Natl. Bur. Std.* (*U.S.*), *Rept.* No. 9449.
116. Furukawa, G. T., Reilly, M. L., Mitchell, G. D., and Armstrong, G. T. (1967). *Natl. Bur. Std.* (*U.S.*), *Rept.* No. 9501.
117. Gehloff, G. (1921). *Z. Physik. Chem.* (*Leipzig*) **98**, 252.
118. George, P., Phillips, S. J., R. C., and Rutman, R. J. (1963). *Biochemistry* **2**, 508.
119. George, P., and Rutman, R. J. (1963). *Proc. 5th Intern. Congr. Biochem., Moscow, 1961* **5**.
120. George, P., and Rutman, R. J. (1964). *Biopolymers, Symp.* **1**, 189–208.

121. Gibson, G. E., and Giauque, W. F. (1923). *J. Am. Chem. Soc.* **45**, 93.
122. Gibson, G. E., Latimer, W. M., and Parks, G. S. (1920). *J. Am. Chem. Soc.* **43**, 1533.
123. Gieskes, J. M. T. M. (1965). *Can. J. Chem.* **43**, 2448.
124. Ging, N. S., and Sturtevant, J. M. (1954). *J. Am. Chem. Soc.* **76**, 2087.
125. Ginsberg (1923). Ph.D. Thesis, Univ. of Braunschweig, Germany. [Quoted in Kharasch, M. S. (1925). *J. Res. Natl. Std.* **2**, 259].
126. Glasstone, S., and Pound, A. (1925). *J. Chem. Soc.* **127**, 2660.
127. Goller, H., and Wicke, E. (1947). *Angew Chem.* **19B**, 117.
128. Gorin, G. (1956). *J. Am. Chem. Soc.* **78**, 767.
129. Gorin, G., and Clary, C. W. (1960). *Arch. Biochem. Biophys.* **90**, 40.
130. Grafius, M. A., and Neilans, J. B. (1955). *J. Am. Chem. Soc.* **77**, 3389.
131. Green, J. H. S. (1961). *Quart. Rev. (London)* **15**, 125.
132. Greenstein, J. P. (1931). *J. Biol. Chem.* **93**, 479.
133. Greenwald, I., Redish, J., and Kibrick, A. C. (1940). *J. Biol. Chem.* **135**, 65.
134. Grzybowski, A. K., and Datta, S. P. (1964). *J. Chem. Soc.* p. 187.
135. Gucker, F. T., Jr., and Allen, T. W. (1942). *J. Am. Chem. Soc.* **64**, 191.
136. Gucker, F. T., Jr., and Ayres, F. D. (1937a). *J. Am. Chem. Soc.* **59**, 2152.
137. Gucker, F. T., Jr., and Ayres, F. D. (1937b). *J. Am. Chem. Soc.* **59**, 447.
138. Gucker, F. T., Jr., and Ford, W. L. (1941). *J. Phys. Chem.* **45**, 309.
140. Gucker, F. T., Jr., Ford, W. L., and Moser, C. E. (1939a). *J. Phys. Chem.* **43**, 153.
141. Gucker, F. T., Jr., and Pickard, H. B. (1940). *J. Am. Chem. Soc.* **62**, 1464.
142. Gucker, F. T., Jr., Pickard, H. B., and Ford, W. L. (1940). *J. Am. Chem. Soc.* **62**, 2698.
143. Gucker, F. T., Jr., Pickard, H. B., and Planck, R. W. (1939b). *J. Am. Chem. Soc.* **61**, 459.
144. Guinchant, J. (1918). *Ann. Chim. (Paris)* **10**, 59.
145. Gunn, S. R. (1965). *J. Phys. Chem.* **69**, 2902.
146. Hahn, A., and Fasold, H. (1925). *Z. Biol.* **82**, 473.
147. Hale, J. D., Izatt, R. M., and Christensen, J. J. (1963). *J. Phys. Chem.* **67**, 2605.
148. Hanes, C. S. (1940). *Proc. Roy. Soc. (London)* **B129**, 174.
149. Hansen, R. S., Miller, F. A., and Christian, S. D. (1955). *J. Phys. Chem.* **59**, 391.
150. Harary, I., Korey, S. R., and Ochoa, S. (1953). *J. Biol. Chem.* **203**, 595.
151. Harned, H. S. (1939). *J. Phys. Chem.* **43**, 275.
152. Harned, H. S., and Owen, B. B. (1958). "The Physical Chemistry of Electrolyte Solutions," 3rd Ed. Reinhold, New York.
153. Harned, H. S., and Pfanstiel, R. (1922). *J. Am. Chem. Soc.* **44**, 2193.
154. Hastings, A. B., and Van Slyke, D. V. (1922). *J. Biol. Chem.* **53**, 269.
155. Hele, P. (1954). *J. Biol. Chem.* **206**, 671.
156. Hendricks, B. C., Dorsey, J. H., LeRoy, R., and Moseley, A. G., Jr. (1930). *J. Phys. Chem.* **34**, 418.
157. Hendricks, B. C., and Steinbach, W. H., Jr. (1938). *J. Phys. Chem.* **42**, 335
158. Hendricks, B. C., Steinbach, W. H., Jr., LeRoy, R. H., and Moseley, A. G., Jr. (1934). *J. Am. Chem. Soc.* **56**, 99.
159. Higbie, H., and Stegeman, G. (1950). *J. Am. Chem. Soc.* **72**, 3799.
160. Hill, T. L. (1944). *J. Phys. Chem.* **48**, 101.
161. Hitchock, D. I. (1958). *J. Phys. Chem.* **62**, 1337.
162. Holmes, W. S. (1962). *Trans. Faraday Soc.* **58**, 1916.
163. Hoskins, W. M., Randall, M., and Schmidt, C. L. A. (1930). *J. Biol. Chem.* **88**, 215.
164. Hudson, C. S. (1904). *J. Am. Chem. Soc.* **26**, 1065.
165. Hudson, C. S. (1908). *J. Am. Chem. Soc.* **30**, 1767.
166. Hudson, C. S., and Brown, F. C. (1908). *J. Am. Chem. Soc.* **30**, 960.

167. Huffman, H. M. (1938). *J. Am. Chem. Soc.* **60**, 1171.
168. Huffman, H. M. (1940). *J. Am. Chem. Soc.* **62**, 1009.
169. Huffman, H. M. (1941). *J. Am. Chem. Soc.* **63**, 688.
170. Huffman, H. M. (1942). *J. Phys. Chem.* **46**, 885.
171. Huffman, H. M., and Borsook, H. (1932). *J. Am. Chem. Soc.* **54**, 4297.
172. Huffman, H. M., and Ellis, E. L. (1935a). *J. Am. Chem. Soc.* **57**, 41.
173. Huffman, H. M., and Ellis, E. L. (1935b). *J. Am. Chem. Soc.* **57**, 46.
174. Huffman, H. M., and Ellis, E. L. (1937). *J. Am. Chem. Soc.* **59**, 2150.
175. Huffman, H. M., Ellis, E. L., and Borsook, H. (1940). *J. Am. Chem. Soc.* **62**, 297.
176. Huffman, H. M., Ellis, E. L., and Fox, S. W. (1936). *J. Am. Chem. Soc.* **58**, 1728.
177. Huffman, H. M., and Fox, S. (1938). *J. Am. Chem. Soc.* **60**, 1400.
178. Huffman, H. M., and Fox, S. (1940). *J. Am. Chem. Soc.* **62**, 3464.
179. Huffman, H. M., Fox, S., and Ellis, E. L. (1937). *J. Am. Chem. Soc.* **59**, 2144.
180. Hutchens, J. O., Cole, A. G., and Stout, J. W. (1960). *J. Am. Chem. Soc.* **82**, 4813.
181. Hutchens, J. O., Cole, A. G., and Stout, J. W. (1963a). *J. Phys. Chem.* **67**, 1128.
182. Hutchens, J. O., Cole, A. G., and Stout, J. W. (1964a). *J. Biol. Chem.* **239**, 591.
183. Hutchens, J. O., Cole, A. G., and Stout, J. W. (1964c). *J. Biol. Chem.* **239**, 4194.
184. Hutchens, J. O., Cole, A. G., Robie, R. A., and Stout, J. W. (1963b). *J. Biol. Chem.* **238**, 2407.
185. Hutchens, J. O., Figlio, K. M., and Granta, S. M. (1963c). *J. Biol. Chem.* **238**, 1419.
186. "International Critical Tables" (1929). Vol. V. McGraw-Hill, New York.
187. Izatt, R. M., and Christensen, J. J. (1962). *J. Phys. Chem.* **66**, 358.
188. Izatt, R. M., Hansen, L. D., Rytting, J. H., and Christensen, J. J. (1965). *J. Am. Chem. Soc.* **87**, 2760.
189. Izatt, R. M., Rytting, J. H., Hansen, L. D., and Christensen, J. J. (1966). *J. Am. Chem. Soc.* **88**, 2641.
190. "JANAF Thermochemical Tables" (1966). Dow Chem. Co., Midland, Michigan.
191. Jencks, W. P., and Gilchrist, M. (1964). *J. Am. Chem. Soc.* **86**, 4651.
192. Jones, M. E. (1953). *Federation Proc.* **12**, 708.
193. Jones, W. J., and Lapworth, A. (1911). *J. Chem. Soc.* **99**, 1427.
194. Jones, I., and Soper, F. G. (1936). *J. Chem. Soc.* p. 133.
195. Jordan, J., and Dumbaugh, W. H., Jr. (1959). *Bull. Thermodyn. Thermochem.* **2**, 9.
196. Kabayama, M. A., Patterson, D., and Piche, L. (1958). *Can. J. Chem.* **36**, 557.
197. Karrier, P., and Fioroni, W. (1923). *Helv. Chim. Acta* **6**, 396.
198. Katchalsky, A., and Curran, P. F. (1965). "Nonequilibrium Thermodynamics in Biophysics," Harvard Univ. Press, Cambridge, Massachusetts.
199. Keffler, L. J. P. (1930). *J. Phys. Chem.* **34**, 1319.
200. Keith, W. A., and Mackle, H. (1958). *Trans. Faraday Soc.* **54**, 353.
201. Kelley, K. K. (1929). *J. Am. Chem. Soc.* **51**, 1145.
202. King, E. J. (1957). *J. Am. Chem. Soc.* **79**, 6151.
203. Kirgintsev, A. N., and Luk'yanov, A. V. (1962). *Izv. Akad. Nauk. SSSR, Otd. Khim. Nauk*, p. 1479.
204. Kitzinger, C., and Benzinger, T. (1955). *Z. Naturforsch.* **10b**, 375.
205. Kitzinger, C., and Hems, R. (1959). *Biochem. J.* **71**, 395.
206. Klotz, C. E., and Benson, B. B. (1963). *J. Phys. Chem.* **67**, 933.
207. Krebs, H. A. (1953a). *Biochem. J.* **54**, 78.
208. Krebs, H. A. (1953b). *Biochem. J.* **54**, 82.
209. Krebs, H. A., and Eggleston, L. V. (1943). *Biochem. J.* **37**, 334.
210. Krebs, H. A., Smyth, D. H., and Evans, E. A. (1940). *Biochem. J.* **34**, 1041.
211. Kresheck, G. C., and Benjamin, L. (1964). *J. Phys. Chem.* **68**, 2476.

212. Kresheck, G. C., Schneider, H., and Scheraga, H. A. (1965). *J. Phys. Chem.* **69**, 3132.
213. Kubowitz, F., and Ott, P. (1943). *Biochem. Z.* **314**, 94.
214. Lama, R. F., and Lu, B. C.-Y. (1965). *J. Chem. Eng. Data* **10**. 316.
215. Lambert, S. M., and Watters, J. I. (1957). *J. Am. Chem. Soc.* **79**, 4262.
216. "Landolt-Börnstein Zahlenwerte und Funktionen aus Physik, Chemie, Astronomie, Geophysik und Technik" (1961). Vol. II, Pt. 4. Springer, Berlin.
217. Lange, E., and Möhring, K. (1953). *Z. Elektrochem.* **57**, 660.
218. Lange, N. A., and Sinks, M. H. (1930). *J. Am. Chem. Soc.* **52**, 2602.
219. Larson, W. D., and Tomsicek, W. S. (1939). *J. Am. Chem. Soc.* **61**, 65.
220. Larsson, E. (1926). *Z. Anorg. Allgem. Chem.* **155**, 247.
221. Larsson, E. (1932). *Z. Physik. Chem.* (*Leipsig*) **A166**, 241.
222. Levedeba, N. D. (1964). *Zh. Fiz. Khim.* **38**, 2648.
223. Levene, P. A., Bass, L. W., and Simms, H. S. (1926). *J. Biol. Chem.* **70**, 243.
224. Levene, P. A., Simms, H. S., and Pfaltz, M. H. (1924). *J. Biol. Chem.* **61**, 445.
225. Levene, P. A., and Simms, H. S. (1925). *J. Biol. Chem.* **65**, 519.
226. Levien, B. J. (1955). *J. Phys. Chem.* **59**, 640.
227. Levintow, L., and Meister, A. (1954). *J. Biol. Chem.* **209**, 265.
228. Lewin, S., and Barnes, M. A. (1966). *J. Chem. Soc.* **B**, 478.
229. Lewis, G. N., and Burrows, G. H. (1912). *J. Am. Chem. Soc.* **34**, 1515.
230. Llopis, J., and Ordonez, D. (1963). *J. Electroanal. Chem.* **5**, 129.
231. Loeffler, M. C., and Moore, W. J. (1948). *J. Am. Chem. Soc.* **70**, 3650.
232. Loewenstein, A., and Roberts, J. D. (1960). *J. Am. Chem. Soc.* **82**, 2705.
233. Louguinine, W. (1888). *Compt. Rend.* **106**, 1289.
234. Louguinine, W. (1891). *Ann. Chim. Phys.* **23**, 179.
235. Mahler, H. R., Wakil, S. J., and Bock, R. M. (1953). *J. Biol. Chem.* **204**, 453.
236. Martell, A. E., and Schwarzenbach, G. (1956). *Helv. Chim. Acta* **39**, 653.
237. Martin, A. W., and Tartar, H. V. (1937). *J. Am. Chem. Soc.* **59**, 2672.
238. Mason, L. S., Offutt, W. F., and Robinson, A. L. (1949). *J. Am. Chem. Soc.* **71**, 1463.
239. Melchior, N. C. (1954). *J. Biol. Chem.* **208**, 615.
240. Melia, T. P. (1964). *Trans. Faraday Soc.* **60**, 1286.
241. Melia, T. P. (1965). *Trans. Faraday Soc.* **61**, 594.
242. Merckel, J. H. C. (1937). *Rec. Trav. Chim.* **56**, 811.
243. Merriman, R. W. (1913). *J. Chem. Soc.* **103**, 1774.
244. Meyerhof, O. (1922). *Biochem. Z.* **129**, 594.
245. Meyerhof, O., and Green, H. (1949). *J. Biol. Chem.* **178**, 655.
246. Miles, C. B., and Hunt, H. (1941). *J. Phys. Chem.* **45**, 1346.
247. Minnick, L. J., and Kilpatrick, M. (1939). *J. Phys. Chem.* **43**, 259.
248. Möbius, H. H. (1955). *J. Prakt. Chem.* **2**, 95.
249. Möller, K. D. (1960). *Compt. Rend.* **250**, 3977.
250. Monod, J., and Torriani, A. M. (1950). *Ann. Inst. Pasteur* **78**, 65.
251. Morrison, J. F., O'Sullivan, W. J., and Ogston, A. G. (1961). *Biochim. Biophys. Acta* **52**, 82.
252. Nanninga, L. B. (1957). *J. Phys. Chem.* **61**, 1144.
253. Nanninga, L. B. (1961a). *Biochim. Biophys. Acta* **52**, 338.
254. Nanninga, L. B. (1961b). *Biochim. Biophys. Acta* **54**, 330.
255. Nelander, L. (1964). *Acta Chem. Scand.* **18**, 973.
256. Nelson, E. W., and Newton, R. F. (1941). *J. Am. Chem. Soc.* **63**, 2178.
257. Nims, L. F. (1936). *J. Am. Chem. Soc.* **58**, 987.
258. Nims, L. F., and Smith, P. K. (1933). *J. Biol. Chem.* **101**, 401.
259. Ochoa, S. (1948). *J. Biol. Chem.* **174**, 133.

260. Ogston, A. G. (1936). *J. Chem. Soc.* p. 1713.
261. Oka, Y. (1944a). *Nippon Seirigaku Zasshi* **9**, 359.
262. Oka, Y. (1944b). *Nippon Seirigaku Zasshi* **9**, 365.
263. Olson, J. A., and Anfinsen, C. B. (1953). *J. Biol. Chem.* **202**, 841.
264. Omel'chenko, F. S. (1962). *Izv. Vysshikh Uchebn. Zavedenii Pishchevaya Tekhnol.* p. 37.
265. Ono, S., Hiromi, K., and Takahashi, K. (1965). *J. Biochem.* (*Tokyo*) **57**, 799.
266. O'Sullivan, W. J., and Perrin, D. D. (1961). *Biochim. Biophys. Acta* **52**, 612.
267. O'Sullivan, W. J., and Perrin, D. D. (1964). *Biochemistry* **3**, 18.
268. Owen, B. B. (1934). *J. Am. Chem. Soc.* **56**, 24.
269. Paoletti, P., Stern, J. H., and Vacca, A. (1965). *J. Phys. Chem.* **69**, 3759.
270. Parker, V. B. (1965). "Thermal Properties of Aqueous Uni-univalent Electrolytes," Natl. Std. Ref. Data Ser., Natl. Bur. Std. No. 2.
271. Parks, G. S., and Anderson, T. (1926). *J. Am. Chem. Soc.* **48**, 1506.
272. Parks, G. S., and Huffman, H. M. (1930). *J. Am. Chem. Soc.* **52**, 4381.
273. Parks, G. S., and Kelley, K. K. (1925). *J. Am. Chem. Soc.* **47**, 2089.
274. Parks, G. S., and Kelley, K. K. (1928). *J. Phys. Chem.* **32**, 734.
275. Parks, G. S., and Thomas, S. B. (1934). *J. Am. Chem. Soc.* **56**, 1423.
276. Parks, G. S., Kelley, K. K., and Huffman, H. M. (1929). *J. Am. Chem. Soc.* **51**, 1969.
277. Parks, G. S., Huffman, H. M., and Barmore, M. (1933). *J. Am. Chem. Soc.* **55**, 2733.
278. Parks, G. S., Thomas, S. B., and Light, D. W. (1936). *J. Chem. Phys.* **4**, 64.
279. Parks, G. S., West, T. J., Naylor, B. F., Fujii, P. S., and McLaine, L. A. (1946). *J. Am. Chem. Soc.* **68**, 2524.
280. Pedersen, K. J. (1952). *Acta Chem. Scand.* **6**, 243.
281. Perman, E. P., and Lovett, T. (1926). *Trans. Faraday Soc.* **22**, 1.
282. Pfeiffer, P., and Angern, O. (1924). *Z. Physiol. Chem.* **133**, 180.
283. Pfeiffer, P., and Wurgler, J. (1916). *Z. Physiol. Chem.* **97**, 128.
284. Phillips, S. J., R. C. (1966). *Chem. Rev.* **66**, 501.
285. Phillips, S. J., R. C., George, P., and Rutman, R. J. (1963). *Biochemistry* **2**, 501.
286. Phillips, S. J., R. C., Eisenberg, P., George, P., and Rutman, R. J. (1965). *J. Biol. Chem.* **240**, 4393.
287. Phillips, S. J., R. C., George, P., and Rutman, R. J. (1966). *J. Am. Chem. Soc.* **88**, 2631.
288. Pihl, A., and Eldjarn, L. (1957). *Acta Chem. Scand.* **11**, 1083.
289. Pilcher, G., and Sutton, L. E. (1955). *Phil. Trans. Roy. Soc. London* **A248**, 23.
290. Pinching, G. D., and Bates, R. (1950a). *J. Res. Natl. Bur. Std.* **45**, 322.
291. Pinching, G. D., and Bates, R. (1950b). *J. Res. Natl. Bur. Std.* **45**, 444.
292. Pitzer, K. S. (1937). *J. Am. Chem. Soc.* **59**, 2365.
293. Pitzer, K. S., and Brewer, L. (1961). Revision of G. N. Lewis and M. Randall, "Thermodynamics", McGraw-Hill, New York.
294. Podalsky, R. J., and Morales, M. F. (1956). *J. Biol. Chem.* **218**, 945.
295. Prigogine, I. (1955). "Introduction to Thermodynamics of Irreversible Processes." Thomas, Springfield, Illinois.
296. Racker, E. (1950). *J. Biol. Chem.* **184**, 313.
297. Ralston, A. W., and Hoerr, C. W. (1942). *J. Org. Chem.* **7**, 546.
298. Rawitscher, M., and Sturtevant, J. M. (1960). *J. Am. Chem. Soc.* **82**, 3739.
299. Richards, M. M. (1938). *J. Biol. Chem.* **122**, 727.
300. Richards, T. W., and Davis, H. S. (1920). *J. Am. Chem. Soc.* **42**, 1599.
301. Richards, T. W., and Gucker, F. T., Jr. (1925). *J. Am. Chem. Soc.* **47**, 1876.
302. Richards, T. W., and Gucker, F. T., Jr. (1929). *J. Am. Chem. Soc.* **51**, 712.
303. Richards, T. W., and Mair, B. J. (1929a). *J. Am. Chem. Soc.* **51**, 737.
304. Richards, T. W., and Mair, B. J. (1926b). *J. Am. Chem. Soc.* **51**, 740.

305. Robbins, E. A., and Boyer, P. D. (1957). *J. Biol. Chem.* **224**, 121.
306. Robinson, R. A., Smith, P. K., and Smith, E. R. B. (1942). *Trans. Faraday Soc.* **38**, 63.
307. Robinson, R. A., and Stokes, R. H. (1961). *J. Phys. Chem.* **65**, 1954.
308. Rodkey, F. L. (1954). *Federation Proc.* **13**, 282.
309. Rodkey, F. L., and Ball, E. G. (1952). *Proc. Natl. Acad. Sci. U.S.* **38**, 396.
310. Rossini, F. D., Wagman, D. D., Evans, W. H., Levine, S., and Jaffe, I. (1952). "Selected Values of Chemical Thermodynamic Properties," *Natl. Bur. Std.* (*U.S.*) *Circ.* No. 500.
311. Roth, W. A. (1923). Landolt-Börnstein Tabellen. [Quoted by Kharsch, M. S. (1929). *J. Res. Natl. Bur. Std.* **2**, 359].
312. Roth, W. A., and Becker, G. (1937). *Z. Physik. Chem.* (*Leipzig*) **A179**, 450.
313. Roth, W. A., and Müller, F. (1927). Landolt-Börnstein Tabellelen. [Quoted by Kharsch, M. S. (1929). *J. Res. Natl. Bur. Std.* **2**, 359].
314. Ruehrwein, R. A., and Huffman, H. M. (1946). *J. Am. Chem. Soc.* **68**, 1759.
315. Rutman, R. J., and George, P. (1961). *Proc. Natl. Acad. Sci. U.S.* **47**, 1094.
316. Saville, G., and Gundry, H. A. (1959). *Trans. Faraday Soc.* **55**, 2036.
317. Saxton, B., and Darken, L. S. (1940). *J. Am. Chem. Soc.* **62**, 846.
318. Scatchard, G., and Prentiss, S. S. (1934). *J. Am. Chem. Soc.* **56**, 1486.
319. Scatchard, G., Hamer, W. J., and Wood, S. E. (1938). *J. Am. Chem. Soc.* **60**, 3061.
320. Schlenk, F., Hellström, H., and von Euler, H. (1938). *Ber. Deut. Chem. Ges.* **71**, 1471.
321. Schmidt, C. L. A., Appelman, W. K., and Kirk, P. L. (1929). *J. Biol. Chem.* **81**, 323.
322. Schmidt, C. L. A., Appelman, W. K., and Kirk, P. L. (1930). *J. Biol. Chem.* **88**, 285.
323. Schwabe, K., and Wagner, W. (1958). *Chem. Ber.* **91**, 686.
324. Scott, E. M., and Powell, R. (1948). *J. Am. Chem. Soc.* **70**, 1104.
325. "Selected Values of Properties of Chemical Compounds" (1966). Thermodyn. Res. Center Data Proj., Texas A & M Univ., College Station, Texas.
326. Simms, H. S. (1928). *J. Phys. Chem.* **32**, 1121.
327. Smith, R. M., and Alberty, R. A. (1956a). *J. Phys. Chem.* **60**, 180.
328. Smith, R. M., and Alberty, R. A. (1956b). *J. Am. Chem. Soc.* **78**, 2376.
329. Smith, E. R. B., and Smith, P. K. (1937). *J. Biol. Chem.* **117**, 209.
330. Smith, E. R. B., and Smith, P. K. (1942). *J. Biol. Chem.* **146**, 187.
331. Smith, R. A., Stamer, J. R., and Gunsalus, I. C. (1956). *Biochim. Biophys. Acta* **19**, 567.
332. Smith, P. K., Taylor, A. C., and Smith, E. R. B. (1937). *J. Biol. Chem.* **122**, 109.
333. Sneathlage, H. C. S. (1952). *Rec. Trav. Chim.* **71**, 699.
334. Stern, J. R., Ochoa, S., and Lynen, F. (1952). *J. Biol. Chem.* **198**, 313.
335. Stehler, R. D., and Huffman, H. M. (1935a). *J. Am. Chem. Soc.* **57**, 1734.
336. Stehler, R. D., and Huffman, H. M. (1935b). *J. Am. Chem. Soc.* **57**, 1741.
337. Stohmann, F. (1890). *Z. Physik. Chem.* (*Leipsig*) **6**, 334.
338. Stohmann, F. (1892). *Z. Physik. Chem.* (*Leipsig*) **10**, 410.
339. Stohmann, F., Kleber, C., and Langbein, H. (1889a). *J. Prakt. Chem.* **40**, 202.
340. Stohmann, F., Kleber, C., and Langbein, H. (1889b). *J. Prakt. Chem.* **40**, 341.
341. Stohmann, F., and Langbein, H. (1890). *J. Prakt. Chem.* **42**, 361.
342. Stohmann, F., and Langbein, H. (1891). *J. Prakt. Chem.* **44**, 336.
343. Stohmann, F., and Langbein, H. (1892). *J. Prakt. Chem.* **45**, 305.
344. Stokes, R. H. (1954). *Trans. Faraday Soc.* **50**, 565.
345. Stokes, R. H. (1966). *J. Phys. Chem.* **70**, 1199.
346. Sturtevant, J. M. (1937). *J. Am. Chem. Soc.* **59**, 1528.
347. Sturtevant, J. M. (1940). *J. Am. Chem. Soc.* **62**, 1879.
348. Sturtevant, J. M. (1941a). *J. Am. Chem. Soc.* **63**, 88.

349. Sturtevant, J. M. (1941b). *J. Phys. Chem.* **45**, 127.
350. Sturtevant, J. J. (1942). *J. Am. Chem. Soc.* **64**, 762.
351. Sturtevant, J. M. (1955). *J. Am. Chem. Soc.* **77**, 255.
352. Swain, H. A., Silbert, L. S., and Miller, J. G. (1964). *J. Am. Chem. Soc.* **86**, 2562.
353. Tabor, H., and Hastings, A. B. (1943). *J. Biol. Chem.* **148**, 627.
354. Taqui Kahn, M. M., and Martell, A. E. (1962). *J. Phys. Chem.* **66**, 10.
355. Taqui Kahn, M. M., and Martell, A. E. (1966). *J. Am. Chem. Soc.* **88**, 668.
356. Taylor, H. F. W. (1948). *J. Chem. Soc.* p. 765.
357. Taylor, J., Hall, C. R. L., and Thomas, H. (1947). *J. Phys. Chem.* **51**, 580.
358. Taylor, J. B., and Rowlinson, J. S. (1955). *Trans. Faraday Soc.* **51**, 1183.
359. Trevelyan, W. E., Mann, P. F. E., and Harrison, J. S. (1952). *Arch. Biochem. Biophys.* **39**, 419.
360. Trimble, H. M., and Richardson, E. L. (1940). *J. Am. Chem. Soc.* **62**, 1018.
361. Ts'o, P. O. P., Melvin, I. S., and Olson, A. C. (1963). *J. Am. Chem. Soc.* **85**, 1289.
362. Tsuzki, T., and Hunt, H. (1957). *J. Phys. Chem.* **61**, 1668.
363. Tsuzki, T., Harper, D. O., and Hunt, H. (1958). *J. Phys. Chem.* **62**, 1594.
364. Vanderzee, C. E., and Swanson, J. A. (1963). *J. Phys. Chem.* **67**, 2608.
365. Verkade, P. E., Hartman, H., and Coops, J. (1926). *Rec. Trav. Chim.* **45**, 373.
366. Wadsö, T. (1962). *Acta Chem. Scand.* **16**, 479.
367. Wagman, D. D., Evans, W. H., Halow, I., Parker, V. P., Bailey, S. M., and Schumm, R. H. (1965). *Natl. Bur. Std. (U.S.), Tech. Note* No. 270–1.
368. Walaas, E. (1957). *Acta Chem. Scand.* **11**, 1082.
369. Walaas, E. (1958). *Acta Chem. Scand.* **12**, 528.
370. Walde, A. W. (1939). *J. Phys. Chem.* **43**, 431.
371. Wallace, W. E., Offutt, W. L., and Robinson, A. L. (1943). *J. Am. Chem. Soc.* **65**, 347.
372. Wasserman, A. (1930). *Z. Physik. Chem. (Leipsig)* **146A**, 418.
373. Weiss, J. M., and Cowns, C. R. (1923). *J. Am. Chem. Soc.* **45**, 1003.
374. White, C. M. (1936). *J. Am. Chem. Soc.* **58**, 1620.
375. Wilhoit, R. C. (1965). Unpublished observations, preliminary values.
376. Wilhoit, R. C., and Lei, I. (1965). *J. Chem. Eng. Data* **10**, 166.
377. Wilhoit, R. C., and Shiao, D. (1964). *J. Chem. Eng. Data* **9**, 595.
378. Williamson, A. T. (1944). *Trans. Faraday Soc.* **40**, 421.
379. Wirth, H. E., Droege, J. W., and Wood, J. H. (1956). *J. Phys. Chem.* **60**, 917.
380. Wrathall, D. P., Izatt, R. M., and Christensen, J. J. (1964). *J. Am. Chem. Soc.* **86**, 4779.
381. Wurmser, R., and Filitti-Wurmser, S. (1936). *J. Chim. Phys.* **33**, 577.
382. Wurmser, R., and Mayer-Reich, N. (1933). *Compt. Rend.* **196**, 612.
383. Zittle, C. A., and Schmidt, C. L. A. (1935). *J. Biol. Chem.* **108**, 161.

CHAPTER III

Experimental Approach and Desired Accuracy in Biochemical Thermochemistry

INGEMAR WADSÖ

I. Calorimeters as Physicochemical Instruments in Biochemistry

A. Introduction

In recent years there has been a very marked trend toward quantitative determination of experimental parameters in various branches of biochemical research. This has been achieved by the use of physicochemical instruments.

Calorimetric measurements offer one approach in this respect. There are, however, two rather different reasons for making calorimetric investigations. The calorimeter may be used either as a general analytical tool or as an instrument for the determination of thermodynamic data.

The importance of calorimetry as an analytical tool is quite obvious: Practically all chemical or physical processes are accompanied by a change in heat content that may be observed calorimetrically. Since the evolution of heat is proportional to the extent of the process, quantitative analytical information thus can be gained from it. Some analytical applications of calorimetry are treated in Chapter VII. Here the discussion will be focused on the use of calorimeters as instruments in experimental thermodynamics.

Why are chemical thermodynamic investigations of interest in biochemistry? In my opinion the greatest value of experimentally determined thermodynamic values lies in discussions concerning structure (including "medium effects") and reaction mechanisms, i.e., using the same thermodynamic arguments as in general chemistry.

Calorimeters form part of the available physicochemical apparatus by which we can determine various thermodynamic quantities such as free energy, enthalpy and entropy changes, and values for partial molar heat capacities and volumes. Enthalpy and heat capacity data usually are best determined by direct calorimetric methods; also, in some favorable cases, free energy data can be calculated from calorimetric experiments. Usually free energy data are determined by various types of equilibrium measurements. Entropy values mostly are obtained as a difference between corresponding free energies and enthalpies ($\Delta G = \Delta H - T\Delta S$).

B. The Field of Biochemical Calorimetry

A schematic summary of the field of biochemical calorimetry is shown in Fig. 1. Starting from the physiological side, we note studies of biological objects retaining life functions that are more or less intact. From a physicochemical point of view such systems are not well known and are very badly defined. Obtained thermochemical data, therefore, can be expected to have a very limited thermodynamic interest. Calorimetric investigations in this area mainly have the character of analytical experiments.

If biochemical compounds are isolated and purified there is more chance that calorimetric studies will give meaningful thermodynamic data. For high molecular weight compounds like proteins and nucleic acids, the purities as

Biological objects	Animals, plants, microorganisms, tissues, tissue fractions	Ill-defined systems
Isolated biochemical compounds		
High molecular compounds	Proteins, nucleic acids	Purity? structure?
Low molecular compounds	Phosphates, thiol esters, peptides, aminoacids, sugars, etc.	Well-defined systems
Model compounds	Simple organic compounds	Well-defined systems

Fig. 1. Biochemical calorimetry.

well as the verified identities cannot be brought to the high standard usually required in general thermochemistry. Sometimes calorimetric results may be ambiguous, even if the calorimetric measurements are made with a high accuracy. Furthermore, as these compounds are very complex and their structures to a large extent are often unknown, it is evident that discussion of the thermochemical parameters obtained sometimes must be rather speculative in nature. Nevertheless, the fact that these systems are very complex and are not well known makes it an important task to place thermodynamic parameters on processes where these compounds are involved. As will be indicated later, there are a number of calorimetric approaches to the study of these compounds that call for a very precise calorimetric technique. The attitude frequently put forward, that these systems do not lend themselves to precise physicochemical investigations, is therefore somewhat shortsighted.

With regard to investigations on simple biochemical compounds, the requirements on purities and definition of states usually can be kept at a level equal to or approaching that in general thermochemistry. In the past, considerable effort has been expended on the determination of thermodynamic properties of these compounds, the intention being to map out quantitatively the energy changes associated with the different reaction steps in various metabolic pathways. These studies have been concerned mainly with determinations of free energy changes through equilibrium measurements. In many instances, however, determination of enthalpy changes has been an important step in arriving at the free energy value.

These studies have given us a picture of the economy of energy transformations in biological systems. It also has demonstrated the thermodynamic requirement of an intimate coupling between endergonic and exergonic reaction steps, and we have found evidence of the energy linking role of key substances like adenosine triphosphate and creatine phosphate.

Unfortunately, we can only calculate energy steps for rather arbitrarily defined states as we have little analytical information about concentrations and activity coefficients in the localities where the various reactions actually do occur. Our knowledge of these energy steps is reasonably complete and cannot, at least at the present time, be expected to provide much more information even if data involved were known with a higher accuracy. In this connection we should recognize that living systems are not equilibrium systems and that life processes should be discussed in terms of nonequilibrium thermodynamics (see, e.g., Katchalsky and Curran, 1965). It therefore seems that continued study to elucidate the energies of metabolic reaction steps will not be very worthwhile. This rather negative attitude does not mean that thermodynamic investigations on simple biochemical compounds should be without biochemical importance. On the contrary, it is believed that a more general approach would be considerably more rewarding when making, for instance, calorimetric investigations on these compounds. This point of view rests on the belief that the main value of experimental thermodynamics in chemistry is that determined data might provide a better understanding of chemical properties.

In general thermochemistry one is usually investigating series of related compounds, and discussions of the results obtained are based on the observed differences in their thermochemical data. Biochemical compounds do not usually appear in series of closely related compounds, and thermochemical results (as well as other physicochemical data) on these compounds therefore tend to be "isolated."

An obvious approach toward a better understanding of thermodynamic data for biochemical compounds is to study nonbiochemical compounds that are similar in structure or have one or more structural features in common with the biochemical prototype. Such nonbiochemical compounds are usually called "model compounds." Within a series of model compounds the structure may be varied systematically, and it is often possible to correlate the data obtained with structural details. Obviously a great many of the simple organic compounds (including complexes between metal ions and organic ligands) can be regarded as biochemical model compounds, and there is, of course, no dividing line between general thermochemistry and thermochemical investigations on biochemical models. This fact doubtlessly can be looked upon as rather encouraging insurance! If results from a study on a biochemical model system do not turn out to be of any immediately discernible biochemical importance, they no doubt will form a valuable contribution to the field of general thermochemistry—providing the experiment is sensibly designed and the measurements are made with acceptable accuracy. It is therefore desirable that any calorimetric experiment on a well-defined system be made with the highest accuracy even if, for a specific biochemical problem (which might have formed the incitement to the study), it would be sufficient to determine a rather crude value.

In the ensuing sections different calorimetric approaches in studies of model compounds and simple biochemical compounds are discussed. In the last part of this chapter aspects on calorimetric investigations performed on more complex biochemical systems are considered. Examples of some recent measurements are given but they are not meant to constitute a complete coverage of the various fields.

II. Calorimetry on Simple Biochemical Compounds and on Model Compounds

A. Heat of Combustion

Results from heat-of-combustion measurements on pure compounds are one of the pillars of thermochemistry. A heat of combustion value, ΔH_c, is the enthalpy change when a compound is oxidized to give CO_2, H_2O, N_2, H_2SO_4, etc. (the reaction products being in strictly defined states). From ΔH_c data the heat of formation of the compound from its elements. ΔH_f, may be calculated. A great number of such ΔH_f values are found in thermochemical compilations. By a combination of these data a standard state* heat of reaction, ΔH°, can be calculated for a real or a hypothetical process. In order to derive values for the heat of reaction in a given solution, the corresponding heat of solution values have to be known.

Heat-of-combustion methods are quite general, and only slightly different experimental procedures have to be applied to different types of compounds. Highly accurate data for compounds containing C, H, O, and N atoms may be obtained by burning the compound in a classical static-bomb calorimeter, whereas compounds containing, for instance, S, P, or a metal atom require a rotating-bomb calorimeter (see, e.g., Waddington, 1962).

Heat-of-combustion values are large numbers compared with heat-of-reaction values. ΔH_c-values, therefore, must be determined with a very high precision in order to give useful heat-of-reaction data. This is particularly true for compounds with a relatively high molecular weight. Unfortunately, highly accuratc combustion measurements are difficult to obtain. There is, in addition, the requirement of exceptionally high purity of the compounds, and great emphasis must be placed on purification work and on purity control. The present practical limit of accuracy is about 0.01%, which may lead to an uncertainty of about ± 0.1 kcal mole^{-1} for a derived heat of reaction value (with molecular weights in the order of 100). It should be noted that there are at present very few laboratories where such an accuracy can be obtained.

The importance of combustion calorimetry in biochemical connections has very much declined with the development of calorimeters suitable for direct determinations of reaction heats; "reaction calorimeters." Still, in some fields related to biochemistry, combustion calorimetry always will retain its importance as there are a number of reactions that will hardly ever be suitable for direct calorimetric measurements.

Examples of recent combustion calorimetric measurements on simple biochemical substances are those by Ponomarev and co-workers (1962) on a number of peptides and by Wilhoit and co-workers (Wilhoit and Shiao, 1964; Wilhoit and Lei, 1965) on Krebs cycle acids and related compounds.

B. Heat of Reaction

1. *Reaction Calorimetry*

Reaction calorimetry lacks the generality of combustion calorimetry. It is required that the particular reaction is specific (or can be analyzed so that proper corrections can be made if necessary) and that it can be made to proceed

* In the "thermochemical standard state" each compound is in its pure form and in its stable aggregation state at the standard temperature (25°C) and at the standard pressure (1 atm).

at a suitable rate. The large variety of reaction conditions in chemistry have called for a great number of differently designed reaction calorimeters. (It has been argued, however that many unnecessary variations have been designed; Sunner and Wadsö, 1966.) Conditions such as reaction rate, temperature, pressure, viscosity, magnitude of the heat effect, corrosion problems, and evaporation effects govern the principle and the practical design of the reaction calorimeter.

Reaction calorimetry on biochemical model compounds and on simple biochemical compounds is primarily concerned with reactions performed in the region of room temperature, at atmospheric pressure, and in low viscosity solutions. Also, heat effects are usually sizable (say 0.01 cal ml^{-1} of calorimetric liquid or larger) and the compound to be studied often can be used in conveniently large quantities. These general conditions are favorable toward making reaction calorimetry attractively simple, and therefore data can be determined with a high degree of accuracy.

Thermodynamic properties of a compound or data for a chemical reaction may vary significantly with aggregation states and with solution media. Heat of solution and heat of dilution measurements therefore form an important complement to heat-of-combustion and heat-of-reaction measurements. These processes are not, from a calorimetric point of view, different from chemical reactions, and measurements are usually made by means of "reaction calorimeters."

At this point it may be of interest to go into some detail by using a specific example. Diacylimides RCO-N(R)-COR′ have been suggested as possible intermediates in biochemical transacylation reactions. To discover some relevant thermodynamic properties of this group of compounds, the heat of hydrolysis was determined for a few diacetylimides, $RNAc_2$ (Wadsö, 1965). In these hydrolysis experiments a sample of the pure imide was mixed with the calorimetric liquid, which consisted of a water–ethanol–sodium hydroxide solution.

For the calorimetrically observed hydrolysis process we may write, for R = Bu,

$$BuNAc_2(l) + H_2O(sol'n) \rightarrow BuNHAc(sol'n) + HOAc(sol'n) \qquad (1)$$

Eq. (1) indicates the pure liquid compound, and (sol'n) is an expression for the composition of the reaction solution. The enthalpy value obtained, $\Delta H = -20.94$ kcal $mole^{-1}$, is a gross value and includes solution and neutralization enthalpies for the reaction components. Sodium hydroxide was in large excess, and the acetic acid produced was accordingly in the acetate form. Furthermore, since amide hydrogens have weakly acidic properties it is possible that the NH group is ionized to some extent in the strongly basic reaction solution.

The enthalpy value as measured directly is obviously rather complex and as such is not very useful in a discussion. It is, therefore, beneficial to transform it to the hypothetical standard state reaction

$$BuNAc_2(l) + H_2O(l) \rightarrow RNHAc(l) + HOAc(l) \qquad (2)$$

The enthalpy change for this process, $\Delta H_2°$, is calculated by performing a series of heat-of-solution measurements that may be written in abbeviated notation as

$$BuNHAc(l) \rightarrow BuNHAc(sol'n) \qquad \Delta H_3 = 0.16 \qquad (3)$$

$$H_2O(l) \rightarrow H_2O(sol'n) \qquad \Delta H_4 = -0.20 \qquad (4)$$

$$HOAc(l) \rightarrow HOAc(sol'n) \qquad \Delta H_5 = -11.62 \qquad (5)$$

If enthalpy values from processes (1) and (3)–(5) are combined, the standard state enthalpy change is obtained, $\Delta H_2° = \Delta H_1 - \Delta H_3 + \Delta H_4 - \Delta H_5 = -9.68$ kcal mole^{-1}. $\Delta H_2°$ is a value which is suitable for compilation and as a starting point for calculating the heat of reaction in a particular solution medium or in the gaseous state. Also, as more standard-state data (heat of formation and heat of reaction data) are accumulated, it will become increasingly likely that combination of such data will lead to values for processes that have not been studied (and which, perhaps, cannot be easily investigated experimentally).

For purposes of discussion, it may be desirable to know the heat of reaction in dilute aqueous solution or perhaps in a nonpolar reaction medium such as a hydrocarbon solution. To derive such values, a further series of heat of solution measurements have to be made. For instance, in finding the heat of hydrolysis value for $BuNAc_2$ in dilute aqueous solution, ΔH_{aq}, the following heat of solution experiments were made. (Data refer to aqueous solutions with ionic strengths of 0.2. In the case of HOAc, ionization was suppressed by the presence of HCl.)

$$RNAc_2(l) \rightarrow RNAc_2(aq) \qquad \Delta H_6 = -1.45 \text{ kcal mole}^{-1} \qquad (6)$$

$$RNHAc(l) \rightarrow RNHAc(aq) \qquad \Delta H_7 = -3.51 \text{ kcal mole}^{-1} \qquad (7)$$

$$HOAc(l) \rightarrow HOAc(aq) \qquad \Delta H_8 = -0.33 \text{ kcal mole}^{-1} \qquad (8)$$

A combination of these values will lead to the standard state value

$$\Delta H_{aq} = \Delta H_2° - \Delta H_6 + \Delta H_7 + \Delta H_8 = -12.04 \text{ kcal mole}^{-1}$$

referring to the process

$$RNAc_2(aq) + H_2O(aq) \rightarrow RNHAc(aq) + HOAc(aq, \text{un-ionized}) \qquad (9)$$

Similarly, a combination of $\Delta H_2°$ with heat of vaporization values for the reaction components will lead to the heat of reaction for the gaseous state, $\Delta H_g° = -4.9$ kcal mole^{-1}.

It is very likely that in many cases we may determine the enthalpy change of a reaction by direct measurement under reaction conditions that are necessary for a particular consideration. From a general thermochemical point of view, however, as well as from a biochemical standpoint, more useful information is derivable from the process if standard-state data plus proper auxiliary data (heats of solution, dilution, ionization, and vaporization) are known.

Sometimes it is not experimentally suitable or possible to use a pure compound as a starting point for the measurement. The reaction, for instance, may be preferably initiated by mixing a solution of the compound with another solution containing a proper catalyst (e.g., an acidic or basic medium or an enzyme solution). Also, in such instances it is usually desirable to transform the directly obtained heat of reaction value to an idealized value. As an example we may look upon the determination of heat of hydrolysis of a peptide performed under enzymatic catalysis in dilute solution and at a suitable pH value as controlled by a buffer system, $BH \rightleftharpoons B^- + H^+$. (For specific examples see Sturtevant, 1962.) Such a process may be represented by the equation

$$\begin{aligned} &\mathrm{RCONHR'(aq) + H_2O(aq) + \alpha B^-(aq) \rightarrow RCOO^-(aq)} \\ &\qquad + (1-\alpha)\mathrm{R'NH_3^+(aq)} + \alpha \mathrm{R'NH_2(aq)} + \alpha \mathrm{BH(aq)} \end{aligned} \tag{10}$$

where α is dependent upon the acidity of the ammonium group and the pH value. (The pH is considered high enough for the carboxylic group to be completely ionized.) To calculate the value for an idealized hydrolysis reaction,

$$\mathrm{RCONHR'(aq) + H_2O(aq) \rightarrow RCOO^-(aq) + RNH_3^+(aq)} \tag{11}$$

heats of ionization have to be known for the buffer substance and for the amine group involved.

2. *Calorimetric Determination of Equilibrium Constants*

For reactions that do not go to "completion" but form an equilibrium mixture, the equilibrium constant can be evaluated by a series of reaction calorimetric experiments (see, e.g., Sturtevant, 1962; Benzinger and Kitzinger, 1963; Bolles and Drago, 1965; Christensen *et al.*, 1966; Paoletti *et al.*, 1966; Nelander, 1966). The equilibrium may be obtained from either side, or its position may be adjusted (and thus the heat value recorded) by varying the concentrations of the reactants. With a precise calorimetric technique it is possible to determine very large K values; for a reaction of the type $A + B \rightarrow C$ one ought to be able to study equilibria with K values of about 10^7, providing reaction calorimetric measurements can be made with a precision of 0.01% (Sunner, 1966).

3. *Heat of Hydrolysis*

Heat of hydrolysis measurements form an important type of reaction calorimetric experiment. Hydrolysis reactions are often characterized by the absence of side reactions and by a convenient (and adjustable) reaction rate that makes these reactions favorable for accurate measurements.

Several heat-of-hydrolysis studies on peptides and phosphate esters have been made and were reviewed by Sturtevant (1962). A more recent study is that by Pin (1965) on phosphagens. Other heat-of-hydrolysis studies involve a comparison between thiol esters and *O*-esters in aqueous solution and in the gaseous state and measurements on *N*-acetylimidazole (Wadsö, 1962a, b) and diacetylimides (Wadsö, 1965).

Ono and his colleagues have made several studies on heats of hydrolysis of glycosides (Ono *et al.*, 1965; Takahashi *et al.*, 1965a, b).

4. *Heat of Ionization*

Biochemical compounds are often in an ionized form, and heat of ionization (ΔH_i) measurements are very important in biochemical thermochemistry. These data are needed as auxiliary information in reaction calorimetric studies, and they are also of great general interest, not the least in connection with current discussions about solvent structure and medium effects.

Among recent calorimetric studies of biochemical interest are included a large number of measurements by Izatt and Christensen and their co-workers on ribonucleotides and related compounds (Christensen and Izatt, 1962; Izatt *et al.*, 1965, 1966) and cysteine (Wrathall *et al.*, 1964) and on pentoses and hexoses (Izatt *et al.*, 1966). They use a very elegant calorimetric titration technique that allows pK_a values and thus also ΔS values to be calculated from the observed heat data. (These workers have given their procedure the name "entropy titration.")

Other recent studies of biochemical interest include measurements on thiols by Irving *et al.* (1964) and by Hepler and his co-workers on pyridine carboxylic acids (Millero *et al.*, 1964) and on taurine (Hopkins *et al.*, 1965). Öjelund and Wadsö (1967a) recently studied ionization reactions for α-ketoacids and could confirm, for instance, that unionized puruvic acid is largely in the hydrated form.

5. *Other Types of Reaction Calorimetric Experiments*

Other reaction types studied recently include investigations on methyl group transfer and the mutarotation of sugars (Chapter IV).

Hydrogenation and aminolysis reactions, oxidation of thiol groups, and formation of metal-ion complexes are examples of important types of biochemical reactions that are often suitable for calorimetric studies. No recent reaction calorimetric work related to biochemistry, however, seems to have been done in these fields.

C. Heat of Solution and Heat of Transfer

Heat-of-solution data reflect interactions between solute and solvent—"medium effects." These effects are of major importance to the energetics of a process in solution, particularly if the solvent is water or a water mixture. From measurements of heat of solution of a compound in different solvents the heat of transfer for the compound between the media can be calculated. It is equal to the difference between the heat-of-solution values. Such experiments often are very simple to perform, and it is believed that much information on medium effects can be obtained from these studies, especially if they are combined with measurements of other properties such as activity coefficients (which will, in combination with enthalpy measurements, lead to entropy values), heat capacities, and volume changes.

An example of this type of work is the extensive study by Kresheck and Benjamin (1964) on the heat of solution of models for protein constituents (carboxylic acids, amino acids, and peptides) in water and in 6 *M* urea. They demonstrated specific interactions between urea and the polar parts of solute molecules, including the peptide bond.

D. Heat of Dilution

Heats of dilution are, like heats of solution, often measured in order to establish necessary auxiliary data in reaction calorimetric investigations. Also, dilution of a solution containing a dissociable complex may cause a significant change in equilibrium position, and in favorable cases heats of dissociation as well as equilibrium constants may be calculated. An example of such a study is the recent measurement by Stoesser and Gill (1967) on the self-association of aqueous purine solutions. These workers used a novel flow microcalorimetric method. Other suitable calorimeters for dilution studies are those designed for "enthalpy titrations" as well as many general reaction calorimeters.

E. Heat of Vaporization

The interpretation of thermochemical results in terms of structural and theoretical considerations (e.g., steric and inductive effects and resonance energies) is most relevant when the compounds are in the gaseous state, where no contribution from medium effects occurs. Vaporization calorimetry, therefore, is a very important method of investigation in thermochemistry. Most biochemical compounds are too involatile to lend themselves to gas phase studies. Often, however, suitable models can be studied, and properties of particular structural details can be illustrated from gas phase data.

At present, few experimentally determined heat-of-vaporization values exist, a fact presumably due to the dubious reliability of results from indirect methods (Clausius–Clapeyron's equation) and because calorimeters suitable for accurate measurements at room temperature on small samples have not been available until very recently (Morawetz and Sunner, 1963; Morawetz, 1968; Wadsö, 1966).

In addition to using data from the gaseous state for studying enthalpy changes in chemical reactions, it may be noted that the gaseous state serves as the perfect reference state for the solvation process. The difference between the heat of solution and the heat of vaporization is called the heat of solvation. This value is thus the enthalpy change for the transfer of a compound from the gaseous phase to the particular state in solution and forms a measure for the interaction between solute and solvent. It should be noted that heat-of-solution values for pure compounds describe a change in medium effect. For a compound in a liquid state the molecules are solvated by their neighbor molecules; for crystalline compounds there are more specific interactions (lattice enthalpies).

F. Heat Capacity Measurements

In general, thermochemistry heat-capacity measurements form a major subject. By far the majority of studies are concerned with measurements on pure compounds, often in connection with determinations of third-law entropies.

Recent examples of such work on biochemical compounds are provided by entropy determinations on a number of amino acids (Hutchens *et al.*, 1960, 1963a, b; Cole *et al.*, 1963) and on citric acid hydrate (Evans *et al.*, 1962).

Concerning heat-capacity measurements on biochemical compounds and on models, the prime interests at present are measurements on solutions, determination of apparent heat capacity ${}^{\phi}C_p$ (or the derived partial molar heat capacity $\bar{C}_p$), or corresponding ΔC_p values for chemical reactions and transfer processes.

An interesting feature of heat-capacity data is that they sometimes reflect structural properties and interactions between components in solution (particularly in aqueous solution). They therefore are of great importance in discussing medium effects. For instance, it has been found that nonpolar compounds or groups have much higher apparent heat-capacity values in aqueous solution than in nonpolar solvents. This has been explained in terms of a structure-promoting effect exerted on the water by the nonpolar solute. These structures (sometimes called "icebergs") are not very stable and tend to "melt" at an increased temperature. An apparent heat-capacity value for a solute therefore includes contributions from the heat of "melting" of these structures. These ideas have a very prominent role in current discussions concerning protein structure and functions. It seems that heat-capacity data for model compounds and for the actual macromolecules are among the most elucidative properties that can be studied.

In principle, measurements can be made with any precise reaction calorimeter (with the exception of true isothermal calorimeters). Usually the heat capacity of the solution is determined relative to that of the pure solvent for a small temperature interval (about 1°), and from the difference between the specific heats of the solvent and the solution, the apparent heat capacity of the solute may be calculated. The most interesting values are usually those for dilute solutions (e.g., $\leqslant 1\%$). The heat capacity of such a solution is very similar to that of the pure solvent. The actual measurements therefore must be made with high precision, preferably in the order of 0.01% or better.

Calorimeters designed specially for the determination of solute heat capacities are often constructed as twin calorimeters (see, e.g., Sturtevant, 1959). Using these calorimeters (where one vessel contains the solution under investigation and the other the pure solvent) the apparent heat capacity of the solute may be obtained directly.

For a chemical reaction or another process, $\Delta^{\phi}C_p$ values* can be obtained by measurements of the enthalpy change of the process at different temperatures. Since $\Delta^{\phi}C_p$ equals the temperature coefficient for the enthalpy change, $\Delta^{\phi}C_p = [\Delta(\Delta H)]/\Delta T$. Also, in these determinations the measurements must be made with high precision, particularly if ΔT is kept small, which is usually desirable.

Alternatively $\Delta^{\phi}C_p$ values may be calculated from very precise calibration experiments before and after the reaction has taken place (Öjelund and Wadsö, 1967b).

White and Benson (1960) have measured the partial molar heat capacity for octanoate solutions at different concentrations and temperatures and discussed the data obtained for this micell-forming substance in terms of water-solute interactions.

Kresheck and Benjamin (1964) performed measurements on some models for protein constituents in aqueous solution and in 6 *M* urea. The authors produced evidence for an increased ordering of water around the nonpolar parts of the

* $^{\phi}C_p$ = the sum of the apparent heat capacities for reactants and reaction products, respectively. For a pure compound the apparent heat capacity is equal to the true heat capacity.

solute molecules and demonstrated that part of this ordering is removed by urea, which thus acts as a structure breaker.

III. Calorimetry on Complex Biochemical Compounds

A. General Considerations

The various equilibrium studies (including data from titriometric experiments) probably have been most important in experimental thermodynamic studies on proteins, nucleic acids, and other complex biochemical compounds. Comparatively few calorimetric investigations have been made so far, probably due to the difficulties involved in producing suitable apparatus. With the introduction of a number of suitable calorimetric methods, however, there are strong indications that reaction and heat capacity calorimetry (in particular in combination with measurements of other physicochemical properties) will play an important role in forming our conception of the function of these compounds.

In this field there are a number of special difficulties involved. Molecular weights are very high, and a "local" chemical process therefore will result in a very small heat effect per unit of weight. For a variety of reasons it is desirable to work with dilute solutions (1% by weight is usually considered a high concentration), and calorimeters accordingly must be very sensitive.

For studies on, e.g., model compounds a suitable volume of the reaction vessel is often in the order of 100 ml. In the calorimetry of proteins and nucleic acids such a volume is sometimes prohibitively large owing to the cost of purified samples of these compounds. The desire to be able to work on small volumes adds to the calorimetric difficulties.

The compounds are usually unstable, and it is difficult to achieve an acceptable state of purity. This fact, in addition to that of high molecular weights, makes a combustion calorimetric approach of very little value for studying these compounds.

Proteins and nucleic acids, from a chemical point of view, are multifunctional compounds. It is therefore difficult to perform a reaction calorimetric experiment where only a selected functional group is involved. The most valuable type of reaction calorimetric experiment seems to be, therefore, that in which the compound itself shows a high specificity relative to the reagent, e.g., coupling to an active center of an enzyme or an acid base reaction.

B. Denaturation Processes

A process in which the comparatively ordered structure of a protein or a nucleic acid is changed to a more random state is usually called "denaturation." Such a transition might have various causes, such as changes in pH, treatment with urea, or a temperature change. Sometimes these transformations are reversible, and we may conclude that the denaturated, or unfolded, form can be considered as a reasonably well-defined state of equilibrium. For such processes several calorimetric studies have been made that have led to conclusions or suggestions concerning the nature of the structural change.

It is usually not possible to analyze an enthalpy value for a denaturation process in any greater detail, and it thus seems as if the accuracy of measurement may be kept rather low. This is probably a relevant standpoint in some cases, particularly if the process under investigation is not fully reversible.

For a thermally induced denaturation process the enthalpy change may be obtained from measurements of the heat capacity of the solution as a function of temperature ("differential thermal analysis"). This technique was first demonstrated by Ackermann and Rüterjans (1964a,b) and by Karasz *et al.* (1964) and since has been used in a number of studies.

In connection with the discussion on calorimetry of model compounds, the great interest in heat-capacity values for compounds in solution was pointed out. This also is highly relevant in the field of biomacromolecules, where heat-capacity measurements (and ΔC_p measurements) are possibly the most promising calorimetric approach. As for model compound calorimetry, it is required that very precise calorimetric equipment is available, as solutions of interest are dilute and their heat capacities are essentially the same as the pure solvent. In addition, it is often essential that the measurements can be made on small volumes. Recently the design of heat-capacity calorimeters for precise measurements on about 2 ml has been reported by Privalov *et al.* (1965) and by Sturtevant and his co-workers (Danforth *et al.*, 1967). These calorimeters are particularly suitable for "differential thermal analysis."

Examples of calorimetric investigations on denaturation processes have been summarized by Sturtevant (1962). More recently work has been reported by Beck *et al.* (1965) and by Danforth *et al.* (1967) on the thermal denaturation of ribonuclease and by Kresheck and Scheraga (1966) on the acid denaturation of the same enzyme. Denaturation reactions for DNA have been reported by Privalov *et al.* (1965) and by Bunville *et al.* (1965).

C. Specific Coupling Reactions between Macromolecules

There are a number of very specific coupling reactions between biochemical macromolecules that have been studied by calorimetric methods. Several recent studies have been made on coupling reactions between polyriboadenylic acid and polyribouridylic acid (see Steiner and Kitzinger, 1963; Rawitscher *et al.*, 1963; Ross and Scruggs, 1965).

The coupling between an antigen and its specific antibody has been investigated calorimetrically (see Sturtevant, 1962), but surprisingly no recent studies of this kind seem to have been done.

D. Enzyme-Substrate Interactions

Canady and Laidler (1958) performed a most interesting type of calorimetric experiment when they measured the heat of coupling between α-chymotrypsin and the inhibitor hydrocinnamic acid. This study is apparently the only one of its kind at the present time. It is believed, however, that this is a field where calorimetric measurements might be very rewarding if they can be made with high enough precision. The structure of inhibitors, supposedly bound to the active center of the enzyme, could be changed systematically, and correlations could be made between the heat effect and conceptions about the mode of binding. It would also be of great interest to determine ΔC_p values for such specific coupling processes, as they possibly may give information concerning configurational changes of the enzyme as a result of substrate binding. For successful experiments of this kind, however, we may need more precise calorimetric techniques than those available today.

E. Transfer Enthalpies

Considering the great interest at present in medium factors for governing the stability of proteins and nucleic acids, it is surprising that calorimetric studies involving the heat of solution into different media have not been made. From such measurements the heat of transfer between media could be obtained. It would seem to be of interest to compare such data with corresponding values for model compounds.

REFERENCES

Ackermann, T., and Rüterjans, H. (1964a). *Ber. Bunsenges. Physik. Chem.* **68**, 850.
Ackermann, T., and Rüterjans, H. (1964b). *Z. Physik. Chem.* (*Frankfurt*) **41**, 116.
Beck, K., Gill, S. J., and Downing, M. (1965). *J. Am. Chem. Soc.* **87**, 901.
Benzinger, T. H., and Kitzinger, C. (1963). *In* "Temperature—Its Measurements and Control in Science and Industry" (I. D. Hardy, ed.) Vol. 3, Pt. 3. Reinhold, New York.
Bolles, T. F., and Drago, R. S. (1965). *J. Am. Chem. Soc.* **87**, 5015.
Bunville, L. G., Geiduscheck, E. P., Rawitscher, M. A., and Sturtevant, J. M. (1965). *Biopolymers* **3**, 213.
Canady, W. J., and Laidler, K. J. (1958). *Can. J. Chem.* **36**, 1289.
Christensen, J. J., and Izatt, R. M. (1962). *J. Phys. Chem.* **66**, 1030.
Christensen, J. J., Izatt, R. M., and Partridge, J. A. (1966). *J. Phys. Chem.* **70**, 2003.
Cole, A. G., Hutchens, J. O., and Stout, J. W. (1963). *J. Phys. Chem.* **67**, 1852.
Danforth, R., Krakauer, H., and Sturtevant, J. M. (1967). *Rev. Sci. Instr.* **38**, 484.
Evans, D. M., Hoare, F. E., and Melia, T. P. (1962). *Trans. Faraday Soc.* **58**, 1511.
Hopkins, H. P., Jr., Wu, C. H., and Hepler, L. G. (1965). *J. Phys. Chem.* **69**, 2244.
Hutchens, J. O., Cole, A. G., and Stout, J. W. (1960). *J. Am. Chem. Soc.* **82**, 4813.
Hutchens, J. O., Cole, A. G., and Stout, J. W. (1963a). *J. Phys. Chem.* **67**, 1128.
Hutchens, J. O., Cole, A. G., Robie, R. A., and Stout, J. W. (1963b). *J. Biol. Chem.* **238**, 2407.

Irving, R. J., Nelander, L., and Wadsö, I. (1964). *Acta Chem. Scand.* **18**, 769.
Izatt, R. M., Hausen, L. D., Rytting, J. H., and Christensen, J. J. (1965). *J. Am. Chem. Soc.* **87**, 2760.
Izatt, R. M., Rytting, J. H., Hausen, L. D., and Christensen, J. J. (1966). *J. Am. Chem. Soc.* **88**, 2641.
Karasz, F. E., O'Reilly, J. M., and Bair, H. E. (1964). *Nature* **202**, 693.
Katchalsky, A., and Curran, P. F. (1965). "Nonequilibrium Thermodynamics in Biophysics." Harvard Univ. Press, Cambridge, Massachusetts.
Kresheck, G. C., and Benjamin, L. (1964). *J. Phys. Chem.* **68**, 2476.
Kresheck, G. C., and Scheraga, H. A. (1966). *J. Am. Chem. Soc.* **88**, 4588.
Millero, F. J., Ahluwalia, J. C., and Hepler, L. G. (1964). *J. Phys. Chem.* **68**, 3435.
Morawetz, E. (1968). *Acta Chem. Scand.* **22**, 1509.
Morawetz, E., and Sunner, S. (1963). *Acta Chem. Scand.* **17**, 473.
Nelander, B. (1966). *Acta Chem. Scand.* **20**, 2289.
Öjelund, G., and Wadsö, I. (1967a). *Acta Chem. Scand.* **21**, 1408.
Öjelund, G., and Wadsö, I. (1967b). *Acta Chem. Scand.* **21**, 1838.
Ono, S., Huomi, K., and Takahashi, K. (1965). *J. Biochem. (Tokyo)* **57**, 799.
Paoletti, P., Vacca, A., and Avenare, D. (1966). *J. Phys. Chem.* **70**, 193.
Pin, P. (1965). *J. Chim. Phys.* **62**, 591.
Ponomarev, V. V., Alekseeva, T. A., and Akimova, L. N. (1962). *Zh. Fiz. Khim.* **36**, 872.
Privalov, P. L., Kafiani, K. A., and Monasilidze, D. R. (1965). *Biofizika* **10**, 393.
Rawitscher, M. A., Ross, P. D., and Sturtevant, J. M. (1963). *J. Am. Chem. Soc.* **85**, 1915.
Ross, P. D., and Scruggs, R. L. (1965). *Biopolymers* **3**, 491.
Steiner, R. F., and Kitzinger, C. (1963). *Nature* **85**, 706.
Stoesser, P. R., and Gill, S. J. (1967). *Rev. Sci. Instr.* **38**, 422.
Sturtevant, J. M. (1959). *In* "Technique of Organic Chemistry" (A. Weissberger, ed.), 3rd Ed., Vol. I, Pt. 1, p. 575. Wiley (Interscience), New York.
Sturtevant, J. M. (1962). *In* "Experimental Thermochemistry" (H. A. Skinner, ed.), Vol. II, pp. 427–442. Wiley (Interscience), New York.
Sunner, S. (1966). Report to the 21st Calorimetry Conf., Boulder, Colorado.
Sunner, S., and Wadsö, I. (1966). *Sci. Tools* **13**, 1.
Takashashi, K., Yoshikawa, Y., Hiromi, K., and Ono, S. (1965a). *J. Biochem. (Tokyo)* **58**, 251.
Takahashi, K., Hiromi, K., and Ono, S. (1965*b*). *J. Biochem. (Tokyo)* **58**, 255.
Waddington, G. (1962). *In* "Experimental Thermochemistry" (H. A. Skinner, ed.), Vol. II, pp. 1–14. Wiley (Interscience), New York.
Wadsö, I. (1962a). *Acta Chem. Scand.* **16**, 487.
Wadsö, I. (1962b). *Acta Chem. Scand.* **16**, 479.
Wadsö, I. (1965). *Acta Chem. Scand.* **19**, 1079.
Wadsö, I. (1966). *Acta Chem. Scand.* **20**, 536.
White, P., and Benson, G. C. (1960). *J. Phys. Chem.* **64**, 599.
Wilhoit, R. C., and Lei, I. (1965). *J. Chem. Eng. Data* **10**, 166.
Wilhoit, R. C., and Shiao, D. (1964). *J. Chem. Eng. Data* **9**, 595.
Wrathall, D. P., Izatt, R. M., and Christensen, J. J. (1964). *J. Am. Chem. Soc.* **86**, 4779.

CHAPTER IV

Chemical Structure and Reaction of Carbohydrates

SÔZABURO ONO AND KATSUTADA TAKAHASHI

Carbohydrates are present in every organism and serve as fuels or as structural elements. Their thermodynamics is of importance when we consider their roles in biochemical processes. Therefore, we shall deal with the enthalpy changes and the thermodynamic properties of reactions involving carbohydrates and shall discuss the application of calorimetry to the analytical study of carbohydrates and the kinetics of enzyme catalyzed reactions in which carbohydrates are the substrates.

I. Energetics of Pyranose Rings

The conformation of pyranose rings is often compared with that of cyclohexane. Because of the great structural resemblance of pyranose rings to cyclohexane, the conformational stability of the pyranose rings is assumed to be almost identical with that of the cyclohexane ring. The chair form of the cyclohexane series was found to be more stable than the boat form by 4500 cal mole^{-1} at ordinary temperatures (Margrave *et al.*, 1963).

Though there seem to exist eight possible strainless conformations of pyranose rings, i.e., 1C, C1, 1B, B1, 2B, B2, 3B, and B3, it has been established that the chair-form conformations, C1 and 1C, are essentially favored for most sugars both in solid state and in aqueous solution (Reeves, 1950; Rao and Foster, 1965; Lenz and Heeschen, 1961).

The pyranose ring involves a hydroxyl group on each C atom from the 1 to 4 position and a carbinol group or a hydrogen atom on C-5 atom. It seems reasonable to suppose that the conformation of the sugars and hence the chemical as well as physical properties vary from one to another because of the changes in the relative position of these substituents. A recent conformational study using NMR revealed that all sugars are not in the perfect chair form but are in more or less distorted conformations (Lenz and Heeschen, 1961).

Conformational stability of pyranose rings has been considered by Reeves (1951) upon the basis of substituent effect. He concludes that any substituent other than hydrogen, oriented perpendicular to the pyranose ring, introduces an instability into the conformation. The predominant anomer in an equilibrium mixture can be deduced from a weighting of the instability factors for each anomer pair. However, changes in the relative position of substituents has not yet been related, quantitatively, to the stability in energy terms.

For some sugars it is possible to consider the stability of α- and β-anomers in terms of the energy difference between them that may be obtained from the direct calorimetric measurement of heat changes caused by mutarotation. In this section the data of heats of isomerization of sugars* is reviewed and the energy relationship among some sugars are considered. Heats of isomerization of sugars reported by several authors are summarized in Table I.

An accurate measurement of heat of isomerization was made by Sturtevant (1941) on the mutarotation of D-glucose. With a precise adiabatic calorimeter both the heats of anomerization and of solution of α- and β-D-glucose were determined within the error of 0.3%. It was found that the enthalpy difference between the α- and β- anomers in solution is very small, while the difference in heats of solution is relatively large, indicating a significant difference in lattice energies. The energy of α-D-glucose was found to be 278 cal $mole^{-1}$ and 299 cal $mole^{-1}$ greater than that of β-D-glucose in solution at 25° and 35°C, respectively.

Kabayama *et al.* (1958) determined the heats of isomerization (anomerization) of D-glucose, D-xylose, lactose, maltose, and cellobiose by observations of their mutarotation using a microcalorimeter of the Tian–Calvet type. Since

* Here the term "heats of isomerization" includes the heats of anomerization, epimerization, and any conversion between the sugar isomers. For the heat of anomerization that corresponds to the enthalpy change for the anomeric conversion $\alpha \rightarrow \beta$ obtained from the mutarotational heat, the term "heat of mutarotation" should not be used since the name "mutarotation" is coined for a change in optical rotation to give an anomeric equilibrium.

the mutarotations studied are all single first-order reactions, a plot of logarithmic temperature change against time was extrapolated to zero time and the total heat evolved by mutarotation was obtained from the intercept $t = 0$. The measurements were made only on either of the α- or β-anomer of each sugar, and the heats of anomerization for the total conversion from α to β were

TABLE I

HEATS OF ISOMERIZATION OF SUGARS

Isomerization	Temperature (°C)	ΔH(cal mole^{-1}), precision in %	Reference
α-D-Glucose → β-D-glucose	25	−278 ± 0.3[a]	Sturtevant (1941)
α-D-Glucose → β-D-glucose	35	−299 ± 0.3[a]	Sturtevant (1941)
α-D-Glucose → β-D-glucose	25	−270 ± 1.4	Kabayama *et al.* (1958)
α-D-Xylose → β-D-xylose	25	−535 ± 3.9	Kabayama *et al.* (1958)
α-D-Mannose → β-D-mannose	25	+455 ± 4.8[b]	Takahashi and Ono (1966b)
α-D-Galactose → β-D-galactose	25	−320	Takahashi and Ono (1966b)
α-Lactose → β-lactose	25	−270 ± 1.4	Kabayama *et al.* (1958)
α-Maltose → β-maltose	25	−126 ± 7.3	Kabayama *et al.* (1958)
α-Cellobiose → β-cellobiose	25	−438 ± 2.0	Kabayama *et al.* (1958)
β-D-Fructopyranose → β-D-fructofuranose	25	+2950 ± 3[b]	Andersen and Grønlund (1965)
eq-D-Glucose → eq-D-fructose	25–70	+2220	Takasaki (1967)
eq-D-Mannose → eq-D-fructose	1–40	0	Takasaki (1967)
eq-D-Glucose → eq-D-mannose	25	+2220	Takasaki (1967)
α-D-Glucose → α-D-mannose	25	+1900	Takahashi and Ono (1966b)
β-D-Glucose → β-D-mannose	25	+2600	Takahashi and Ono (1966b)

[a] Average deviation.
[b] Standard deviation.

calculated using the values of α/β in equilibrium mixture drawn from literature. Although the sugars studied are identical in the neighborhood of the so-called anomeric carbon, the ΔH values thus obtained were found to vary over a wide range from one to another, while ΔG is nearly constant for these sugars, indicating the large variation in ΔS. These variations in ΔH and ΔS among the sugars observed were interpreted in terms of water structure and of the changes in the hydration of the reducing hydroxyl group on the basis of possible conformations of the sugars in aqueous solution.

A calorimetric study of the mutarotation of D-fructose was made by Andersen and Grønlund (1965) with a Dewar vessel-type calorimeter, and the ΔH for the conversion from β-D-fructopyranose to β-D-fructofuranose was obtained. Recently Takahashi and Ono (1966b) measured the mutarotational heats of D-mannose and D-galactose with a simple Dewar vessel-type calorimeter and made an energetic consideration of the pyranose ring. Since the rates of mutarotation for these sugars are dependent upon pH, the measurements were carried out in a calorimeter with a buffered solution at pH 4.5. At this pH the mutarotation of the sugars obeys first-order kinetics with the minimum rate constant. Because of the difficulty in obtaining the pure β-anomers of both sugars, the measurements were made only for the reaction $\alpha \rightarrow$ eq. Total heat evolved by mutarotation was obtained by a kinetic treatment on the temperature–time curves using a rate constant determined separately by an autorecording polarimeter.

The result obtained showed that the α-anomer of D-mannose is 455 cal $mole^{-1}$ more stable than the β-anomer, while the β-anomer of D-galactose is 320 cal $mole^{-1}$ more stable than the α-anomer. The percentages of β in equilibrium mixtures are known to be 32.61 and 69.63 for D-mannose and D-galactose, respectively. The corresponding ΔG values are +423 and −492 cal $mole^{-1}$ for the change $\alpha \rightarrow \beta$. It is obviously expected that β-D-mannose and α-D-galactose, both being the minor component in the equilibrium compositions of respective sugar, are energetically unfavorable because of the repulsion between the two hydroxyl groups on C-1 and C-2 of the pyranose ring. This also is suggested by the results of Sturtevant (1941) and Kabayama *et al.* (1958). These authors have reported that the more stable anomer of D-glucose is the β-form, in which the relative position of C-1—C-2 is identical with that of β-D-galactose. This indicates that the situation of C-2 plays a major role in determining the favorable structure of anomeric forms.

Judging from the reported values, the magnitude of enthalpy change for the anomeric conversion, i.e., conversion at C-1, of various sugars seems to be within the order of a few hundred calories per mole.

Now some data on the enthalpy relationship between some sugar isomers are available for further consideration. Recently Takasaki (1967) studied the temperature dependence of equilibrium constants for the isomerizations eq-D-glucose $\rightleftarrows$ eq-D-fructose and eq-D-mannose $\rightleftarrows$ eq-D-fructose by using glucose–isomerase and mannose–isomerase, respectively, and obtained the thermodynamic quantities. Furthermore, by using these values the thermodynamic quantities of a hypothetical reaction, eq-D-glucose $\rightleftarrows$ eq-D-mannose, were evaluated, and it was found that the enthalpy of eq-D-mannose is 2220 cal $mole^{-1}$ greater than that of eq-D-glucose. These values can be used to obtain the energy differences between α-D-glucose and α-D-mannose and between β-D-glucose and β-D-mannose using the known composition of equilibrium mixture and the heats of anomerization reported. From this treatment the energies required for the conversion of a hydroxyl group on C-2 of α-D-glucose and β-D-glucose from equatorial to axial in chair-1 form (and from axial to equatorial in 1-chair form) were estimated to be 1900 and 2600 cal $mole^{-1}$, respectively. The enthalpy relationship among the sugars thus obtained are schematically illustrated in Fig. 1.

More interesting information concerning the substituent effects on stability could be obtained by the comparison of the heats of isomerization observed for various sugars if more data were available.

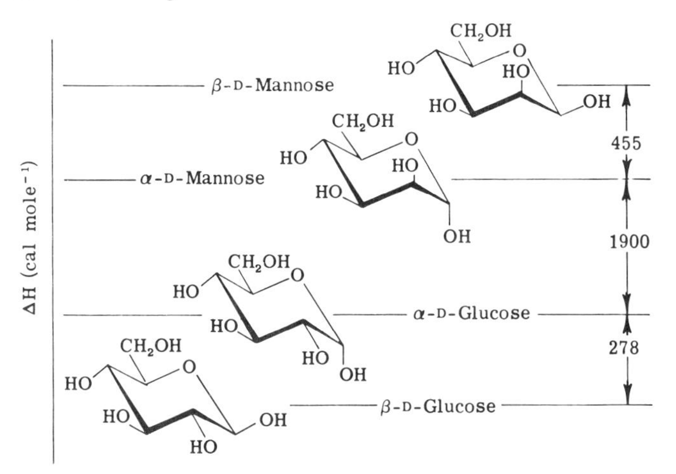

FIG. 1. Energy correlation between D-glucose and D-mannose in aqueous solution (Takahashi and Ono, 1966b).

II. Hydrolysis of Glucosidic Linkages

A study of the formation of the glucosidic linkages to give oligosaccharides and polysaccharides seems to be of special biochemical interest. While α-1,4 glucosidic linkages are involved in starches and many other polysaccharides to form long chains of glucose units, α-1,6 glucosidic linkages participate in cross-linking of the individual glucose chains to give a ramified structure (Fig. 2).

Thus the roles of both linkages in the formation of polysaccharides are quite different. This fact suggests that there would be a considerable difference in thermodynamic properties of both the linkages. Recently Ono *et al.* (1965b) and Takahashi *et al.* (1965a, b) made a calorimetric study of the hydrolyses of these two linkages. The present discussion refers to these results and presents a thermodynamic consideration of the formation of the linkages.

Acid or alkaline hydrolyses of the α-1,4 and α-1,6 glucosidic linkages of starch were not useful for a reaction calorimetry study. Recent progress in the isolation of starch hydrolyzing enzymes, however, has made possible calorimetric measurement of such hydrolyses. Gluc-amylase, obtained in crystalline form by Tsujisaka *et al.* (1958) is an exo-amylase that splits glucose from the

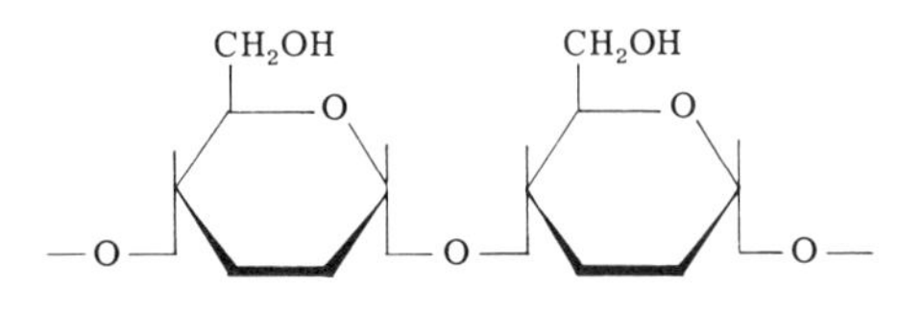

α-1, 4-Glucosidic linkage

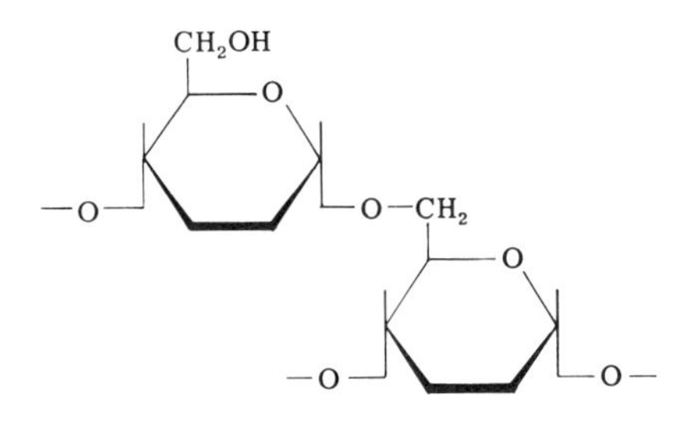

α-1, 6-Glucosidic linkage

FIG. 2. α-Glucosidic linkages.

nonreducing end of starch with an excellent hydrolytic activity upon both the α-1,4 and α-1,6 glucosidic linkages and seems to be most suitable for the present purpose. The authors performed the measurement of the heats of reaction of the hydrolyses of various sugars having the two linkages by using gluc-amylase with a simple Dewar vessel-type calorimeter (Fig. 3).

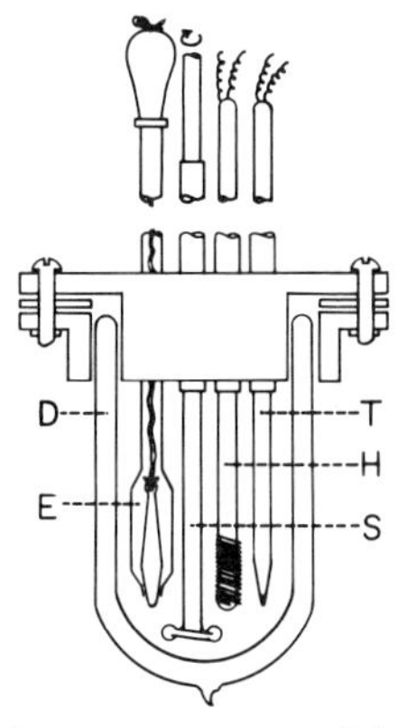

FIG. 3. Calorimeter used for the measurements of heats of hydrolysis of α-glucosidic linkages (Ono *et al.*, 1965b; Takahashi *et al.*, 1965a, b). D = 50 ml Dewar vessel; T = Thermistor TOA-CS-503 (%/°C = −3.8 at 25°C, B = 3400°K, R = 52.41 KΩ at 25°C); H = Platinum heater; S = Glass propeller stirrer; E = Enzyme vessel.

The total heat change produced by the hydrolytic reaction was obtained from the temperature–time curve after correcting for the heat exchange between the calorimeter and the surroundings. At the hydrolysis of glucosidic linkages by gluc-amylase the Walden inversion takes place to yield β-glucose (Ono *et al.*, 1965a; Hamauzu *et al.*, 1965), and the β-glucose formed causes mutarotation accompanied by a slight thermal change. Therefore, the correction for this effect also was made on the temperature–time curve by using the values for mutarotational heats reported by Sturtevant (1941). Heats of hydrolysis of the two linkages obtained for various glucosides are summarized in Table II.

The difference between the heats of hydrolysis of maltose and that of phenyl α-maltoside, both having an α-1,4 glucosidic linkage, was found to be very small. It seems that the phenyl radical combined to the reducing end of maltose molecule slightly lowers the heat of hydrolysis of the α-1,4 glucosidic linkage. This fact suggests that there is no strong intramolecular interaction between the phenyl group and the remote glucose residue.

The heats of hydrolysis of panose, which has an α-1,6 and an α-1,4 glucosidic linkage, and of isomaltose, which has an α-1,6 glucosidic linkage, were found to be $\Delta H = +140$ cal mole^{-1} and $\Delta H = +1300$ cal mole^{-1}, respectively. The heat of hydrolysis of panose is considered to be the sum of the heats of hydrolysis of the two linkages involved since no intramolecular interaction affecting the heat data is present. It has been found that gluc-amylase hydrolyzes the α-1,6 glucosidic linkage of the panose molecule first and subsequently the α-1,4 glucosidic linkage of the remaining part, which corresponds to the maltose molecule (Tsujisaka and Fukumoto, 1956; Ono *et al.*, 1963). Therefore, using the value obtained for maltose previously, the heat of hydrolysis of α-1,6 glucosidic linkage in the panose molecule is calculated as

$$\begin{aligned}\Delta H_{1,6} &= \Delta H_{\text{pan}} - \Delta H_{\text{mal}}\\ &= +140 - (-1100)\\ &= +1240 \text{ cal mole}^{-1}\end{aligned}$$

This value is only 6% lower than that obtained for isomaltose, indicating that there is no strong interaction between the glucose residues in the panose molecule. From this fact it may be assumed that there would be a simple additive property among the heats of hydrolysis of the individual linkage involved in the carbohydrate. Furthermore, the results provide a very interesting and significant finding that the hydrolysis of α-1,6 glucosidic linkages is endothermic, while that of α-1,4 glucosidic linkages is exothermic.

Validity for thermal additivity in the heat of hydrolysis that was seen to hold for panose has been checked by the further study of the heat data of the α-1,4 linkages involved in linear glucosides. For the purpose of obtaining the heat of hydrolysis of α-1,4 glucosidic linkages involved in higher glucosides,

TABLE II

HEATS OF HYDROLYSIS OF α-1,4 AND α-1,6 GLUCOSIDIC LINKAGES IN SEVERAL GLUCOSIDES AT 25°C

Glucoside	Number of glucose residue	Linkages involved	$\Delta H_{1,4}$(cal mole^{-1})		$\Delta H_{1,6}$(cal mole^{-1})		Reference
			1[a]	2[b]	1[a]	2[b]	
Maltose	2	α-1,4	−1100				Ono *et al.* (1965b)
Phenyl α-maltoside	2	α-1,4	−1070				Ono *et al.* (1965b)
Maltotriose	3	α-1,4 × 2	−1100	−1010			Takahashi *et al.* (1965a)
Amylose	about 6000	α-1,4 × 6000		−1030			Takahashi *et al.* (1965a)
Isomaltose	2	α-1,6			+1300		Takahashi *et al.* (1965b)
Panose	3	α-1,4, α-1,6	−1100			+1240	Takahashi *et al.* (1965b)

[a] The case involving no more hydrolyzable linkage.
[b] The case involving a next hydrolyzable linkage.

measurements were made on the heats of hydrolysis of a linear trisaccharide, maltotriose, and of amylose, whose average degree of polymerization is about 6000. The molar heat of hydrolysis of maltotriose was found to be $\Delta H = -2110$ cal mole^{-1}. The heat evolved at the hydrolysis of 1 gm of amylose was found to be 6.35 cal gm^{-1}. Since maltotriose has two α-1,4 linkages, the heat of hydrolysis of the linkage cleaved in the first step (the first linkage) is obtainable by subtracting that of the linkage cleaved in the second step, which corresponds to the heat of hydrolysis of maltose, from the above value.

Thus the heat of hydrolysis of the first linkage in maltotriose was obtained as

$$\begin{aligned}\Delta H_{1,4} &= \Delta H_{\text{tri}} - \Delta H_{\text{mal}} \\ &= -2110 - (-1100) \\ &= -1010 \text{ cal mole}^{-1}\end{aligned}$$

Although the molar heat of hydrolysis of amylose is not obtainable because of its heterogeneity and lack of precise knowledge about the molecular weight, the average heat of hydrolysis of the linkages involved in amylose may be obtained to a good approximation as the heat evolution per mole of the linkage. The average heat of hydrolysis of the linkages was determined to be $\Delta H = -1030$ cal mole^{-1}.

The heat of hydrolysis of the first linkage in maltotriose, $\Delta H = -1010$ cal mole^{-1}, is slightly lower than that obtained for maltose, $\Delta H = -1100$ cal mole^{-1}. It seems that there is a weak interaction between the first and the neighboring glucose residues. This fact resembles the previous finding that the phenyl group attached to the reducing end of the maltose molecule lowers the heat evolution accompanying hydrolysis. The average heat of hydrolysis of α-1,4 glucosidic linkages involved in amylose, $\Delta H = -1030$ cal mole^{-1}, is also slightly lower than that of maltose and is very close to the heat of hydrolysis of the first linkage of maltotriose. These facts indicate that only the linkage adjacent to the reducing end is of exceptional property and that the thermal data of the linkages except for this one are scarcely affected by the chain length.

It is obvious that the heats of hydrolysis of peptides are not simply given as the sum of the heat data of each peptide bond involved because of the occurrence of the intramolecular coulombic interactions among amino acid residues (Dobry and Sturtevant, 1952; Rawitscher *et al.*, 1961). However, in carbohydrates there is no strong intramolecular interaction affecting the thermal data, and hence the molar heats of hydrolysis of oligosaccharides are obtainable as the additive function of the thermal data of the individual linkages to be hydrolyzed. This is a significant property, the recognition of which provides an approach to the quantitative application of thermal data to carbohydrate analysis.

The most remarkable finding in these studies is that the enthalpy changes found for the hydrolysis of the two kinds of glucosidic linkage are approximately equal in their absolute magnitudes but opposite in sign. The hydrolytic reaction of α-glucosidic linkages by gluc-amylase at ordinary concentrations are practically irreversible, and their equilibrium constants and hence standard free energy changes are not readily obtained by chemical analysis of reactants and products. However, Fukumoto *et al.* (1962) reported that when glucose solution of as high a concentration as 40% was incubated with gluc-amylase for a long period, a considerable amount of isomaltose gradually appeared on the paper chromatogram while the formation of maltose was not detectable. This fact indicates that in the dimerization of glucose the equilibrium is inclined toward the formation of the α-1,6 glucosidic linkage rather than toward α-1,4 linkage. Thus the following relation holds for the free energy changes of the formation of two linkages from glucose:

$$\Delta G_{f\text{-}1,6} < \Delta G_{f\text{-}1,4}$$

where the subscripts *f*-1,6 and *f*-1,4 denote the formation of the α-1,6 and α-1,4 glucosidic linkages, respectively.

Numerical data corresponding to the above relation are in the literature. Burton and Krebs (1953) evaluated the free energy change for the formation of α-1,4 linkage in glycogen biosynthesis to be $\Delta G_{f\text{-}1,4} = +4300$ cal mole^{-1} of glucose unit. The free energy of the formation of the α-1,6 linkage was deduced to be $\Delta G_{f\text{-}1,6} = +2000$ cal mole^{-1} or less (Kalckar, 1954). However, these values were indirectly evaluated by combining data for some successively coupled reactions whose free energy changes had been known. Hence they seem to have little quantitative significance. For example, from the result of Fukumoto *et al.* (1962) one may assume that the $\Delta G_{f\text{-}1,6}$ has a negative value, being far less than 2000 cal mole^{-1}. Although no reliable value for the free energy changes of the two linkages is available in the literature, the above relation that the free energy of the formation of α-1,6 linkage is more negative than that of α-1,4 linkage is considered to be reasonable. From the data obtained for the hydrolyses of maltose and of isomaltose, the enthalpy changes for the formation of two linkages may be written as $\Delta H_{f\text{-}1,4} = +1100$ cal mole^{-1} and $\Delta H_{f\text{-}1,6} = -1300$ cal mole^{-1}. Thus the enthalpy changes for the α-1,6 formation is lower by 2400 cal mole^{-1} than that for the α-1,4 formation. This indicates that the enthalpy term contributes to the ease of the formation of α-1,6 linkage rather than that of α-1,4 linkage to give a lower ΔG value for α-1,6 formation than that for α-1,4 formation.

As for the hydrolytic reactions of sugars studied calorimetrically, except for that of the two linkages mentioned above only the inversion of sucrose has been known. Inversion of sucrose is one of the most familiar chemical reactions, and the heat measurements of this reaction have been made by several investigators (Barry, 1920; Kôzaki, 1935; Sturtevant, 1937; Łaźniewski, 1959).

The most accurate determination of the heat of sucrose inversion was made by Sturtevant (1937) in acid-catalyzed systems at various acid concentrations. Since the inversion catalyzed by acids is the complete first-order reaction, the temperature–time curves observed were treated kinetically, taking into account the small changes of rate constants by a temperature rise during the course of reaction. The heat of inversion of sucrose was obtained to be $\Delta H = -3560$ cal mole^{-1}.

III. Other Reactions Involving Carbohydrates

A. Amylose–Iodine Complex Formation

The physicochemical properties of the amylose–iodine complex have been investigated by several workers with respect to the amylose conformation in aqueous solution (Rundle and Baldwin, 1943; Ono *et al.*, 1953, 1965c; Szejtli, 1963). It has been established that the basic structure of the amylose–iodine complex is a helical amylose chain in which the iodine molecules are arranged in a linear array (Rundle and Baldwin, 1943). The enthalpy change of this reaction also has been estimated from the temperature dependence of equilibrium constants obtained by spectrophotometric, potentiometric, or amperometric methods (Holló and Szejtli, 1957; Kuge and Ono, 1960).

Recently Takahashi and Ono (1966c) have studied this reaction by direct calorimetry. In practical measurements the aggregation of the complex formed takes place gradually after the rapid complex formation, and this results in precipitation. If the heat of reaction is determined from the measurement of the total area under the temperature–time curves, the sum of the heat of complex formation and the heat of aggregation, if produced, will be given. In order to avoid this complication, the heat of complex formation was determined from the immediate deflection on the chart by linear extrapolation of the temperature–time curves that was obtained with a thermistor as a sensor.

Although the deflection thus obtained is considered to be almost entirely due to the heat of complex formation, the following device was taken for further verification. Since the rate process of aggregation was known to be proportional to the iodide concentration, the heat measurements were performed at three different KI concentrations and the heat change obtained at each salt concentration from the above method was extrapolated to zero salt concentration. However, with this procedure no variation in the heat data with the change in KI concentrations was observable within the experimental error, and the heat of complex formation was estimated as the average of the total experimental values obtained at different KI concentrations. The heat of amylose–iodine complex formation thus was found to be $\Delta H = -16{,}800$ cal mole^{-1} of I_2. The previous values obtained by the indirect methods (Holló and Szejtli, 1957; Kuge and Ono, 1960) are in good agreement with this value.

B. Bioenergetics in Glycolysis

The anaerobic conversion of glucose or glycogen to pyruvic acid or lactic acid in living cells or various tissues is one of the important processes for organisms to obtain biological energies. Since almost all the reactions involved in glycolysis are reversible, their free energy values were calculated from the equilibrium constants. In this way the energetics of glycolytic reactions have been illustrated in free energy terms (Burton and Krebs, 1953).

However, the published evaluations of the enthalpy change of these reactions by means of direct calorimetry are few. Attempts were made only by Meyerhof and Schulz (1935) on the enzymatic formation of lactic acid from phospho-pyruvate. They also studied the enzymatic aldol condensation of glyceraldehyde and dihydroxyacetone phosphate to give ketohexose-1-phosphate (Meyerhof and Schulz, 1936). These measurements were carried out in rather impure biochemical systems, and reinvestigation of the values may be needed. However, it is worth noting that these investigations were the first calorimetric attempt in the field of bioenergetics.

In 1960 the enthalpy change of isomerization from glucose-6-phosphate to fructose-6-phosphate by phosphoglucoisomerase was shown to be $\Delta H = -2300$ cal mole^{-1} from the temperature dependence of the equilibrium constants (Kahana *et al.*, 1960).

C. Cellulose

Although cellulose is an important substance present in many organisms as a structural element, no attempt has been made to investigate its biochemistry calorimetrically. Physical properties of cellulose during wetting by water, alcohol, and other organic reagents, however, have been investigated by many workers. The heats of wetting by various solvents have been measured in consequence of the industrial significance of these phenomena (Rees, 1948; Dumanskii and Nekryach, 1953; Mischenko *et al.*, 1959; Dymarchuk and Mischenko, 1959; Wahba and Aziz, 1959; Usmanov and Khakimov, 1962; Balcerzyk and Hempel, 1964).

IV. Application

A. Calorimetric Determination of α-1,4 Glucosidic Linkage Content in Starches and Glycogens

Starches and glycogens are high polymers of glucose in which the α-1,4 glucosidic linkages form long chains of glucose unit and the α-1,6 glucosidic linkages participate in cross-linking of the individual glucose chains to give a ramified structure. Therefore, as in starch chemistry, the content of the two kinds of linkage in such polyglucosides is one of the important parameters that describes their structural features.

Earlier in this chapter it was shown that the heats of hydrolysis of α-1,4 and α-1,6 glucosidic linkages are approximately equal in their absolute magnitudes but opposite in sign for the two linkages. It also was shown that the heat of hydrolysis of sugar compounds is obtainable as an additive function of the thermal data of the individual linkages to be hydrolyzed.

If one obtains quantitatively the total heat liberated upon the complete hydrolysis of both the linkages involved in starches or glycogens, the content of the linkages is given by a simple mathematical treatment using the thermal properties mentioned above. An attempt has been made by Takahashi and Ono (1966a) to determine calorimetrically the α-1,4 glucosidic linkage content in some starches and glycogens. These calorimetric measurements have been made on enzymatic hydrolysis of starches from various sources; potato, corn, sweet potato, waxy rice, and of animal- and phyto-glycogens. The reactions were carried out in a Dewar vessel-type calorimeter using gluc-amylase. From the thermal data obtained, the α-1,4 glucosidic linkage content has been estimated for all the samples studied.

If we define $n_{1,4}$ and $n_{1,6}$ as the average numbers of α-1,4 and α-1,6 glucosidic linkages involved in 1 mole of a sample and $\Delta H_{1,4}$ and $\Delta H_{1,6}$ as the heats of hydrolysis of the respective linkages, the total molar heat of hydrolysis of the sample is given by Eq. (1).

$$n_{1,4} \cdot \Delta H_{1,4} + n_{1,6} \cdot \Delta H_{1,6} \tag{1}$$

The heat evolution per gram of the sample, q cal gm^{-1}, is then given by Eq. (2),

$$-q = \frac{n_{1,4} \cdot \Delta H_{1,4} + n_{1,6} \cdot \Delta H_{1,6}}{M_s} \tag{2}$$

where M_s is the average molecular weight of sample, which is given by the following equation with a reasonable approximation that $n_{1,4}$ and $n_{1,6}$ are sufficiently large compared to unity:

$$\begin{aligned} M_s &= (n_{1,4} + n_{1,6}) \times (M_g - M_w) \\ &= 162 \cdot (n_{1,4} + n_{1,6}) \end{aligned} \tag{3}$$

where M_g and M_w are the molecular weights of glucose and water, respectively.

Substituting Eq. (3) into Eq. (2), α-1,4 glucosidic linkage content, α-1,4 %, is given as Eq. (4). Since q is obtainable from the calorimetry of the hydrolytic

$$\begin{aligned} \alpha\text{-}1,4\ \% &= \frac{n_{1,4}}{n_{1,4} + n_{1,6}} \times 100 \\ &= \frac{-\Delta H_{1,6} - 162q}{\Delta H_{1,4} - \Delta H_{1,6}} \times 100 \end{aligned} \tag{4}$$

processes, α-1,4 linkage content, α-1,4 %, is calculated from Eq. (4) by employing the values of $\Delta H_{1,4}$ and $\Delta H_{1,6}$ previously obtained.

The values of $\Delta H_{1,4}$ and $\Delta H_{1,6}$ in several glucosides are summarized in Table II. For calculation of the present purpose it would be more reasonable to use the value obtained for amylose and panose, since both the values correspond to the heats of hydrolysis of the linkages in the presence of the neighboring hydrolyzable linkages.

Therefore, employing the values of $\Delta H_{1,4} = -1030$ cal mole^{-1} and $\Delta H_{1,6} = +1240$ cal mole^{-1}, Eq. (4) becomes Eq. (5), which was used to calculate

$$\alpha\text{-}1{,}4\,\% = \frac{-1240 - 162q}{-1030 - 1240} \times 100$$

$$= \frac{1240 + 162q}{2270} \times 100 \qquad (5)$$

the α-1,4 glucosidic linkage content. Values obtained are listed in Table III.

TABLE III

α-1,4 GLUCOSIDIC LINKAGE CONTENT AND AVERAGE LENGTH OF UNIT CHAIN OBSERVED FOR STARCHES AND GLYCOGENS[a]

Sample	q_{ob} (cal gm^{-1})	q_{corr} (cal gm^{-1})	Linkage content (α-1,4 %)	Amylose content (Am %)	Average length of unit chain (r)
Potato	4.94_9	5.83_0	96.2 ± 0.6	25.0	19
Corn	5.04_9	5.91_4	96.8 ± 1.2	14.4	26
Sweet potato	4.91_0	5.81_3	96.2 ± 1.5	18.0	21
Waxy rice	4.59_8	5.35_6	92.9 ± 1.2	0	13
Animal-glycogen	4.52_3	5.25_9	92.2 ± 0.6		12
Phyto-glycogen	4.87_7	5.04_2	90.6 ± 0.7		10

[a] From Takahashi and Ono (1966a).

In the second column are the values for the heat observed upon the hydrolysis per gram of the sample (q_{ob}), and in the third column are heat data corrected for the anhydrous pure sample (q_{corr}). Substituting q_{corr} of the third column into Eq. (5), the α-1,4 content, α-1,4 %, was obtained and is shown in the fourth column.

If the amylose content in a whole starch is known, the average length of glucose unit chain of the amylopectin component can be calculated from the above data. The amylose content of the starches was determined from the iodine affinity measurement by the microamperometric titration method.

The ratio of α-1,4 to α-1,6 glucosidic linkages in amylopectin, r, is given by combining the value of α-1,4% obtained calorimetrically for whole starch using Eq. (5) with the value of amylose content determined separately by the above method. The equation for r is (6), where Am% is the amylose content in

$$r = \frac{(\alpha\text{-}1{,}4\,\%) - (\text{Am}\,\%)}{100 - (\alpha\text{-}1{,}4\,\%)} \tag{6}$$

whole starch obtained by the amperometric titration method. The value of r is practically equal to the average unit chain length of amylopectin, which is usually determined from the end-group measurement by means of chemical analysis using the methylation or periodic acid oxidation method (see, e.g., Archibald *et al.*, 1961). The amylose content and the average unit chain length of glucose in amylopectin are given in the fifth and the sixth columns, respectively.

Data for the average unit chain length are in good agreement with values obtained by end-group analysis (Hassid and McCready, 1943; Anderson *et al.*, 1955; Wolff *et al.*, 1955; Brown *et al.*, 1945; Peat *et al.*, 1956). The determination of the average unit chain length of amylopectin from various sources is often important in starch chemistry, especially in the industrial field. Many starch chemists are interested in this problem, and chemical methods are generally used. General use of the calorimetric method involves some technical difficulties. However, it should be noted that it is unique in the sense that the linkage content as well as the average unit chain length can be obtained from the "direct" heat measurement based upon the cleavage of individual linkages.

B. Thermal Analysis of Reaction Process of Panose Hydrolysis by Gluc-amylase

Reaction calorimetry as an analytical tool for the reaction of carbohydrates was employed earlier by Kôzaki (1935) and by Osugi and Hiromi (1952). More recently, the enzymatic hydrolysis of panose by gluc-amylase, a typical consecutive reaction, has been studied by the thermoanalytical method. Panose is a trisaccharide that consists of three glucopyranose units joined together by the two glucosidic linkages of different properties, i.e., an α-1,4 and an α-1,6 linkage. Its time course of hydrolysis by gluc-amylase seems to be of special interest since the two linkages have quite different kinetic properties in the enzymatic hydrolysis. As described earlier, gluc-amylase hydrolyzes first the α-1,6 linkage in the panose molecule to yield maltose as an intermediate material that is subsequently hydrolyzed into the final product glucose (Tsujisaka and Fukumoto, 1956). In order to elucidate the time course of panose, maltose, and glucose in this reaction, Ono *et al.* (1963) studied the same reaction by following the thermal change, the changes in the optical rotatory power, and the increase in reducing power.

If we define the four variables C_P, C_M, C_α, and C_β to be the molar concentrations of eq-panose, eq-maltose, α-glucose, and β-glucose, respectively, in the reaction mixture at a given time, the total heat evolution h of the system at this time may be written by Eq. (7), where $C_P{}^0$ is the initial molar concentration of

$$h = \Delta H_P(C_P - C_P{}^0) + \Delta H_M C_M + \Delta H_\alpha C_\alpha + \Delta H_\beta C_\beta \quad (7)$$

the substrate panose and ΔH_P, ΔH_M, ΔH_α, and ΔH_β are the enthalpy changes for the reactions 3 eq-glucose $\rightarrow$ eq-panose, 2 eq-glucose $\rightarrow$ eq-maltose, eq-glucose $\rightarrow$ α-glucose, and eq-glucose $\rightarrow$ β-glucose, respectively. All the values for ΔH are known from the heats of hydrolysis and the heats of anomerization given in the literature.

Equation (8) holds for the conservation of glucose residues. On the other

$$3\,C_P{}^0 = 3\,C_P + 2\,C_M + (C_\alpha + C_\beta) \quad (8)$$

hand, the optical rotatory power α and the glucose equivalent reducing power R of the reaction mixture at a given time may be given by Eqs. (9) and (10),

$$10\alpha = [M_P]\,C_P + [M_M]\,C_M + [M_\alpha]\,C_\alpha + [M_\beta]\,C_\beta \quad (9)$$

$$R = \rho_P\,C_P + \rho_M\,C_M + \rho_G(C_\alpha + C_\beta) \quad (10)$$

where $[M]$ is the molecular rotation of the respective sugar and ρ is the molar glucose equivalent reducing power of respective sugar ($\rho_G = 1$).

The reaction was carried out in a calorimeter (Fig. 2), and the time course of h was obtained from the observed temperature–time curve after correcting for the heat exchange between the calorimeter and the surroundings and for the mutarotational heat due to the β-glucose produced. The values for α and R were determined separately under the same reaction condition with that in the calorimetry. The change in optical rotatory power with time was recorded by using an autorecording polarimeter. The reducing power determination was made by the modified Somogyi–Nelson method (Hiromi *et al.*, 1963) on each aliquot of reaction mixture that was pipetted out at appropriate time intervals.

Using the values of h, α, and R determined separately in the above manner, the values of C_P, C_M, C_α, and C_β at a given time were obtained by solving Eqs. (7)–(10) simultaneously and were plotted against time (Fig. 4).

The time courses of C_P, C_M, and $(C_\alpha + C_\beta)$ were found to be in reasonable agreement with the theoretical curves, which were obtained on the basis of kinetic equations for consecutive enzyme reactions (Hiromi and Ono, 1963) by using the values of two kinetic parameters, the Michaelis constant, and the maximum velocity, determined separately for panose and maltose, respectively (Hiromi *et al.*, 1966a, b).

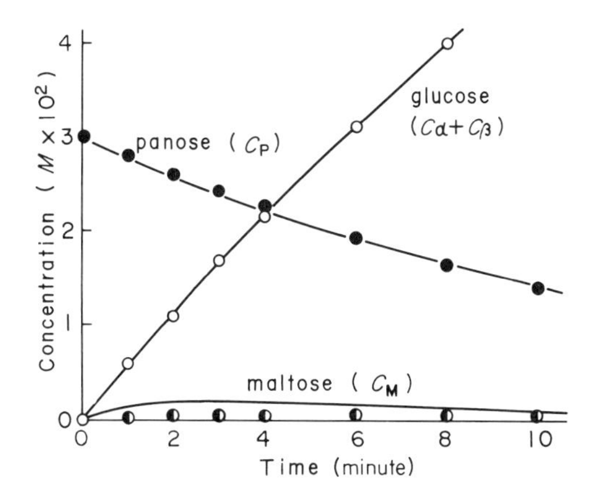

FIG. 4. Time courses of the concentrations of panose, maltose, and glucose for the reaction of panose hydrolysis by enzyme gluc-amylase of Rh. delemar (Ono *et al.*, 1963).

REFERENCES

Andersen, B., and Grønlund, F. (1965). *Acta Chem. Scand.* **19**, 723.

Anderson, D. M. W., Greenwood, C. T., and Hirst, E. L. (1955). *J. Chem. Soc.* p. 225.

Archibald, A. R., Fleming, I. D., Liddle, A. M., Manners, D. J., Mercer, G. A., and Wright, A. (1961). *J. Chem. Soc.* p. 1183.

Balcerzyk, E., and Hempel, K. (1964). *Przeglad Papier.* **20** (10), 309.

Barry, F. (1920). *J. Am. Chem. Soc.* **42**, 1911.

Brown, F., Dunstan, S., Halsall, T. G., Hirst, E. L., and Jones, J. K. N. (1945). *Nature* **156**, 785.

Burton, K., and Krebs, H. A. (1953). *Biochem. J.* **54**, 94.

Dobry, A., and Sturtevant, J. M. (1952). *J. Biol. Chem.* **195**, 141.

Dumanskii, A. V., and Nekryach, E. F. (1953). *Kolloidn. Zh.* **15**, 97.

Dymarchuk, N. P., and Mischenko, K. P. (1959). *Tr. Leningr. Tekhnol. Inst.* No. 7, 115.

Fukumoto, J., Tsujisaka, Y., and Kimoto, K. (1962). *Abstr. 14th Symposium on Enzyme Chemistry, Fukuoka, Japan*, p. 13.

Hamauzu, Z., Hiromi, K., and Ono, S. (1965). *J. Biochem.* (*Tokyo*) **57**, 39.

Hassid, W. Z., and McCreedy, R. M. (1943). *J. Am. Chem. Soc.* **65**, 1157.

Hiromi, K., and Ono, S. (1963). *J. Biochem.* (*Tokyo*) **53**, 164.

Hiromi, K., Takasaki, Y., and Ono, S. (1963). *Bull. Chem. Soc. Japan* **36**, 563.

Hiromi, K., Hamauzu, Z., Takahashi, K., and Ono, S. (1966a). *J. Biochem.* (*Tokyo*) **59**, 411.

Hiromi, K., Takahashi, K., Hamauzu, Z., and Ono, S. (1966b). *J. Biochem.* (*Tokyo*) **59**, 469.

Holló, J., and Szejtli, J. (1957). *Staerke* **9**, 109.

Kabayama, M. A., Patterson, D., and Piche, L. (1958). *Can. J. Chem.* **36**, 557.

Kahana, S. E., Lowry, O. H., Schulz, D. W., Passonneau, J. V., and Crawford, E. J. (1960). *J. Biol. Chem.* **235**, 2178.

Kalckar, H. M. (1954). *In* "The Mechanism of Enzyme Action" (W. D. McElroy and B. Glass, eds.), pp. 675–739. Johns Hopkins Press, Baltimore, Maryland.

Kôzaki, T. (1935). *Rev. Phys. Chem. Japan* **9**, 64.

Kuge, T., and Ono, S. (1960). *Bull. Chem. Soc. Japan* **33**, 1273.

Łaźniewski, M. (1959). *Bull. Acad. Polon. Sci., Sér. Sci., Chim. Géol. Géograph.* **7**, 163.
Lenz, R. W., and Heeschen, J. P. (1961). *J. Polymer Sci.* **51**, 247.
Margrave, J. L., Frisch, M. A., Bautista, R. G., Clarke, R. L., and Johnson, W. S. (1963). *J. Am. Chem. Soc.* **85**, 546.
Meyerhof, O., and Schulz, W. (1935). *Biochem. Z.* **281**, 292.
Meyerhof, O., and Schulz, W. (1936). *Biochem. Z.* **289**, 87.
Mischenko, K. P., Talmud, S. L., and Yakimova, V. I. (1959). *Vysokomolekul. Soedin.* **1**, 662.
Ono, S., Tsuchihashi, S., and Kuge, T. (1953). *J. Am. Chem. Soc.* **75**, 3601.
Ono, S., Hamauzu, Z., Takahashi, K., and Hiromi, K. (1963). *Abstr. 15th Symposium on Enzyme Chemistry, Osaka, Japan*, p. 111.
Ono, S., Hiromi, K., and Hamauzu, Z. (1965a). *J. Biochem. (Tokyo)* **57**, 34.
Ono, S., Hiromi, K., and Takahashi, K. (1965b). *J. Biochem. (Tokyo)* **57**, 799.
Ono, S., Watanabe, T., Ogawa, K., and Okazaki, N. (1965c). *Bull. Chem. Soc. Japan* **38**, 643.
Osugi, J., and Hiromi, K. (1952). *Rev. Phys. Chem. Japan* **22**, 76.
Peat, S., Whelan, W. J., and Turvey, J. R. (1956). *J. Chem. Soc.* p. 2317.
Rao, V. S. R., and Foster, J. F. (1965). *J. Phys. Chem.* **69**, 636.
Rawitscher, M., Wadsö, I., and Sturtevant, J. M. (1961). *J. Am. Chem. Soc.* **83**, 3180.
Rees, W. H. (1948). *J. Textile Inst.* **39**, T351.
Reeves, R. E. (1950). *J. Am. Chem. Soc.* **72**, 1499 (includes additional references).
Reeves, R. E. (1951). *Advan. Carbohydrate Chem.* **6**, 107.
Rundle, R. E., and Baldwin, R. R. (1943). *J. Am. Chem. Soc.* **65**, 554.
Sturtevant, J. M. (1937). *J. Am. Chem. Soc.* **59**, 1528.
Sturtevant, J. M. (1941). *J. Phys. Chem.* **45**, 127.
Szejtli, J. (1963). *Periodica Polytech.* **7**, 259.
Takahashi, K., and Ono, S. (1966a). *J. Biochem. (Tokyo)* **59**, 290.
Takahashi, K., and Ono, S. (1966b). *Abstr. 2nd Japanese Calorimetry Conference, Tokyo.*
Takahashi, K., and Ono, S. (1966c). *Abstr. 19th Ann. Meet. Chem. Soc. Japan*, p. 259.
Takahashi, K., Hiromi, K., and Ono, S. (1965a). *J. Biochem. (Tokyo)* **58**, 255.
Takahashi, K., Yoshikawa, Y., Hiromi, K., and Ono, S. (1965b). *J. Biochem. (Tokyo)* **58**, 251.
Takasaki, Y. (1967). *Agr. Biol. Chem. (Tokyo)* **31**, 309, 435.
Tsujisaka, Y., and Fukumoto, J. (1956). *Kagaku To Kogyo (Osaka)* **30**, 130.
Tsujisaka, Y., Fukumoto, J., and Yamamoto, T. (1958). *Nature* **181**, 770.
Usmanov, K. U., and Khakimov, I. K. (1962). *Dokl. Akad. Nauk Uz. SSR* **19**, No. 3, 42.
Wahba, M., and Aziz, K. (1959). *J. Textile Inst., Trans.* **50**, T558.
Wolff, I. A., Hofreiter, B. T., Watson, P. R., Deatherage, W. L., and McMasters, M. M. (1955). *J. Am. Chem. Soc.* **77**, 1654.

CHAPTER V

Chemical Structure and Reaction Mechanisms: Proteins and Other *N*-Containing Compounds

WILLIAM J. EVANS

I. Heats of Hydrolysis of Amide and Peptide Bonds

Sturtevant and his co-workers have studied by reaction calorimetry the heats of hydrolysis of dipeptides, amino acid amides, and one polypeptide with the purpose of gaining further insight into the biosynthesis of proteins. Another aspect, the definition of state of ionization, recently has been presented (Sturtevant, 1962), but because of its applicability to many biochemical reactions, the importance of this particular point hardly can be over-emphasized. For example, the hydrolysis of benzoyl-L-tyrosyl-glycine amide at approximately neutral pH may be represented as Eq. (1) where RCOOH

$$RCO\text{—}NHCH_2CONH_2(aq) + H_2O = RCOO^-(aq) + (1 - \alpha)^+NH_3CH_2CONH_2(aq) + \alpha H^+(aq) + \alpha NH_2CH_2CONH_2(aq) \qquad \Delta H_1 \quad (1)$$

represents benzoyl-L-tyrosine. If the solution is buffered, the reaction (2) also

$$\alpha BH^{=}(aq) + \alpha H^+(aq) = \alpha H_2B^-(aq) \qquad \alpha\Delta H_2 \quad (2)$$

occurs. The quantity α is estimated from the ionization constant of glycine

$$RCO\text{—}NHCH_2CONH_2(aq) + H_2O = RCOO^-(aq) + {}^+NH_3CH_2CONH_2(aq) \qquad \Delta H_h \quad (3)$$

amide. The heat of reaction (3) is given by Eq. (4), where ΔH_p is the heat of

$$\Delta H_h = \Delta H_{obs} + \alpha(\Delta H_l - \Delta H_p) = \Delta H_1 - \alpha\Delta H_p \quad (4)$$

ionization of glycine amide and is approximately 10,000 cal mole^{-1}. Since ΔH_h is only -1550 cal mole^{-1}, the importance of the definition of the state of ionization in this particular example is immediately obvious. Some of the early calorimetric work has suffered from neglect of this fact and of suitable buffer corrections, a striking example being the determination of the heat of hydrolysis of adenosine triphosphate to form adenosine diphosphate (Meyerhof and Lohmann, 1932). The value obtained later was shown (Podolsky and Sturtevant, 1955) to be approximately twice too large because, among other factors, no allowance was made for heat effects resulting from the addition of protons to the buffer during the reactions.

II. Heats of Denaturation of Proteins

Several calorimetric determinations of the heat of denaturation of methemoglobin and pepsin by alkali and methemoglobin by salicylate have been carried out by Kistiakowsky and his co-workers (Conn *et al.*, 1940, 1941; Roberts, 1942). The alkaline denaturation of methemoglobin was studied at constant alkali concentration and at constant pH, though the former was emphasized due to the difficulty of obtaining stable pH readings. At constant alkali concentration the following equations hold:

$$(\text{Native protein} + n\text{KOH}) + m\text{KOH} \rightarrow [\text{denatured protein} + (n+m)\text{KOH}] \quad \Delta H_1$$

$$(\text{Denatured protein} + n\text{KOH}) + m\text{KOH} \rightarrow [\text{denatured protein} + (n+m)\text{KOH}] \quad \Delta H_2$$

The heat of these two reactions was measured in the calorimeter, and the difference between them was taken as the heat of denaturation. A value of 138 kcal mole^{-1} was found for the denaturation reaction.

Sturtevant and his associates (Bender and Sturtevant, 1947; Buzzell and Sturtevant, 1952; Sturtevant, 1954) studied the alkaline denaturation of pepsin, finding the heat of reaction to be a complex variation of pH and highly dependent upon temperature. Forrest and Sturtevant (1960) conducted a calorimetric study of the acid denaturation of horse ferrihemoglobin and found ΔH values independent of pH over a moderate pH range but strongly dependent upon temperature. An interesting consequence of this work was the observation that the calorimetric data were more consistent with protons adding predominately to carboxylate ions rather than imidazole groups in the denatured protein. It has been previously proposed on the basis of tirtration studies that of the approximately 36 groups that become available to tritration during the denaturation process, 22 are somehow masked in the native protein and are most likely to be imidazole groups of histidine residues (Beychok and Steinhardt, 1959). In the calorimetric experiments on ferrihemoglobin, the denaturation was found to be strongly exothermic at 15° and weakly endothermic at 25°. Comparison of the heat values obtained at these temperatures indicated that for the denaturation process at 20°, ΔC_p was +9850 cal mole^{-1} deg^{-1}. The only previously reported ΔC_p (Bro and Sturtevant, 1958) of this magnitude was +8000 cal mole^{-1} deg^{-1}, which value was obtained for bovine serum albumin in solution when the pH was lowered from 5 to 3. The similarity of these figures suggests that the two processes may be fundamentally similar and that the enormous changes in heat capacity are indicative of the complexity of the processes.

Hermans and Rialdi (1965) recently made a study of the heat of ionization and denaturation of sperm-whale myoglobin over the pH range 2–12.5. For the acid denaturation at about pH 4.5, a heat of approximately 40 kcal mole^{-1} had to be assumed, which balanced the heat of uptake of the hydrogen ion by the six abnormal histidine side chains existing in the native protein. Denaturation occurred at pH values greater than 11.5 accompanied by a heat uptake of about 30 kcal mole^{-1}. The calorimetric results were in agreement with those obtained by spectrophotometric determinations of the unfolding equilibrium. A rather interesting fact that is sometimes overlooked was pointed out in this work: The heat of denaturation can be compared with values obtained from the temperature dependence of equilibrium, which nearly always can be studied by other techniques. However, the two methods of obtaining ΔH denaturation need not give identical results. For example, for a protein molecule consisting of two identical halves that can unfold independently, the value of the heat of denaturation obtained calorimetrically will be exactly twice as large as the value determined using the van't Hoff equation. Calorimetric measurement, therefore, may be able in many cases to provide information not otherwise available.

III. Aspects of Calorimetric Study of Antigen–Antibody Reactions

The first attempt to measure the heat of an antigen–antibody reaction by calorimetric means was that of Bayne-Jones (1925), who obtained a value of -1.5×10^9 cal mole^{-1} as ΔH of diphtheria toxin–antitoxin interaction. Later Boyd and co-workers (1941) concluded that this value was greatly in error and that the source of error was probably mechanical, i.e., due to a heat of stirring attributed to the reaction proper. In the work of Boyd *et al.*, the heat of interaction of hemocyanin with horse antihemocyanin was measured and the results indicated a value of about 40,000 cal mole^{-1} of antibody. Assuming what was considered a reasonable value of the change in free energy, −10,000 cal in this instance, they estimated a change in entropy of approximately −100 entropy units. More recently Steiner and Kitzinger (1956) have directly measured the heat of combination of human serum albumin with rabbit antibody and obtained a value of $\Delta H = -3500$ cal bond^{-1} under the assumption that all antibodies are bivalent. This value of ΔH was combined with the free energies of association computed from equilibrium studies on the same material to give an entropy of association of about +14 entropy units. It is interesting that the heat of reaction found in this study was much smaller than that found for the hemocyanin–antihemocyanin system. Current evidence indicates that the small, favorable free energy changes for antigen–antibody reactions must be largely attributed to an entropy increase associated with the reactions.

REFERENCES

Bayne-Jones, S. J. (1925). *J. Immunol.* **10**, 663.

Bender, M., and Sturtevant, J. M. (1947). *J. Am. Chem. Soc.* **69**, 607.

Beychok, S., and Steinhardt, J. (1959). *J. Am. Chem. Soc.* **81**, 5679.

Boyd, W. C., Conn, J. B., Gregg, D. C., Kistiakowsky, G. B., and Roberts, R. M. (1941). *J. Biol. Chem.* **139**, 787.

Bro, P., and Sturtevant, J. M. (1958). *J. Am. Chem. Soc.* **80**, 1789.

Buzzell, A., and Sturtevant, J. M. (1951). *J. Am. Chem. Soc.* **73**, 2454.

Conn, J. B., Kistiakowsky, G. B., and Roberts, R. M. (1940). *J. Am. Chem. Soc.* **62**, 1895.

Conn, J. B., Gregg, D. C., Kistiakowsky, G. B., and Roberts, R. M. (1941). *J. Am. Chem. Soc.* **63**, 2080.

Forrest, W. W., and Sturtevant, J. M. (1960). *J. Am. Chem. Soc.* **82**, 585.

Hermans, J., and Rialdi, G. (1965). *Biochemistry* **4**, 1277.

Meyerhof, O., and Lohmann, K. (1932). *Biochem. Z.* **253**, 431.

Podolsky, R. J., and Sturtevant, J. M. (1955). *J. Biol. Chem.* **217**, 603.

Roberts, R. M. (1942). *J. Am. Chem. Soc.* **64**, 1472.

Steiner, R. F., and Kitzinger, C. (1956). *J. Biol. Chem.* **222**, 271.

Sturtevant, J. M. (1954). *J. Phys. Chem.* **58**, 97.

Sturtevant, J. M. (1962). *In* "Experimental Thermochemistry" (H. A. Skinner, Ed.), Vol. 2, Chapt. 19. Wiley (Interscience), New York.

CHAPTER VI

Physical States of Biomolecules: Calorimetric Study of Helix-Random Coil Transitions in Solution

T. ACKERMANN

I. Introduction

A. Physical Properties and the Secondary Structure of Nucleic Acids and Polypeptides in Solution

As shown recently, many biopolymers (nucleic acids and proteins) and their synthetic models have a very high degree of intramolecular ordering—a definite arrangement of groups and distribution of secondary bonds. The helical duplex structure (Watson and Crick, 1953) found for native deoxyribonucleic acid reoccurs for a whole class of polynucleotides of high helical content. Indeed, this kind of secondary structure, stabilized by interbase hydrogen bonding, occupies the same central position in the polynucleotide field as does the α-helix (Pauling *et al.*, 1951) in the polypeptide field. There is, however, a basic difference between the two cases. The α-helix is shown in Fig. 1. The structure is stabilized by hydrogen bonding of the "vertical" type,

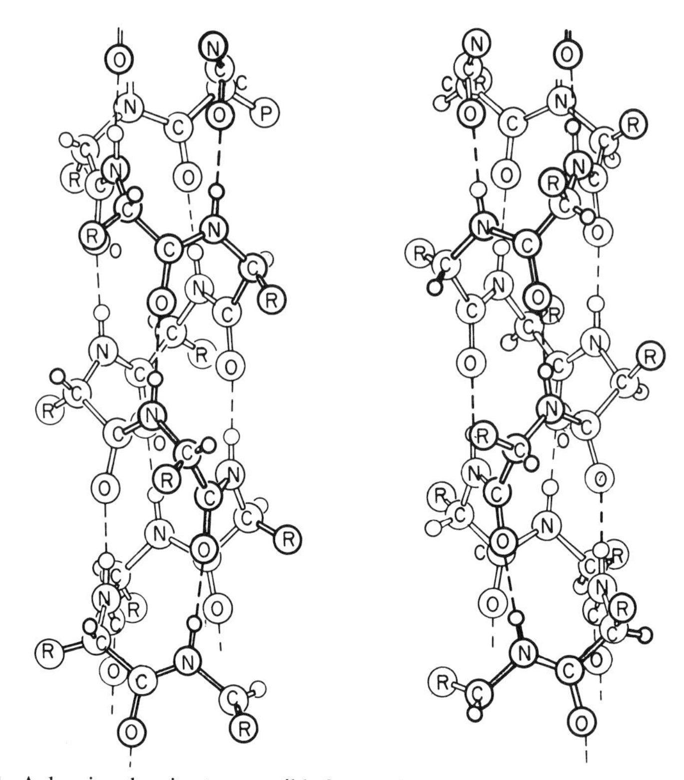

FIG. 1. A drawing showing two possible forms of the α-helix; the one on the left is a left-handed helix, the one on the right is a right-handed helix. In both, the amino acid residues have the L-configuration (Pauling *et al.*, 1951).

for which the hydrogen bonds link groups of the peptide backbone itself and are parallel to the fiber axis. The hydrogen bonding of the polynucleotide helical duplex (Fig. 2) does not usually involve the sugar–phosphate backbone. Furthermore, the bonding in the helical form of the polynucleotides is of the "horizontal" type, with the hydrogen bonds roughly perpendicular to the fiber axis.

Helical forms also have been found to occur for the dissolved polymeric molecules of the polynucleotide and polypeptide type. In solvents that competitively disrupt hydrogen bonding, however, both the magnitudes of the observed physical properties and their dependence upon molecular weight are characteristic of randomly coiled polymer molecules. For example, poly-γ-benzyl-L-glutamate, depending upon the solvent, can exist in either the α-helical or the randomly coiled configuration. Doty and co-workers (1957) found that the hydrodynamic properties of poly-γ-benzyl-L-glutamate are profoundly dependent upon solvent composition. In solvents relatively favourable to the formation of intramolecular hydrogen bonds, such as ethylene dichloride, the molecular dimensions computed from intrinsic viscosity correspond closely to those expected for essentially perfect α-helices.

The determination of the spatial organization of proteins and polypeptides by X-ray diffraction methods is extremely laborious and is limited by the nature of the technique to the crystalline state. Thus it is useful to have secondary

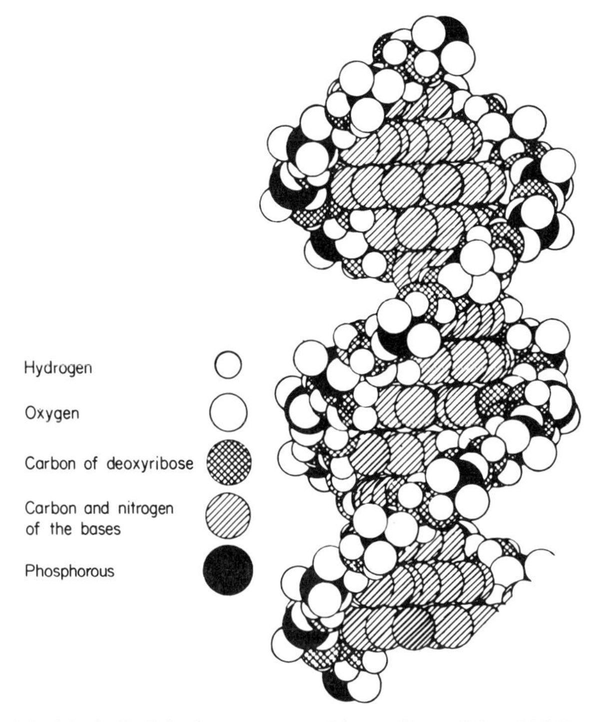

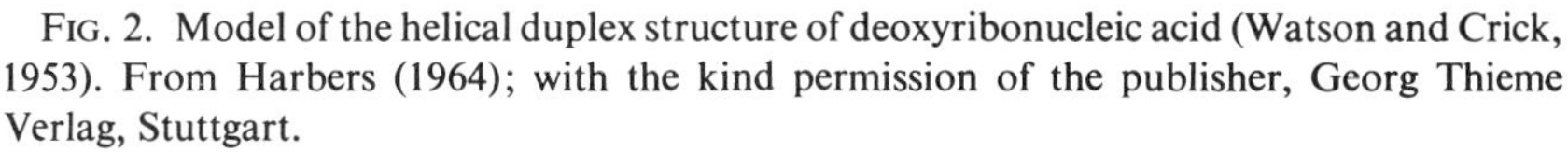

FIG. 2. Model of the helical duplex structure of deoxyribonucleic acid (Watson and Crick, 1953). From Harbers (1964); with the kind permission of the publisher, Georg Thieme Verlag, Stuttgart.

criteria for the helical content of biopolymers and their synthetic models in solution. The limitations of the available techniques should be recognized. They do not attempt to do more than assess the overall fraction of the polymer chain in the helical conformation. In particular, they cannot locate the helical regions within the molecule or differentiate between a single long helix and several shorter helices. Nevertheless, these secondary criteria remain the only available structural probes that can assess the helical content of polypeptides and polynucleotides in solution. Their relative simplicity and applicability to noncrystalline systems has rendered them the only source of information for most proteins and nucleic acids.

Since proteins are formed exclusively from amino acids in the L-configuration, a major contribution to the optical rotation arises from the summed contributions of the individual amino acid residues. This intrinsic optical activity persists under conditions where the spatial organization of the protein is partially, or completely, lost. A similar magnitude of intrinsic optical activity is to be expected for synthetic polypeptides of the poly-γ-benzyl-L-glutamate type. In the case of proteins and polypeptides with an important degree of α-helical content, a second major part of the optical rotation arises from the intrinsic asymmetry of the α-helix itself. The α-helix can exist in either a right-handed or a left-handed form (Fig. 1). The two forms are mirror images and cannot be interconverted by rotating the helices in space. As a consequence of its spatial asymmetry, the α-helix will have a definite optical activity quite distinct from that arising from the summed contributions of the individual amino acids. In the case of helical polypeptides composed of L-amino acids, the two components of the resulting optical activity are comparable in magnitude and opposite in sign. It is not yet possible to relate the specific rotation $[\alpha]$ quantitatively to the α-helical content of a protein or synthetic polypeptide. Certain semiquantitative relationships, however, have emerged from studies upon the synthetic systems. An increase in helical content is (for polymers of L-amino acids) generally accompanied by a decrease in L-rotation. For solutions of poly-γ-benzyl-L-glutamate in mixtures of ethylene dichloride and dichloroacetic acid, the change in optical rotation has been found to be roughly proportional to the fraction of helical content. However, there is definite evidence for the existence of nonspecific solvent effects upon the specific rotation.

A more quantitative basis for the estimation of α-helical content is provided by the wavelength dependence, or dispersion, of optical rotation. Estimates of the fractional helical content obtained from rotatory dispersion are in agreement with those predicted from the more complete structural determination. For detailed information the reader is referred to the monographs listed as "General References" in the reference list of this chapter.

Native deoxyribonucleic acid approaches the extreme case of a completely helical conformation (Fig. 2). In dilute solutions the polynucleotides have been found to include all gradations of secondary structure, from the completely amorphous to the highly helical. Helical forms other than the duplex also have been found to occur for the biosynthetic polymers. In general it probably can be stated that the establishment of the secondary structures of most polynucleotide systems is approaching completion. The achievements in this area have been made possible only by calling upon all of the conventional physicochemical methods for determining the gross size and shape of macromolecules in solution. These include light scattering, ultracentrifugation, viscometry, and streaming birefringence.

In studies upon polynucleotide systems, ultraviolet hypochromism can be used as a secondary criterion for the helical content of the dissolved polymer molecules. The very strong ultraviolet absorption of all the purine and pyrimidine bases and the sensitivity of this observed parameter to structural and configurational changes have made ultraviolet spectroscopy one of the most commonly employed techniques in studying such systems. The change in optical density is not proportional to the helical content of the polynucleotide complex. In general the appearance of changes in optical density appears to parallel the onset of a disruption of the secondary structure, as judged by other criteria. A quantitative treatment of the hypochromatic effect for polynucleotides is not yet available.

B. Methods of Altering the Secondary Structure of the Polymeric Solute in Polypeptide and Nucleic Acid Solutions

It is clear from the discussion of Section I, A that many biopolymers and their synthetic models can undergo a transition from the completely ordered helical state to the entirely amorphous situation of a randomly coiled molecule. This helix-random coil transition can be mediated by a change in a parameter that influences the degree of intramolecular hydrogen bonding in the dissolved polymeric molecule. By varying a single experimental parameter (e.g., temperature, solvent composition, or pH), it is often possible to observe the transition between the two states. The helix-coil transition also may be brought about by varying the pressure applied to the system. For polypeptides, the pressure dependence of the helix content has been studied by Rifkind and Applequist (1964). A characteristic feature of the helix-coil transition for polypeptides and polynucleotides of high molecular weight is the dramatically sharp character it often assumes. In the case of poly-γ-benzyl-L-glutamate of high molecular weight (see Section I, A), the process attains completion over a temperature range of about 5°. In view of the low enthalpy change accompanying the formation of hydrogen bonds, the abruptness of this transition cannot be accounted for by any simple equilibrium treatment.

1. *Variation of the Solvent Composition* (*Effect of pH Changes*)

An example of a transition that is mediated by a change in a solvent parameter is furnished by an equimolar mixture of polyadenylic acid (poly A) and polyuridylic acid (poly U) in 0.10 *M* aqueous potassium chloride solution. The characteristic influence of pH changes on ultraviolet hypochromism is shown in Fig. 3. The absorbency at 259 mμ of the equimolar poly A–poly U complex is constant until a critical alkaline pH is attained, and then, it displays an almost discontinuous increase to a value corresponding to that for the nonassociated components.

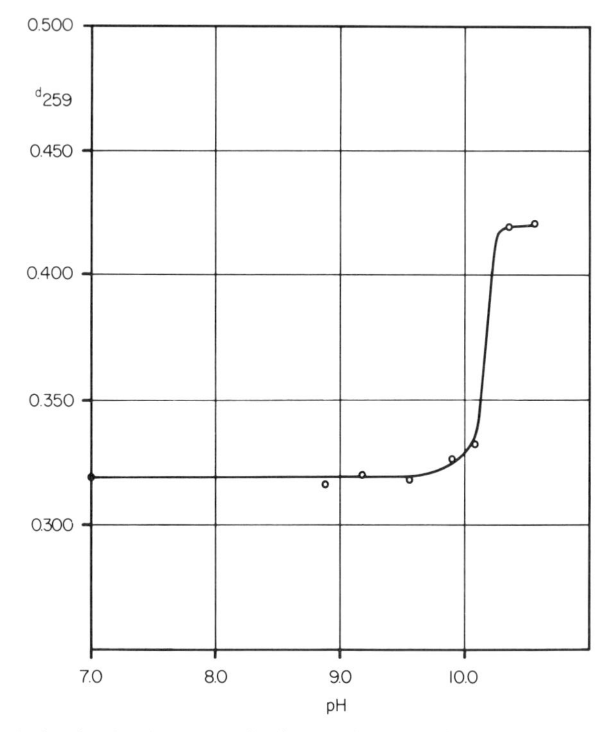

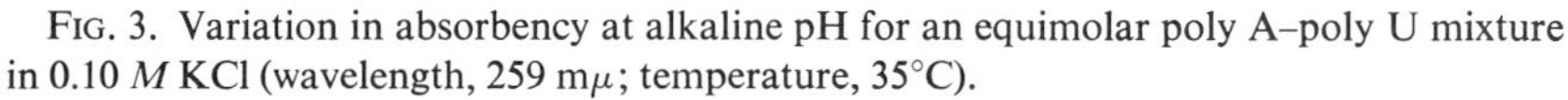

FIG. 3. Variation in absorbency at alkaline pH for an equimolar poly A–poly U mixture in 0.10 *M* KCl (wavelength, 259 mμ; temperature, 35°C).

Analogously, the specific rotation [α] of poly-γ-benzyl-L-glutamate in a mixture of dichloroacetic acid and chloroform is independent of the chloroform content at room temperature until a critical solvent composition is attained, at which point the measured optical rotation increases very sharply to that predicted for the helical form of the dissolved polypeptide.

2. *Variation of Temperature*

A thermal transition of the secondary structure of a polypeptide occurs in the case of poly-γ-benzyl-L-glutamate (Fig. 4) in ethylene dichloride–dichloroacetic acid. Optical rotation is a convenient index of the extent of conversion, the change in rotation being roughly proportional to the fraction of residues that have undergone the transition. In a similar way the reversible dissociation of the poly A–poly U complex can be brought about by varying the temperature. The optical density at 259 mμ of the equimolar polynucleotide mixture is almost independent of temperature at neutral pH until the transition temperature range is attained, at which point a major change in absorbency is observed. The so-called "melting point" (i.e., the apparent half conversion

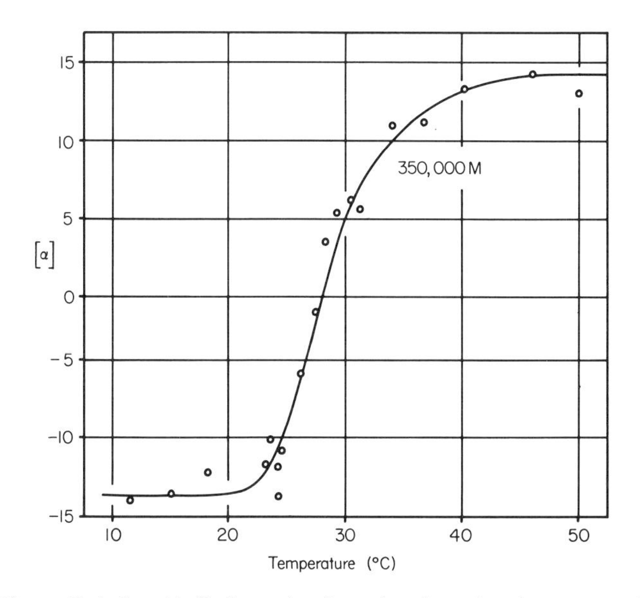

FIG. 4. Thermally induced helix formation for poly-γ-benzyl-L-glutamate, as followed by optical rotation. The solvent is ethylene dichloride–dichloroacetic acid (molecular weight of the sample, 350,000).

temperature) of the ordered helical duplex structure increases with increasing ionic strength (see General References). The helix-coil transition may be regarded as somewhat similar to a phase transition. Strictly speaking, however, the cooperative change in secondary structure is quite different from a first-order phase transition. The imperfect character of the analogy is indicated by the finite breadth of the transition range, as contrasted with the discontinuous nature of a true phase transition. Another example of a thermal helix-coil transition is furnished by native deoxyribonucleic acid (DNA) in aqueous solutions of definite pH and ionic strength. As in the case of the poly A–poly U system, the transition from the ordered helical duplex to the randomly coiled conformation can be observed by measuring the ultraviolet absorbency as a function of temperature. The thermal denaturation of the more complex DNA system, however, is not entirely reversible.

C. Heat (Enthalpy, ΔH) of Helix Formation (Standard States and Definition of Symbols)

Starting from general thermodynamic considerations, one would expect that heat is absorbed in an endothermic transconformation of dissolved macromolecules, as in the melting of solids. From the absorbed heat one could determine the number and energy of the broken hydrogen bonds and also the change of entropy in the helix-random coil transitions. However, in spite of the exceeding importance of information of this type, the experimental data on the enthalpy changes corresponding to helix-coil transitions of biopolymers were lacking until recently. The main cause was the great difficulty of recording weak thermal effects in dilute solutions of macromolecules that are available only in very limited amounts. If we take further account of the fact that the transition temperature range is sometimes broader than in the examples mentioned above and, consequently, at a low heating rate the heat absorption due to the thermal transition is strongly stretched out in time, it becomes obvious that such research encounters technical difficulties. In the meantime most of these difficulties have been overcome by using improved experimental techniques. The enthalpies of transition have been determined for some typical biopolymers and their synthetic models. The present situation in this field is reflected in the survey of ΔH values (Table I), which is discussed in the following section. Usually it is the standard state functions (i.e., at infinite dilution) that are desired. The state of a system (macroscopic state) is determined by its properties just insofar as these properties can be investigated directly or indirectly by experiment. The enthalpy H is a state function of the system. Thus, the change in enthalpy for a given process is determined only by the initial and final states and not by the particular manner in which the process is carried out. If we have a helix-random coil transition involving substances at different pH values, e.g., we have ΔH as the difference between the partial molal enthalpies of the dissolved polymeric substance corresponding to the initial and final pH values of the system. For the definition of the apparent and partial molal quantities, the reader is referred to Lewis and Randall (1961).

If the transition is brought about by a variation in temperature (i.e., at constant solvent composition), starting from an initial state that is equal to that of the example mentioned above, the final state of the system differs from the final state for the transition mediated by varying the solvent composition at constant temperature. Hence the definition of the molal enthalpies requires that the enthalpy change involved in the thermal transition be different from the ΔH value observed for a helix-random coil transition that is mediated by a variation in solvent composition at constant temperature. This difference must be regarded, at least in principle, when a collection of experimental values of transition enthalpies is considered for comparison. Since the exact definition of the property denoted by ΔH is not clearly stated in various papers dealing with direct or indirect determinations of transition enthalpies for helix-random coil transitions, the reader should be careful in looking for additional information on the initial and final states of these transitions.

Furthermore, the concentration dependence of the ΔH values is disregarded, for the most part, in reports of calorimetric work in this field. In other words, the standard state values (i.e., ΔH at infinite dilution) have not been determined. This property, which we denote by ΔH^0, can be obtained only by determining ΔH at various concentrations and extrapolating these values to infinite dilution. In this context the ΔH values determined at a single finite concentration must be regarded as preliminary results that can be used for approximative calculations.

Finally, we must state what we mean by a molal property (enthalpy, heat capacity, etc.) of a polymer molecule. In principle, we may define ΔH as the measured enthalpy change of the system (i.e., the additional enthalpy change involved in the transition) divided by the average number of moles of the dissolved polymer. The average molecular weight of a polymeric substance, however, is less well determined and depends upon the special type of method used in the determination of this quantity. We therefore define ΔH as the measured transition enthalpy divided by the number of moles of the basic stoichiometric unit of the polymeric substance. This basic stoichiometric unit is defined as the base pair in the helical duplex, when results of calorimetric measurements of polynucleotides are reported (see Table I). In the case of polypeptides, the basic stoichiometric unit is simply defined as the amino acid residue.

A somewhat different situation occurs when the molecular weight of a biomolecule is confined to a relatively small unequivocal value. For example, the molecular weight of ribonuclease is a well-determined quantity. In this case, ΔH is defined as the measured enthalpy change involved in the structural transition per mole of the enzyme.

A list of the symbols used in this section is given below:

Θ	Fractional helical content of the polymer.
$[\alpha]$	Specific optical rotation.
pH	Negative logarithm of the hydrogen ion activity in solution.
T	Absolute temperature (Kelvin temperature).
R	Gas constant.
ΔH	Molal enthalpy change involved in a helix-random transition (this symbol requires additional information on the initial and final state of the system).
ΔH^0	Standard state value of ΔH (i.e., at infinite dilution, see ΔH).
C_c	Excess heat capacity in the temperature range of a thermal helix-random coil transition.
σ	Cooperative parameter of the Zimm–Bragg theory (see Zimm and Bragg, 1959).

TABLE I

CALORIMETRIC ΔH VALUES ACCOMPANYING CONFORMATIONAL CHANGES OF MACROMOLECULES IN SOLUTION[a]

Macromolecule	Molecular weight	S_{20}	Solvent	pH[b]	Concentration[b]	Temperature (°C)[c]	Type of transition	Type of measurement	ΔH (kcal)	References[d]
Pepsin	3.5×10^5		0.0 *M* phosphate and about 0.15 *M* KCl	6.41	0.2–0.5%	15	Denaturation	Heat of mixing	22[e] mole^{-1}	(1, 2)
						35			69[e] mole^{-1}	
Trypsin	2.0×10^4		0.1 *M* NaCl	1.4–2.5		25	Denaturation	Heat of mixing	8.0 mole	(3)
Fibrin	3.3×10^5		1.0 *M* NaBr-acetate,	6.08	2.91%	25	Polymerization	Heat of mixing	−19 mole^{-1}	(4)
			phosphate	6.88			Clotting		−44.5 mole^{-1}	
Fibrinogen	3.3×10^5		Phosphate	6–8.5	5 mg ml^{-1}	—	Clotting	Heat of mixing	−44 mole^{-1}	(5)
Mercaptalbumin	6.7×10^4		0.1 *M* NaCl	2.8–4.7	2–3.5%	25	Denaturation	Heat of mixing	1.5–3.4 mole^{-1}	(3, 6)
						15			1.5 mole^{-1}	(6)
Ferrihemoglobin	6.8×10^4		0.02 *M* Sodium formate	3.2–3.8	0.76%	25	Denaturation	Heat of mixing	10±0.3 mole^{-1}	(7)
					1.17%	15			76±1.6 mole^{-1}	
Horse serum albumin	6.9×10^5		0.1 *M* glycine	7.0	2%	55	Denaturation	Heat capacity	90±15 mole^{-1}	(8)
						68			75±10 mole^{-1}	
						76			55±7 mole^{-1}	
Sperm-whale myoglobin	1.78×10^4		0.15 *M* KCl	4.5	3 mg ml^{-1}	39	Denaturation	Heat of mixing	40 mole^{-1}	(9)
Ribonuclease A	1.37×10^4		0.15 *M* KCl	2.8	1.385 and 2.69 wt%	43	Denaturation	Heat capacity	70[f]±1 mole^{-1}	(10)
	1.37×10^4		0.1 *M* KCl	2.2	0.5–1.0 mg ml^{-1}	45	Denaturation	Heat of mixing	109±5 mole^{-1}	(11)
	1.37×10^4		0.15 *M* KCl	2.8	1.5%	44	Denaturation	Heat capacity	86.5×4.4 mole^{-1}	(12)
Procollagen	3.6×10^5		0.1 *M* citrate	4.0	0.15%	37	Denaturation	Heat capacity	2.52×10^6 mole^{-1}	(13)
Lysozyme	1.43×10^4		H_2O-HCl	1.0	3 wt vol%$^{-1}$	45	Denaturation	Heat capacity	55 mole^{-1}	(14)
Poly-γ-benzyl-L-glutamate	2.35×10^5		19 wt% DCE 81 wt% DCA		0.257 m residue^{-1}	32	Coil-helix	Heat capacity	0.43 residue^{-1}	(15, 16)
					0.132 m residue^{-1}				0.68 residue^{-1}	
					0.068 m residue^{-1}				0.81 residue^{-1}	
					C→O				0.95±0.03 residue^{-1}	
	2.7×10^5		25 vol% DCE 75 vol% DCA		0.097 m residue^{-1}	26	Coil-helix	Heat capacity	0.525±0.08 residue^{-1}	(17)
	1.6×10^5		DCA-DCE → 100% DCA		0.007 m residue	30	Coil-helix	Heat of solution	0.70±0.05 residue^{-1}	(18)
	3.5×10^5		DCE–DCA				Coil-helix	Heat capacity		(19)
			82 wt% DCA		0.25 m residue^{-1}	37			0.38 residue^{-1}	
					0.07 m	43			0.79 residue^{-1}	

			83 wt% DCA		0.25 m residue^{-1}	40			0.32 residue^{-1}	
					0.13 m residue^{-1}	44			0.62 residue^{-1}	
					0.07 m residue^{-1}	46			0.78 residue^{-1}	
			85 wt% DCA		0.25 m residue^{-1}	46			0.29 residue^{-1}	
					0.13 m residue^{-1}	50			0.58 residue^{-1}	
					0.07 m residue^{-1}	53			0.76 residue^{-1}	
			88 wt% DCA		0.25 m residue^{-1}				0.26 residue^{-1}	
					0.13 m residue^{-1}				0.55 residue^{-1}	
					0.07 m residue^{-1}				0.74 residue^{-1}	
Poly-γ-benzyl-L-glutamate (deuterated)	2.7×10^5		34 vol% DEC 66 vol% DCA		3%	8.5	Coil-helix	Heat capacity	0.67 ± 0.05 residue^{-1}	(20)
			18 vol% DCE 82 vol% DCA			40			0.38 ± 0.05 residue^{-1}	
Poly-ε-Carbo-benzoxy-L-lysine	2.75×10^5		37 vol% DCA 63 vol% $CHCl_3$		3%	26	Coil-helix	Heat capacity	0.21 ± 0.06 residue^{-1}	(21)
Poly-L-glutamate	$0.4–1.0 \times 10^5$		0.1 M KCl		0.5 mg ml^{-1}	30	Coil-helix	Heat of mixing	1.1 ± 0.2 residue^{-1}	(22)
Poly-β-benzyl-L-aspartate	1.78×10^5		8 vol% DCA 92 vol% $CHCl_3$		3%	10	Coil-helix	Heat capacity	0.050 residue^{-1}	(23)
Salmon DNA		21.7	0.1 M NaCl	2.75	0.15–0.6 mg ml^{-1}	25	Acid denaturation	Heat of mixing	8.31 base pair^{-1}	(24, 25)
Herring Sperm-atozoa DNA			0.015 M NaCl and 0.0015 M citrate		1%	75	Thermal denaturation	Heat capacity	5 base pair^{-1}	(16)
P. Fluorescens DNA		20.1	0.1 M NaCl	2.5	0.15–0.6 mg ml^{-1}	25	Acid denaturation	Heat of mixing	7.83 basic pair^{-1}	(25)
S. Marcescens DNA		17.4	0.1 M NaCl	2.5	0.15–0.6 mg ml^{-1}	25	Acid denaturation	Heat of mixing	7.83 basic pair^{-1}	(25)
Sea urchin DNA		23.3	0.1 M NaCl	2.5	0.15–0.6 mg ml^{-1}	25	Acid denaturation	Heat of mixing	8.03 base pair^{-1}	(25)
Calf thymus DNA	$>10^6$		0.015 M NaCl and 0.0015 M citrate		10 mg ml^{-1}	72	Thermal denaturation	Heat capacity	7.0 ± 0.5 base pair^{-1}	(26)
T_2 phage DNA			0.1 M NaCl and 0.007 M phosphate	7.0	1.1 mg ml^{-1}	78	Thermal denaturation	Heat capacity	8.5 ± 0.4 base pair^{-1}	(13)[g], (27)
Poly(A + U)		4.5–10.0 Poly U 8.0–12.0 Poly A	0.1 M KCl and 0.01 M cacodylate	6.6	2×10^{-4} 5×10^{-5} M nucleotide	25	Poly A + Poly U = Poly (A + U)	Heat of mixing	-5.9 ± 0.2 base pair^{-1}	(28)

TABLE I—*continued*

Macromolecule	Molecular weight	S_{20}	Solvent	pH[b]	Concentration[b]	Temperature (°C)[c]	Type of transition	Type of measurement	ΔH (kcal)	References[d]
			0.5 *M* KCl and 0.01 *M* cacodylate						−5.25±0.2 base pair^{-1}	
			1.0 *M* KCl and 0.01 *M* cacodylate						−4.75±0.3 base pair^{-1}	
		2.1–12.2 Poly A	0.1 *M* KCl and 0.01 *M* cacodylate						−5.95±0.1 base pair^{-1}	
		6.1 Poly A 7.2 Poly U	0.1 *M* KCl and 0.01 *M* cacodylate	7.0	3.5×10^{-4} *M* nucleotide	10		Heat of mixing	−6.29±0.19[h] base pair^{-1}	(29)
						25			−6.97±0.17[h] base pair^{-1}	
						40			−7.72±0.29[h] base pair^{-1}	
			0.1 *M* NaCl and 0.01 *M* cacodylate	6.8	8×10^{-4} *M* nucleotide	24		Heat of mixing	−5.95±0.1 base pair^{-1}	(30)
						37		Heat of mixing	−6.5±0.1 base pair^{-1}	
			0.5 *M* NaCl and 0.01 *M* cacodylate			37			−6.69±0.1 base pair^{-1}	
Poly(A+U)	~10^5		0.01 *M* citrate and 0.057 *M* NaCl	6.8	0.0085 *M* base pairs	49	Helix-coil Poly (A+U)= Poly A+ Poly U	Heat capacity	6.8 base pair^{-1}	(31)
			0.01 *M* citrate and 0.10 *M* NaCl			54.8			7.3 base pair^{-1}	
			0.01 *M* citrate and 0.15 *M* NaCl			58.4			7.7 base pair^{-1}	
						85–90		(extrapolated)	8.5±0.5 base pair^{-1}	
			0.01 *M* citrate and 0.5 *M* NaCl			54.3	2 Poly(A+U)= Poly(A+2U)+ Poly A		3.3 (A+2U)$^{-1}$	
			0.01 *M* citrate and 0.6*M* NaCl			53.9			3.2 (A+2U)$^{-1}$	
						85–90		(extrapolated)	4.5±0.5 (A+2U)$^{-1}$	

			0.01 M citrate and 0.5 M NaCl			72.1	Poly(A+2U)= Poly A+2 Poly U		11.5 (A+2U)	
			0.01 M citrate and 0.6 M NaCl			75.0			12.2 $(A+2U)^{-1}$	
						85–90		(extrapolated)	12.5±0.5 $(A+2U)^{-1}$	
			0.01 M citrate and 0.5 M NaCl			54.3	Poly (A+U)+ Poly U=poly (A+2U)		−4.0±0.1 $(A+2)^{-1}$	
Poly (A+2U)			0.1 M NaCl and 0.01 M cacodylate			24	Poly (A+U)+ Poly U= Poly (A+2U)	Heat of mixing	−3.82±0.1 $(A+2U)^{-1}$	(30)
						37			−3.5±0.5 $(A+2U)^{-1}$	
			0.5 M NaCl and 0.01 M cacodylate			24			−3.80±0.1 $(A+2U)^{-1}$	
						37			−4.09±0.1 $(A+2U)^{-1}$	
Poly A	~10^5		0.01 M citrate and 0.2 M NaCl	6.8	0.0085 M nucleotide	42–45	Helix-coil	Heat capacity	5.5±1.0 $nucleotide^{-1}$	(31)
Poly A	~10^5		0.24 M citrate and 0.12 M NaCl and HCl aq	5.50	0.0132 M nucleotide	31.5	Double helix-coil $[(Poly\ A)H^+]_2$ = 2 poly A+$2H^+$	Heat capacity	1.46 $nucleotide^{-1}$	(33)
				5.30		39.2			1.78 $nucleotide^{-1}$	
				5.06		47.1			2.03 $nucleotide^{-1}$	
				4.89		56.6			2.23 $nucleotide^{-1}$	
				4.70		65.5			2.42 $nucleotide^{-1}$	
Poly A		6.1	0.1 M KCl and 0.01 M cacodylate	4.0		10	Helix-coil	Heat of mixing	1.80±0.25 $nucleotide^{-1}$ [i]	(29)
						25			2.74±0.20 $nucleotide^{-1}$ [i]	
Poly A	~10^5	4.23	0.1 M NaCl and 0.01 M tris	7.30	C → O 0.37–1.34 mg ml^{-1}	35	Helix-coil	Heat capacity	9.4±1.4 $nucleotide^{-1}$	(32)

[a] Compiled by Gordon C. Kresheck.
[b] Final value in mixing experiments.
[c] Transition temperature for heat-capacity experiments.
[d] Numbers refer to those cited under "Table I References," page **148**.
[e] These values depend upon the choice of expressing pepsin concentration and vary greatly with pH.
[f] Value depended upon commercial source of protein.
[g] Also given are data that can be summarized approximately as $[\Delta H_{denat}(cal/gm) = 0.21\ T_m(°C) - 5.5]_{55-88°}$.
[h] Heat change corrected for unfolding poly A before reaction.
[i] These calorimetric data were recalculated by Stevens and Felsenfeld [*Biopolymers* **2**, 293 (1964)] to yield values of 6.5–8.5 kcal/mole-nucleotide.

II. Calorimetric Measurements of Characteristic ΔH-Values in Molecular Biology

A. Summary of Methods Available

1. *Measurement of the Temperature Change Induced by a Measured Variation of pH or Solvent Composition (Calorimeter with a Thermally Insulated Sample Holder in a Constant Temperature Bath)*

In principle, the enthalpy change involved in a helix-random transition can be determined in a simple calorimetric experiment by measuring the temperature change induced by a variation of pH or solvent composition. The elementary calorimetric apparatus, which can be used in measurements of this type, consists of a thermally insulated sample holder in a constant temperature bath. Since the thermal insulation is not perfect, the heat flow from the sample to the temperature bath must be accounted for when ΔH is calculated from the measured change in the temperature of the calorimeter proper. In an improved calorimetric system the liquid bath is furnished with an electric heater and the electric current in the heater wirings is controlled by a regulator, i.e., the difference between the sample temperature and the temperature of the liquid bath is minimized in order to eliminate heat flows. An adiabatic calorimeter of this type has been used by Kresheck and Scheraga (1966) in a study on enthalpy changes accompanying the acid denaturation of ribonuclease. This special type of calorimeter was first described by Benjamin (1963). The calorimeter assembly is shown schematically in Fig. 5. The polished stainless steel calorimeter vessel is suspended by nylon loops and light springs in a cradle made of steel rods. A multiple pin connector collects the leads from the vessel components and fits a socket sealed in the cap of the glass submarine with an epoxy resin. The submarine is constructed from freeze-drying flasks and can be evacuated using the stopcock indicated by the arrow in Fig. 5. The flexible nature of the cradle allows it to be held rigidly in the submarine when the cap is in position, the parts of the cradle coming in contact with the glass being covered with tape or coated with Tygon. The calorimeter vessel is sealed by caps tightening on Neoprene O-rings. The capacity of the vessel is approximately 50 ml. Epoxy resin is used to seal two thermistors (for temperature measurement and control) and a heater in the top cap and fuse connectors in the lower cap. Sample cells are made from standard female tapered joints. The cut ends of the tubes are sealed with a thin cover glass. Sealing of the upper ends of the tubes can be achieved using sections of Neoprene stoppers. The sample cells are aligned with the supporting rods along two grooves in their sides. The coil spring is held in compression by the stainless steel wire across the fuse connectors. This coil spring allows opening of the sample cell by applying a voltage to the fuse wire when the latter is above the liquid level. The submarine, held in a metal clamp around its center, is attached to a shaft passing through a graphitized-nylon bearing in the side of a temperature-controlled bath. The latter is kept as small as possible while still allowing room for rotation of the submarine. A circulating pump provides rapid stirring, and the bath can be heated electrically or cooled by passing cold tap water through

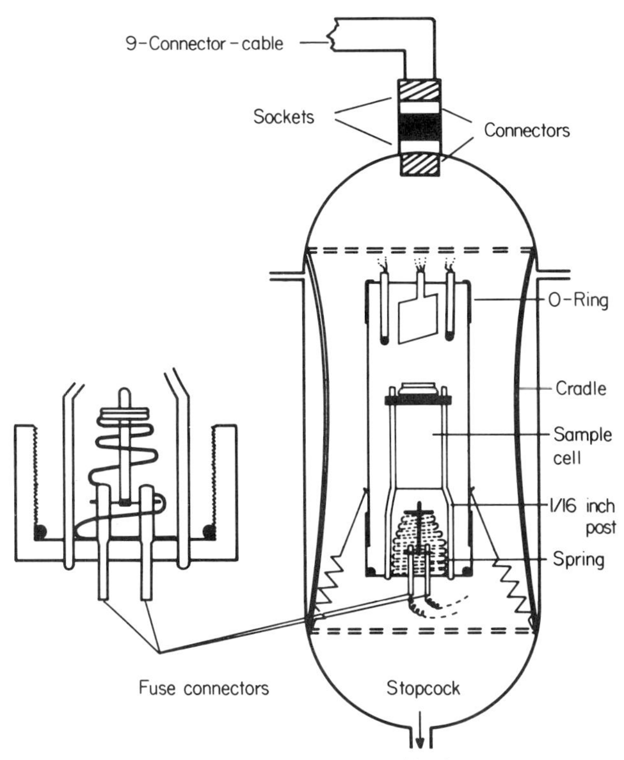

FIG. 5. Diagram of the solution calorimeter assembly for measurements of enthalpy changes accompanying the acid denaturation of proteins. From Benjamin, 1963; with the kind permission of the publisher.

an immersed copper coil. Continuous rocking of the submarine through 180° is achieved using a rack and pinion movement. In this way good stirring of the calorimeter contents by the enclosed air bubble is obtained. All calorimeter components in a laboratory are maintained at a constant temperature and humidity. A regulated power supply is used for operating the calorimeter.

Temperature measurement is achieved by measuring the resistance of one of the thermistors with a Wheatstone bridge, the off-balance signal from which is amplified and recorded. Automatic adiabatic control in the calorimeter can be achieved using a circuit described by Cleland and Harding (1957). Two thermistors, one in the calorimeter vessel and one in the bath, form arms of a resistance bridge, the other arms being a precision resistance and a four-dial decade resistance. A variable capacitance also can be connected in parallel to either of the resistance arms and allows good balancing of the bridge to be achieved without the care otherwise needed in mounting thermistors. An extra amplification stage has been added to the original circuit together with a meter to record the amplified signal. The circuit is stable and actuates the bath heater and the cooling device, giving a uniform sawtooth temperature control.

In a typical experiment, 20–40 mg of ribonuclease was weighed into the sample cell with 3.0 ml of either 0.10 or 0.15 *M* KCl. The solution was brought to thermal equilibrium within about 45 minutes in the calorimeter vessel containing both the cell and about 40 ml of salt solution (containing the desired amount of either HCl or KOH) to be mixed with the protein solution. The cell was broken, and the temperature change was measured. The observed heat changes ranged up to 0.180 cal (corresponding to a measured temperature change of about 0.0025°), which is considerably in excess of the heat change observed in a blank experiment in which the same two solvents were mixed in the absence of protein, the latter exothermic heat change, amounting to 0.048 cal was subtracted from that observed with the protein solution. For calibration, a heating period is carried out during which the calorimeter proper is heated using the electric heater shown in Fig. 5.

A similar apparatus has been used in a preliminary calorimetric study of the helix-random coil transition of poly-γ-benzyl-L-glutamate (Block and Jackson, 1963). Known volumes of two polymer solutions, one from an α-helix solvent (chloroform) and the second from a random-coil solvent (dichloroacetic acid), were mixed in the calorimeter and the temperature change was noted. The measured temperature changes were plotted against the volume fraction of $CHCl_3$, and the occurrence of a transition is indicated by a distinct break in the curves. However, it was not possible to estimate the enthalpy changes of the observed effects. Meanwhile, the transition enthalpies of the dissolved poly-γ-benzyl-L-glutamate have been determined by measuring the temperature course of the heat capacity of the polypeptide solution in the temperature region of the helix-coil transition (see Section II, A, 3). For further applications of this simple calorimetric technique, the reader is referred to Section II, B.

A temperature change is also induced by the heat liberated in mixing equal volumes of poly A and poly U solutions of approximately equal concentrations in one of the two cells of the twin calorimeter described by Buzzell and Sturtevant (1951). This type of calorimeter was discussed briefly in Chapter I. The development of accurate devices for integrating a varying voltage makes it possible to employ the twin calorimetric method with automatic compensation, by means of which the temperatures of the two calorimeter propers are held closely equal, the energy resulting from a reaction proceeding in one of the cells being duplicated by electrical heating. The two cells A and B are each equipped with constant resistance electrical heaters H_A and H_B, and resistance thermometers T_A and T_B. The thermometers are in adjacent arms of a Wheatstone bridge, the output of which is proportional to the temperature difference between the two calorimeter cells. The bridge output is amplified, and the amplified voltage is continuously squared by an analog computer. The squared voltage is integrated by another computer. The squarer also produces an unsquared voltage, which is fed back to the heater of the calorimeter vessel that requires electrical heat. It is evident that under these circumstances there is no need to measure the actual temperature change of the system. The integrator output, which is recorded on a strip-chart recorder, is accurately proportional to the total electrical energy fed back to H_B, i.e., the recorded integrator output is proportional to the heat liberated in the reaction vessel of the apparatus. The reader is referred to the original paper (Buzzell and Sturtevant, 1951) for a more detailed description of the calorimetric equipment. Experience with a wide range of reactions (e.g., acid denaturation reactions of deoxyribonucleic acids) has demonstrated that this apparatus gives satisfactory enthalpy data not only for exothermic reactions but also for various processes in which heat is absorbed by the sample. Since most of the measurements of the enthalpy of denaturation of nucleic acids have been carried out at a constant temperature of the calorimetric system, this application of the apparatus will be discussed further in the following section.

2. *Measurement of the Enthalpy Change Induced by a Measured Variation of pH or Solvent Composition* (*Isothermal and Heat-Flow Calorimeters*)

In classical calorimetry, the various types of reaction calorimeters are classified in two groups:

1. Adiabatic calorimeters. Most of the heat produced in the reaction vessel is retained by the calorimeter container, causing its temperature to rise. The temperature change is measured by a suitable type of calorimeter.
2. Isothermal calorimeters. All the heat produced in the reaction vessel is compensated.

It was pointed out in Chapter I that this classification is not entirely sound. Most of the calorimeters mentioned in Section II,A,1 may be classified as belonging to the first group, but they cannot be regarded as completely adiabatic calorimeters. The second group also includes calorimeters for endothermic processes in which sufficient electrical energy is introduced in the reaction vessel to prevent any temperature change. For exothermic processes, the compensation is achieved by using Peltier cooling devices or by melting a certain weight of solid in a surrounding container. However, for precise thermochemical measurements it is not necessary, nor is it desirable, to totally compensate the heat of the chemical process. The complexity of the automatic devices needed to effect total compensation would, in any case, probably introduce additional sources of error. In the classical Bunsen ice calorimeter a perfect compensation of the heat produced in the reaction vessel may be achieved, but the thermal inertia of this type of apparatus is too large to allow it to be used as an instrument for our purposes. In this section we shall deal with applications of the conduction calorimeter (Tian–Calvet type) in studies on exothermal protein denaturation reactions and with another application of the Buzzell–Sturtevant calorimeter in a series of enthalpy measurements of endothermic denaturation processes of various deoxyribonucleic acids.

The Calvet microcalorimeter was discussed in Chapter I. A description of a Tian–Calvet type calorimeter is given in Chapter XIV. The reader is referred to Calvet and Prat's (1963) text for a detailed description of this particular technique. The apparatus, built by Société D.A.M. (Lyon, France), has been used by Hermans and Rialdi (1965) for measurements of the heat of ionization and denaturation of sperm-whale myoglobin. A schematic diagram of the cell used for the measurements is shown in Fig. 6. The microcalorimeter consists of a thermostated block with two cavities. The cavities contain silver tubes that are in thermal contact with the block, each via 1000 thermocouples, connected in series. In the silver tubes, the stainless steel cells (one the sample and the other the reference) are introduced. The two thermocouple piles are connected so that they oppose each other, and the circuit is closed with a recording galvanometer. It can be shown (Calvet and Prat, 1956) that the integral of the current as a function of the time is proportional to the difference in heat produced in the two cells, and calibration can be performed by producing a known quantity of heat electrically in the sample cell. In conduction microcalorimetry, no effort is made to preserve the difference in temperature produced by the heat of reaction with thermal insulation. Instead, the heat is dissipated at the maximum obtainable velocity. The heat-flow calorimeter might be classified as an isothermal calorimeter, but a small temperature change of the sample is still required as the signal source of the recorder circuit. A similar instrument and its use is described by Kitzinger and Benzinger (1960).

The design of the cell shown in Fig. 6 is adapted to the particular denaturation experiment. The basic cell is a hollow, stainless steel cylinder 2 cm in diameter and about 10 cm long. Its use has the advantage that only a relatively small amount of sample is required for the calorimetric measurements. Teflon is preferred for the screwed-on stopper. The stopper is pierced so that a syringe fits into it and can be connected via a short piece of Teflon to a small glass tube containing acid (II in Fig. 6). Compression of the syringe forces out the mercury that seals the end of this tube during thermal equilibration and permits mixing of the acid or base with the dilute neutral protein solution (I in Fig. 6). When thermal equilibrium is achieved in the starting period of the experiment, the measurement requires 45 minutes to again reach thermal equilibrium after the mixing of the two solutions. A review of ΔH

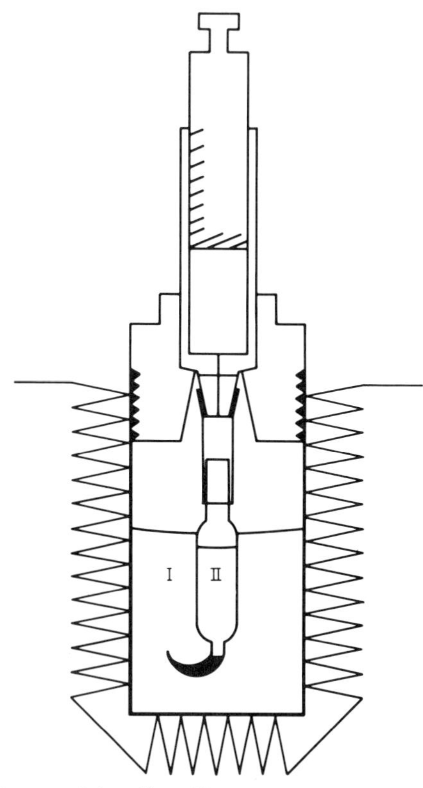

FIG. 6. Schematic diagram of the cell used by Hermans and Rialdi (1965) as an element of a Calvet microcalorimeter in myoglobin denaturation experiments (I, dilute neutral myoglobin solution; II, acid or base solution).

values obtained in a series of measurements with this particular technique is given in Section II,B. The reader is also referred to Chapter IV, Section B (calorimetric studies of proteins), for a description of the measurements.

The Buzzell–Sturtevant reaction twin calorimeter described in Section II,A,1 also has been used in a series of isothermal nucleic acid denaturation experiments performed by Bunville, Geiduschek, Rawitscher, and Sturtevant (1965). Compensation of the heat absorbed during the acid denaturation is achieved by introducing a sufficient amount of electrical energy in the resistance heater H_A (see Section II,A,1) of the calorimeter vessel in which the process takes place. The heat of denaturation is then calculated from the integrator readings, which are recorded during the measurement as in the experiments described in the foregoing section. These acid denaturation experiments can be regarded as a typical example for a calorimetric determination of a transition enthalpy at constant temperature (isothermal calorimetry),. provided that any temperature change in the reaction vessel is prevented by electrical heating. The acid denaturation measurements of this particular type have been extended over a wide range of pH variations using DNA samples from various sources. The results are discussed briefly in Section II, B. A more detailed discussion can be found in the original paper (Bunville *et al.*, 1965). The heat of the acid denaturation of deoxyribonucleic acid also is discussed in connection with calorimetric studies on proteins in Chapter V.

3. *Measurement of the Heat Capacity of a Solution as a Function of Temperature in the Temperature Region of the Helix-Random Coil Transition* (*Adiabatic Calorimeters*)

Recently the experimental investigations concerning biopolymers have been widened by a new method. Nearly at the same time various groups of authors (Privalov *et al.*, 1964; Ackermann and Rüterjans, 1964; Karasz *et al.*, 1964) succeeded in gaining information on the thermal behavior and on the thermodynamics of dissolved polypeptides and nucleic acids in a direct way. The accurate measurement of the temperature course of the heat capacity through the transition region opens a way to determine directly the transition enthalpy ΔH for the solvated biopolymer and, in the case of polypeptides, additionally the cooperative parameter σ. In the statistical model developed by Zimm and Bragg (1959), the term σ is defined as a key parameter, which now is experimentally accessible by a calorimetric method. The measurement of the heat capacity as a function of temperature results in calorimetric transition curves that are characterized by a maximum of the additional heat capacity C_c in the temperature range of the thermal transition. An example, the calorimetric transition curve of a solution of poly-γ-benzyl-L-glutamate in a mixture of dichloroacetic acid and ethylene dichloride, as measured by Ackermann and Rüterjans (1964), is shown in Fig. 7. A linear dependence of the measured heat capacity data on temperature is obtained at temperatures above and below the transition region. The temperature corresponding to the maximum value of the excess heat capacity (observed minus baseline, see Fig. 7) is called T_c, originally defined as the midpoint of the thermal conversion in the polarimetric transition curve. For comparison, the temperature course of the optical rotatory power of the system is plotted in the lower part of the figure. The maximum value,

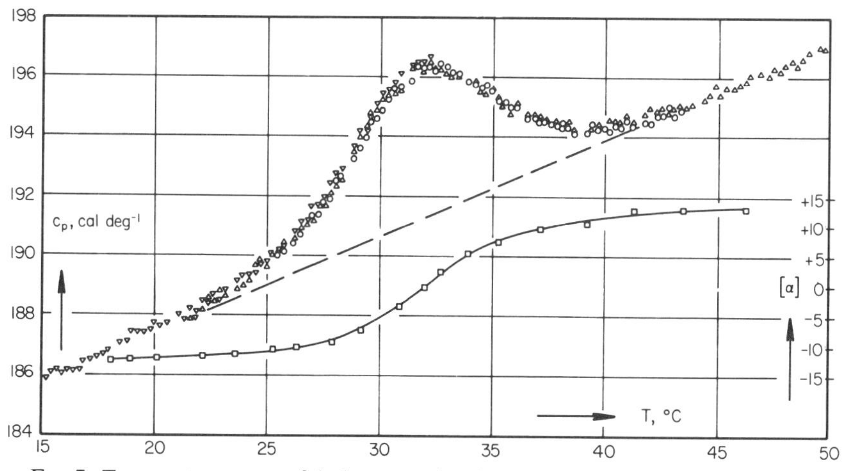

FIG. 7. Temperature course of the heat capacity of a solution of poly-γ-benzyl-L-glutamate in a mixture of dichloroacetic acid and ethylene dichloride (80:20), as measured by Ackermann and Rüterjans (1964). The open circles, triangles, etc. correspond to measured values obtained in successive experiments. ΔH is proportional to the area under the peak limited by the measured curve and the dashed line (the latter corresponding to the limiting case $\Delta H = 0$). For comparison, the [α] vs. T curve is plotted in the lower part of the figure.

$C_{c_{\max}}$, is proportional to the maximum value of the first temperature derivative of the fractional helical content. As demonstrated by Ackermann and Rüterjans (1964), C_c is related to the fractional helical content Θ by a very simple thermodynamic equation, Eq. (1), which can be derived from

$$C_c = \Delta H \frac{\partial \Theta}{\partial T} \tag{1}$$

elementary thermodynamic relationships. Hence, we have Eq. (2) for the

$$C_{c_{\max}} = \Delta H (d\Theta/dT)_{T_c} \tag{2}$$

maximum value of the excess heat capacity at T_c. The transition enthalpy ΔH, i.e., the additional molar enthalpy change of the system, is directly proportional to the area under the peak limited by the measured calorimetric transition curve and the dashed line in Fig. 7 (the latter corresponding to the hypothetical limiting case $\Delta H = 0$). Since an approximative relation between Θ and T is derived by Applequist (1963), we can write

$$(d\Theta/dT)_{T_c} = \Delta H / 4RT_c^{\,2} \sigma^{\frac{1}{2}} \tag{3}$$

or

$$C_{c_{\max}} = (\Delta H)^2 / 4RT^2 \sigma^{\frac{1}{2}} \tag{4}$$

in the case of polypeptides. For a given ΔH value, the sharpness of the thermal transition therefore depends on the cooperative parameter σ. A detailed discussion of this parameter is given in Davidson's text (1962) and in the original papers of Applequist (1963) and Zimm and Bragg (1959). Since $C_{c_{max}}$ is proportional to $\sigma^{-\frac{1}{2}}$, numerical values of the cooperative parameter σ can be calculated from the measured maximum value of the excess molar heat capacity (i.e., from the height of the peak in Fig. 7, e.g.). A complete thermodynamic analysis of the PBG–DCA–EDC system has been carried out by means of this particular calorimetric method (Ackermann and Neumann, 1967). The results are summarized together with other calorimetric data in Table I.

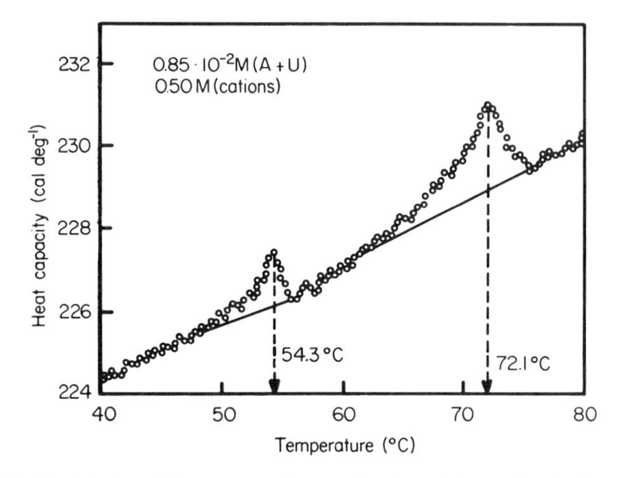

FIG. 8. Calorimetric transition curve of an equimolar mixture of poly A and poly U in aqueous salt solution. The total concentration of cations was 0.50 M; 0.01 M citrate buffer, pH 6.8.

Calorimetric transition curves also have been determined in a series of measurements on solutions of polynucleotides. A typical example for a calorimetric transition curve, showing two maximum values of the additional heat capacity in a temperature range from 40 to 80°C, is presented in Fig. 8. The curve is obtained in a single experiment, using a recording adiabatic calorimeter that is described in Chapter XIII, Section II. The sample was an equimolar mixture of polyribouridylic and polyriboadenylic acid in an aqueous salt solution (the total concentration of cations was 0.50 M; 0.01 M citrate buffer, pH 6.8). According to the phase diagram given by Stevens and Felsenfeld (1964), the first peak of the heat capacity vs. T curve in Fig. 8 refers to the conversion of poly (A + U) to poly (A + 2U) and poly (A); the second one is caused by the dissociation of the poly (A + 2U) complex. Again, the areas under the peaks are proportional to the enthalpy changes involved in the transitions. The measured transition enthalpies, however, are reduced ΔH

values, since poly A forms an intramolecular secondary structure in the course of these reactions. Therefore, an extrapolation to about 95°C is necessary, where poly A is assumed to exist completely in the randomly coiled conformation (see Leng and Felsenfeld, 1966). The extrapolated values are given in the review of measured transition enthalpies (Table I). The method of measuring the heat capacity of the sample in the temperature region of the helix-random coil transition is particularly suitable for the analysis of thermal helix-coil transitions. Its use has the advantage that the composition of the sample is not changed during the experiment. In the case of reversible transitions, the reproducibility of the results therefore can be checked without changing the sample. Furthermore, by integrating Eq. (1) the transition curve (i.e., the temperature course of Θ) can be evaluated from the results of the calorimetric measurements. This independent method has been proved to give correct values for the fractional helical content in the Θ vs. T curves, in good agreement with those obtained in measurements of the optical rotatory power (Ackermann and Rüterjans, 1964).

4. *Indirect Methods*

An indirect method for the determination of ΔH values involved in thermal transitions of the secondary structure of polypeptides has been proposed by Karasz and O'Reilly (1967). This method is based on the assumption that the thermal helix-random coil transition can be treated as a first-order phase transition. An experimental proof of the method has been worked out by the authors for the PBG–DCA–EDC system. In this case it is assumed that a single amino acid residue of the polypeptide is bonded on the average to one dichloroacetic acid (DCA) molecule when it is in the coiled form, and bonded intramolecularly (not, of course, to the adjacent residue) when in the helical conformation. The role of the ethylene dichloride (EDC) is that of an "inert" solvent in this first-order approximation. As is well known, the transition temperature T_c can be changed by suitably altering the composition of the solvent (see Section I,A). In the PBG–DCA–EDC system, lowering the temperature favors the formation of the disordered conformation in the polymer. Below T_c, therefore, the DCA bound to the polypeptide has lost its translational freedom and is therefore assumed to be in a "solid" phase. Above T_c the "released" DCA, in contrast, is in the liquid state. The transition temperature for the formation of the polypeptide helix therefore may be regarded equally correctly as the melting point of the bound DCA. The ethylene dichloride has a profound thermodynamic effect on the DCA in that its presence lowers the chemical potential of the latter relative to the pure state. Therefore, the observed lowering in T_c on increasing the relative concentration of EDC in the solvent phase can be treated as a diluent-induced depression of the DCA melting point. In the ideal solution approximation the usual type of equation for the melting point depression, Eq. (5), is applied,

$$d(\ln x_{DCA})/dT_c = \Delta H/RT_c^2 \tag{5}$$

where x_{DCA} is the mole fraction of the dichloroacetic acid in the solvent mixture and ΔH is the overall heat, per mole of DCA, associated with the helix formation. The fact that only a small fraction of the DCA undergoes a phase change is neglected, since it is irrelevant to the thermodynamic analysis. The authors have further assumed that there is a one-to-one molar equivalence between the bound DCA molecules and the polypeptide residues and that ΔH can be identified with the calorimetric heat of transition. In effect the "smeared" character of the transition is disregarded in this first-order approximation. The numerical values of the property ΔH, calculated from Eq. (5), exceed those found in the heat-capacity measurements. The difference (200 cal mole^{-1}, i.e., 20% of the calorimetric value) reflects the approximative character of this simplified theoretical treatment.

ΔH values have also been calculated from potentiometric titration curves in a study on poly-L-glutamic acid (Hermans, 1966). This special type of a non-calorimetric method, however, is beyond the scope of a book on micro-calorimetry. The reader therefore is referred to the original paper for a detailed description of the measurements.

B. Review of Measured ΔH Values of Helix-Random Coil Transitions in Solution

1. *Collection of Measured Values (Comparison of Results from Different Methods)*

A general picture of the field as it presently exists is reflected in the table of calorimetric ΔH values accompanying conformational changes of macro-molecules in solution (Table I).* The contents of the table are divided into two groups. The first group includes values for enzymes, proteins, synthetic polypeptides, and polyamino acids. The data for nucleic acids and poly-nucleotides are collected in the second group. Since most of the various measurements have been performed at different solvent compositions, pH values, concentrations, and temperatures, the numerical values of these parameters are listed in the fourth, fifth, sixth, and seventh columns, respectively. The type of transition and the type of experiment is indicated in the eighth and ninth columns. Additional information on the molecular weight or sedimentation coefficient of the sample is provided in the second and third columns. Dr. Kresheck numbered and collected in a separate list the

* The author is indebted to Dr. Gordon C. Kresheck for communicating part of the data of Table I prior to publication (*Handbook of Biochemistry*, in preparation),

references to the original papers (see Special References to Table I, page 148). At present a comparison of ΔH values that are determined by means of different experimental methods is not possible, except for a few characteristic model systems. For example, the enthalpy changes accompanying the formation of the poly (A + U) complex have been determined by Rawitscher, Ross, and Sturtevant (1963) in a series of mixing experiments. The results are in fair agreement with those derived from heat-capacity measurements* (Neumann and Ackermann, 1968). In most cases, however, the measured ΔH values should be regarded as preliminary results that can be used for approximative calculations. As stated in Section I,C, the influence of concentration and solvent composition has been neglected for most of the calorimetric measurements in this field. The various results obtained in a calorimetric analysis of the PBG–DCA–EDC system (Ackermann and Neumann, 1967) cannot be interpreted if ΔH is assumed to be a constant value over the whole range of concentration and solvent composition. Additional measurements at various concentrations and solvent compositions therefore are required.

2. *The Physical Meaning of the Results* (*Discussion of ΔH Values with an Emphasis upon the Probable Future*)

The thermodynamic parameters associated with conformational transitions in polypeptides and polynucleotides are of interest for at least three reasons: (1) as a quantitative means of studying changes in molecular interactions that occur during the transitions, (2) as a method of deciding between postulated mechanisms, particularly in proteins, and (3) in providing data with which the validity of the various theoretical treatments of the helix-coil transition may be tested. Any detailed interpretation of the measured enthalpy changes of course will depend on the particular mechanisms of the elementary processes involved in the conformational transition. As pointed out in Section II,A,4, poly-γ-benzyl-L-glutamate in ethylene dichloride–dichloroacetic acid (DCA) mixtures is stable as a helix at high temperature and as a coil at low temperature. This may be interpreted by saying that the reaction is really

$$\text{Coil (DCA)}_N \rightleftarrows \text{helix} + N \cdot \text{DCA}$$

The reaction is endothermic because of the hydrogen bonds formed between DCA and the amide groups in the compound on the left-hand side of the equation; the entropy change is positive because of the release of the DCA molecules into the solvent on forming the helix. In other words, the negative

* In any case, the elementary differences between the two methods must be regarded. The negative sign of the values reported (Rawitscher *et al.*, 1963) is due to the fact that the complex formation is exothermic.

enthalpy change resulting from the formation of the intramolecular helical hydrogen bonds is superimposed by a positive contribution to ΔH corresponding to the disruption of the intermolecular hydrogen bonds between DCA and the amide groups of the polypeptide. An additional negative enthalpy change is due to the dimerization of the released DCA molecules. The fundamental thermodynamic property that is a measure of the stability of the intramolecular hydrogen bonds in the α-helix, therefore, is not accessible to direct calorimetric measurement in this particular case. Additional information concerning the enthalpy changes involved in the solvation and dimerization reactions is necessary.

For aqueous salt solutions of polynucleotides, the influence of interactions between solvent molecules and the dissolved polymer is less important. It is generally assumed in a first-order approximation that the heat absorbed in the disruption of the hydrogen bonds of the helical duplex is compensated by the heat produced during the formation of hydrogen bonds between water molecules and polar groups of the single-stranded polynucleotides. The calorimetric transition enthalpy, therefore, may be regarded as a measure for the relative strength of the interactions between the stacked bases in the helical duplex structure (stacking enthalpy). It is well known that the transition temperature T_c of nucleic acids from various sources depends on the relative content of the guanine–cytosine base pair in the double-stranded polynucleotide. The nature of the forces responsible for the maintenance of the helical structure of DNA is not as yet clearly understood in quantitative detail. A systematic analysis of calorimetric data including all possible combinations of base pairs will provide us with the information required for a theoretical discussion of this phenomenon.

In various theories of elementary processes in molecular biology the assumption is made that so-called "loops" can occur in the secondary structure of polynucleotides. These loops may differ in thermal stability from the main part of the helical duplex structure. Measurements of the heat capacity of appropriate samples, therefore, will be useful as a method with which the validity of the particular theoretical treatment may be checked.

Since a relatively large number of papers dealing with calorimetric measurements on solutions of biopolymers and related model substances has been published during the last years, it can be said without undue optimism that calorimetry is becoming a standard laboratory method in molecular biology. There is, however, a serious problem that leaves us with the requirement for an improved calorimetric apparatus. In most cases the amount of biochemical samples is too small compared with the amount of sample required for a precise calorimetric measurement. Thus calorimeter designers must focus upon the construction of calorimeters giving a maximum of precision for a reduced amount of sample.

REFERENCES

General References

Davidson, J. N. (1965). "The Biochemistry of the Nucleic Acids," 5th Ed. Wiley, New York.
Lewis, G. N., and Randall, M. (1961). "Thermodynamics," 2nd Ed. McGraw-Hill, New York.
Scheraga, H. A. (1961). "Protein Structure." Academic Press, New York.
Scheraga, H. A. (1963). *In* "The Proteins" (H. Neurath, ed.), Vol. I, pp. 477–594. Academic Press, New York.
Stahmann, M. A., ed. (1962). "Polyamino Acids, Polypeptides and Proteins." Univ. of Wisconsin Press, Madison, Wisconsin.
Steiner, R. F. (1965). "The Chemical Foundations of Molecular Biology." Van Nostrand, Princeton, New Jersey.
Steiner, R. F., and Beers, R. F., Jr. (1961). "Polynucleotides." Elsevier, Amsterdam.

Publications Cited

Ackermann, T., and Neumann, E. (1967). *Biopolymers* **5**, 649.
Ackermann, T., and Rüterjans, H. (1964). *Ber. Bunsenges. Physik. Chem.* **68**, 850.
Applequist, J. (1963). *J. Chem. Phys.* **38**, 934.
Benjamin, L. (1963). *Can. J. Chem.* **41**, 2210.
Block, H., and Jackson, J. B. (1963). *Proc. Chem. Soc.* p. 381.
Bunville, L. G., Geiduschek, E. P., Rawitscher, M. A., and Sturtevant, J. M. (1965). *Biopolymers* **3**, 213.
Buzzell, A., and Sturtevant, J. M. (1951). *J. Am. Chem. Soc.* **73**, 2454.
Calvet, M., and Prat, H. (1956). "Microcalorimetrie." Masson, Paris.
Calvet, M., and Prat, H. (1963). "Recent Progress in Microcalorimetry." Macmillan (Pergamon), New York.
Cleland, W. W., and Harding, R. S. (1957). *Rev. Sci. Instr.* **28**, 696.
Davidson, N. (1962). "Statistical Mechanics," pp. 385–393. McGraw-Hill, New York.
Doty, P., Wada, A., Yang, J. T., and Blout, E. R. (1957). *J. Polymer Sci.* **23**, 851.
Harbers, E. (1964). "Die Nucleinsäuren," Georg Thieme Verlag, Stuttgart.
Hermans, J., Jr. (1966). *J. Phys. Chem.* **70**, 510.
Hermans, J., Jr., and Rialdi, G. (1965). *Biochemistry* **4**, 1277.
Karasz, F. E., and O'Reilly, J. M. (1967). *Biopolymers* **5**, 27.
Karasz, F. E., O'Reilly, J. M., and Bair, H. E. (1964). *Nature* **202**, 693.
Kitzinger, C., and Benzinger, T. H. (1960). *Methods Biochem. Analysis*, D. Glick, Editor, Interscience, New York, 1960, Vol. VIII.
Kresheck, G. C., and Scheraga, H. A. (1966). *J. Am. Chem. Soc.* **88**, 4588.
Leng, M., and Felsenfeld, G. (1966). *J. Mol. Biol.* **15**, 455.
Lewis, G. N., and Randall, M. (1961). "Thermodynamics," 2nd Ed. McGraw-Hill, New York.
Neumann, E., and Ackermann, T. (1968). *J. Phys. Chem.* (in press).
Pauling, L., Corey, R., and Branson, H. (1951). *Proc. Natl. Acad. Sci. U.S.* **37**, 205.
Privalov, P. L., Kafiani, K. A., and Monaselidze, D. R. (1964). *Dokl. Akad. Nauk SSSR* **156**, 951.
Rawitscher, M., Ross, P., and Sturtevant, J. M. (1963). *J. Am. Chem. Soc.* **85**, 1915.
Rifkind, J., and Applequist, J. (1964). *J. Am. Chem. Soc.* **86**, 4207.
Stevens, C. L., and Felsenfeld, G. (1964). *Biopolymers* **2**, 293.
Watson, J. D., and Crick, F. H. C. (1953). *Nature* **171**, 737, 964.
Zimm, B. H., and Bragg, J. K. (1959). *J. Chem. Phys.* **31**, 526.
Zimm, B. H., Doty, P., and Iso, K. (1959). *Proc. Natl. Acad. Sci. U.S.* **45**, 1601.

Table I References

1. Buzzell, A., and Sturtevant, J. M. (1952). *J. Am. Chem. Soc.* **74**, 1983.
2. Sturtevant, J. M. (1954). *J. Phys. Chem.* **58**, 97.
3. Gutfreund, H., and Sturtevant, J. M. (1953). *J. Am. Chem. Soc.* **75**, 5447.
4. Sturtevant, J. M., Laskowski, M., Laskowski, M., Jr., Donnelly, T., and Scheraga, H. A. (1955). *J. Am. Chem. Soc.* **77**, 6168.
5. Laki, K., and Kitzinger, C. (1956). *Nature* **178**, 985.
6. Bro, P., and Sturtevant, J. M. (1958). *J. Am. Chem. Soc.* **80**, 1789.
7. Forrest, W., and Sturtevant, J. M. (1960). *J. Am. Chem. Soc.* **82**, 585.
8. Privalov, P. L., and Monaselidze, D. R. (1963). *Biofizika* **8**, 420.
9. Hermans, J., Jr., and Rialdi, G. (1965). *Biochemistry* **4**, 1277.
10. Beck, K., Gill, S., and Downing, T. (1965). *J. Am. Chem. Soc.* **87**, 901.
11. Kresheck, G. C., and Scheraga, H. A., (1966). *J. Am. Chem. Soc.* **88**, 4588.
12. Danforth, R., Krakauer, H., and Sturtevant, J. M. (1967). *Rev. Sci. Instr.* **38**, 484.
13. Privalov, P. L., Monaselidze, D. R., Mrevlishvili, G., and Magaldadze, V. (1965). *Soviet Phys. JETP (English Transl.)* **20**, 1393.
14. O'Reilly, J. M., Bair, H. E., and Karasz, F. E. (1968). *J. Am. Chem. Soc.* Submitted for publication.
15. Ackermann, T., and Rüterjans, H. (1964). *Z. Physik. Chem. (Frankfurt)* **41**, 116.
16. Ackermann, T., and Rüterjans, H. (1964). *Ber. Bunsenges. Physik. Chem.* **68**, 850.
17. Karasz, F. E., O'Reilly, J. M., and Bair, H. E. (1964). *Nature* **202**, 693.
18. Giacometti, G., and Turolla, A. (1966). *Z. Physik. Chem. (Frankfurt)* **51**, 108.
19. Ackermann, T., and Neumann, E. (1968). *Biopolymers* (in press).
20. Karasz, F. E., and O'Reilly, J. M. (1966). *Biopolymers* **4**, 1015.
21. Karasz, F. E., O'Reilly, J. M., and Bair, H. E. (1965). *Biopolymers* **3**, 241.
22. Rialdi, G., and Hermans, J., Jr. (1966). *J. Am. Chem. Soc.* **88**, 5719.
23. O'Reilly, J. M., Bair, H. E., and Karasz, F. E. (1968). *Biopolymers* (in press).
24. Sturtevant, J. M., and Geiduschek, E. P. (1958). *J. Am. Chem. Soc.* **80**, 2911.
25. Bunville, L. G., Geiduschek, E. P., Rawitscher, M. A., and Sturtevant, J. M. (1965). *Biopolymers*, **3**, 213.
26. Rüterjans, H. (1965). Ph.D. Thesis, Univ. of Münster, Germany.
27. Privalov, P. L., Kafiani, K. A., and Monaselidze, D. R. (1965). *Biofizika* **10**, 393.
28. Steiner, R. F., and Kitzinger, C. (1962). *Nature* **194**, 1172.
29. Rawitscher, M. A., Ross, P., and Sturtevant, J. M. (1963). *J. Am. Chem. Soc.* **85**, 1915.
30. Ross, P., and Scruggs, R. (1965). *Biopolymers* **3**, 491.
31. Neumann, E., and Ackermann, T. (1968). *J. Phys. Chem.* (in press).
32. Epand, G., and Scheraga, H. A. (1968). *J. Am. Chem. Soc.* (in press).
33. Klump, H., Neumann, E., and Ackermann, T. (1968). *Biopolymers* (in press).

CHAPTER VII

Calorimetry of Enzyme-Catalyzed Reactions

HARRY DARROW BROWN

I. Introduction

Development of direct-reaction calorimetry as an enzymologic tool has followed a general recognition of the need for thermodynamic data for chemical reactions that occur in biological systems. In enzyme-catalyzed reactions, to a far greater extent than in other areas of chemistry, pertinent energy data refers to reactions in solution under rigidly specified physical conditions. Because the components of these reactions are seldom available with sufficient purity to allow estimation of enthalpy changes from combustion heats in the absence of additional thermodynamic data, bomb calorimetry has been of only infrequent utility. The rarity of components of biochemical systems is so great that development here ordinarily has implied *micro*calorimetric instrumentation.

Paradoxically, the resulting development of instrument designs offering extreme sensitivity and reasonably facile use has made it attractive also to consider the microcalorimeter as an analytic tool entirely divorced from the need for thermodynamic information. Calvet (1963; Calvet and Prat, 1956), writing some years ago, looked forward to the day when the microcalorimeter would be a conventional laboratory instrument, as common on the bench as the spectrophotometer. The rational base for this expectation derives from the ubiquitousness of heat changes in chemical interaction. Although exceptions exist, almost all chemical reactions involve measurable heat changes. Thus we may expect that in a reaction in which an enzyme attacks a substrate, a heat change will characterize the interaction. Moreover, though the technique itself is nonspecific, if either the enzyme or the substrate can be specified, even though present in a complex mixture, the reaction heat approximately will correspond to the extent of the specified reaction. Not only is the total heat generally proportionate to the extent of the transformation, but the course of the reaction may be followed. For this reason, in many cases the calorimeter can provide an observation usefully interpreted in terms of reaction kinetics. In favorable reactions it has been found possible to establish equilibria with thermal data alone and hence possible to obtain ΔG with two or a small series of direct-reaction calorimetric measurements. Even molecular weight has been approximated in some cases.

Published reports for the most part have dealt with hydrolytic reactions. However, the methyl-transferase catalyzed transfer of a methyl group from dimethyl acetothetin to homocysteine (Dobry and Sturtevant, 1952; Durell *et al.*, 1962) and the fumarate $\rightleftharpoons$ malate isomerization catalyzed by fumarase are examples of other classes of reaction that have been described. The fumarase isomerization has been used as the basis for the development of a calorimetrically determined equilibrium useful in the determination of ΔG (Benzinger and Kitzinger, 1963) of reversible reactions.

Most of the existing literature illustrates that the historic motivation of calorimetric studies has been the quest for thermodynamic values. Unfortunately, however, many of the older reports (prior to 1952) tabulate heat values that include heat of buffer protonation. Often this significant contribution was neglected. Though the need for thermodynamic data is great, the useful literature is small. Use of the microcalorimeter as an analytic tool for enzymatic reactions has resulted, unfortunately, in even fewer reports, and indeed this "discipline" only barely has seen a beginning of what, many calorimetrists feel confident, shall be a long and respectable useage.

An outline of enzyme-catalyzed reactions that have been studied calorimetrically is given in Table I. Calculated thermodynamic values for several enzymes are given in Table II.

II. Oxidoreductases

Katz (1955) used an isothermally jacketed calorimeter to measure the temperature change for the reaction of pyruvate with reduced NAD. The reaction,

$$H^+ + \text{pyruvate} + \text{NADH} \xrightarrow{\text{lactate dehydrogenase}} \text{lactate} + \text{NAD}^+ \qquad (1)$$

was chosen for study because the standard free energy, ΔG^0, was known. Thus the availability of measured heat values made it possible to calculate the

ENZYME CATALYZED REACTIONS STUDIED CALORIMETRICALLY

Enzyme commission number	Systematic name	Trivial name	Source	Reaction pH	Substrate	$-\Delta H$ (cal mole^{-1})	Reference
I. Oxidoreductases							
1.1.1.47	B-D-Glucose: NAD(P) oxidoreductase	Glucose dehydrogenase	Yeast	7.2	Glucose	[a]	Brown *et al.* (1965a)
1.1.1.27	L-Lactate: NAD oxidoreductase	Lactate dehydrogenase		7.3	Na pyruvate	10,600 ± 365	Katz (1955)
II. Transferases							
2.1.1.5	Betaine: L-homocysteine S-methyltransferase	Betaine homocysteine methyltransferase		7.1	Betaine	8,800 ± 500	Durell *et al.* (1962); Durell and Sturtevant (1957)
					Dimethyl-acetothetin	11,300	Durell *et al.* (1962); Durell and Sturtevant (1957)
					s-Methyl methionine sulfonium bromide	7,700	Durrell *et al.* (1962)
2.7.4.3	ATP: AMP phosphotransferase	Adenylate kinase			Inosine diphosphate	25,800	Meyerhof and Lohmann (1932); Meyerhof and Schulz (1935, 1936)
III. Hydrolases							
3.1.1.7	Acetylcholine-hydrolase	Acetylcholine-esterase			Acetylcholine	[b, c]	Brown *et al.* (1968); This work, Wadsö, Chapter III
3.1.3		Phosphatases			AMP	1,550	Ohlmeyer (1945)
					ATP	24,250	Ohlmeyer (1945)
3.1.3.2	Orthophosphoric monoester phosphohydrolase	Acid phosphatase	Prostate	4.5 + 0.80	*p*-Nitro phenyl phosphate	6,280 ± 100	Sturtevant (1955a)
					ATP	5,800	Sturtevant (1962)
					Pyrophosphate	5,800	Sturtevant (1962)

TABLE I—*continued*

Enzyme commission number	Systematic name	Trivial name	Source	Reaction pH	Substrate	$-\Delta H$ (cal mole^{-1})	Reference
			Muscle		3-Phosphoglycerate	8,250	Meyerhof and Schulz (1935, 1936)
					Glucose phosphate	8,450	Meyerhof and Schulz (1935, 1936)
					Creatine phosphate	3,200	Meyerhof and Schulz (1935, 1936)
					Arginine phosphate	0 ± 300^{d}	Meyerhof and Schulz (1935, 1936)
3.1.4.5	Deoxyribonucleate oligonucleotidy-hydrolase	Deoxyribonuclease	Bovine pancreas	7.2	DNA		Brown *et al.* (1968)
3.2.1	Glycoside hydrolases			4.5	Maltose	1,100	Ono *et al.* (1965)
					Phenyl α^{-} maltoside	1,070	Ono *et al.* (1965)
					Maltotriose	2,110	Takahashi *et al.* (1965a)
					Amylose	$1,030 \pm 15$	Takahashi *et al.* (1965a)
					Isomaltose	$1,300 \pm 30$	Takahashi *et al.* (1965b)
					Panose	200	Takahashi *et al.* (1965b)
3.4.1.1	L-Leucylpeptide hydrolase	Leucine amino peptidase	Hog kidney	8.3 ± 0.3	L-Tyrosyl-glycinamide	$1,300 \pm 150$	Poe *et al.* (1967)
					Benzoyl-L-tyrosylglycin-amide	$1,550 \pm 100$	Sturtevant (1953)
3.4.2.1	Peptidyl-L-amino acid hydrolase	Carboxypeptidase A	Bovine pancreas	7.0–8.0	Benzoyl-L-tyrosylglycine	$1,330 \pm 90$	Poe *et al.* (1967)
				7.05–8.05	Benzoyl-L-tyrsine	$1,980 \pm 100$	Poe *et al.* (1967)
				7.3 ± 0.6	Carbobenzoxy-glycyl-L-phenylalanine	$2,550 \pm 50$	Dobry and Sturtevant (1952); Sturtevant (1953)

				7.0 ± 0.5	Carbobenzoxy-glycyl-L-leucine	2,110 ± 50	Ohlmeyer (1945); Sturtevant (1962)
3.4.4.4		Trypsin	Bovine pancreas	7.6	Poly-L-lysine (hydro bromide)	1,240	Poe *et al.* (1967); Sturtevant (1955b)
				6.9 ± 0.4	Benzoyl-L-argininamide	6,650 ± 200	Forrest *et al.* (1956); Sturtevant (1953)
3.4.4.5		Chymotrypsin	Bovine pancreas	6.6 ± 0.6	Benzoyl-L-tyrosinamide	5,840 ± 220	Dobry and Sturtevant (1952); Sturtevant (1953)
				6.6 ± 0.3	Benzoyl-L-tyrosyl-glycinamide	1,550 ± 100	Dobry *et al.* (1952); Sturtevant (1953)
3.4.4.9		Cathepsin C		5.1 ± 0.5	Glycyl-L-phenyl-alanylamide	6,220 ± 150	Ohlmeyer (1945); Sturtevant (1953)
3.5.1.1	L-Asparagine amidohydrolase	Asparaginase	Pig serum	7.0	Asparagine	5,710 ± 100	Benzinger *et al.* (1959); Kitzinger and Benzinger (1960); Kitzinger and Hems (1959)
3.5.1.2	L-Glutamine amidohydrolase	Glutaminase	E. coli Cl. welchii	7.0	Glutamine	5,160 ± 70	Benzinger and Hems (1956); Benzinger *et al.* (1959); Kitzinger and Benzinger (1960); Kitzinger and Hems (1959)
3.5.1.5	Urea amidohydro-lase	Urease	Jack bean	7.0	Urea	1,570	Brown *et al.* (1968)
3.6.1.1	Pyrophosphate phospho-hydrolase	Inorganic pyrophos-phatase	Baker's yeast	7.2	Na pyrophosphate	8,950	Ohlmeyer and Shatas (1952)
				7.4 ± 0.2	Inorganic pyrophosphate	7,310 ± 180[b]	Ging and Sturtevant (1954)
				7.3		5.810 ± 130[g]	Ging and Sturtevant (1954)
3.6.1.2	Trimetaphosphate hydrolase	Trimetaphosphatase	Baker's yeast	6.95 ± 05	Trisodium trimeta phosphate	19,310 ± 900	Meyerhof *et al.* (1953)
3.6.1.3	ATP phospho-hydrolase	ATPase	Myosin	8.0	ATP	5,400 ± 700	Podolsky and Morales (1956); Podolsky and Sturtevant (1955)
3.6.1.5	ATP diphospho-hydrolase	Apyrase	Potato	7.2	ATP	4,700 ± 700[h]	Brown *et al.* (1964a, b, c)

TABLE I—*continued*

Enzyme commission number	Systematic name	Trivial name	Source	Reaction pH	Substrate	$-\Delta H$ (cal $mole^{-1}$)	Reference
3.6.1.9	Dinucleotide nucleotido-hydrolase	Nucleotide pyrophosphatase		7.0	NAD	13,045[i,j]	Ohlmeyer and Shatas (1952)
						11,610[k]	
		Intestinal phosphatase		9.6	p-Nitrophenyl phosphate	4,530	Ohlmeyer and Shatas (1952)
4.1.1		Carboxylating enzyme		8.0	Ribulose diphosphate	4,800 ± 500	Kitzinger *et al.* (1956)
4.2.1.1		Carbonic anhydrase				[l]	This work, Berger, Chapter XII
4.2.1.2	L-Malate hydro-lyase	Fumerase			Fumerate	3,700	Kitzinger and Benzinger (1960); Ohlmeyer (1945)
	NADH dehydrogenase	Electron transport particles (ETP)			NADH	61,600 ± 1300[m]	Poe *et al.* (1967)
	Succinate dehydrogenase	Electron transport particles (ETP)			Succinate	36,200 ± 1200	Poe *et al.* (1967)

[a] Reaction used as an indirect indicator of membrane transport (highly exothermic).
[b] Data not yet published, see comment by Wadsö, Chapter III.
[c] Analytic procedure for the measurement of an inhibitor, dimethydichlorovinyl phosphate.
[d] Thermoneutral reaction.
[e] Analytic procedure.
[f] Veronal buffer.
[g] Phosphate buffer.
[h] Average value for β and γ phosphates.
[i] Schwartz cozymase.
[j] NAD → NMN + AMP.
[k] Sigma "65" cozymase.
[l] Used as calibrating reaction coupled to a color change.

TABLE II

THERMODYNAMIC QUANTITIES FOR SOME ENZYMIC REACTIONS[a]

Enzyme	Substrate	$\Delta F_{\alpha\ddagger}$ (kcal mole^{-1})	$\Delta H_{\alpha\ddagger}$ (kcal mole^{-1})	$\Delta S_{\alpha\ddagger}$ (eu mole^{-1})	$\Delta F_{\beta\ddagger}$ (kcal mole^{-1})	$\Delta H_{\beta\ddagger}$ (kcal mole^{-1})	$\Delta S_{\alpha\ddagger}$ (eu mole^{-1})	$\Delta F_{\alpha/\beta}$ (kcal mole^{-1})	$\Delta H_{\alpha/\beta}$ (eu mole^{-1})	$\Delta S_{\alpha/\beta}$ (eu mole^{-1})
Chymotrypsin[b]	Methyl hydrocinnamate	17.8	10.9	−23.2	19.7	16.2	−11.8	−1.9	−5.3	−11.4
	Methyl-DL-α-chloro-β-phenylpropionate	16.1	6.3	−33.0	18.7	14.8	−13.2	−2.6	−8.5	−19.8
	Methyl-D-β-phenyllactate	16.6	2.5	−47.2	18.7	14.5	−14.2	−2.2	−12.0	−33.0
	Methyl-L-β-phenyllactate	14.7	3.2	−38.5	17.5	10.5	−23.4	−2.8	−7.3	−15.1
	Benzoyl-L-tyrosine ethyl ester	10.7	0.2	−38.5	15.0	8.6	−21.4	−3.3	−8.4	−17.1
	Benzoyl-L-tyrosine amide	15.9	3.1	−43.0	17.9	14.0	−13.0	0.9	−9.9	−30.0
Acetylcholin-esterase[b]	Acetylcholine	2.5 (av)	(14–19)	(34–52)	8 (av)	(14–19)	(16–34)	−5.5	0	18.5
	Dimethylaminoethyl acetate	5.5 (av)	(6.7–8)	(4–8)	9.9 (av)	(6.7–8)	−(6.5–10.5)	−4.4	0	14.6
	Methylaminoethyl acetate	7.8	8	0.7	10.7	8	−9	−2.9	0	9.7
	Aminoethyl acetate	9.7	9.5	−0.6	12.2	9.5	−9	−2.5	0	8.4
	"Acyl enzyme hydrolysis"	—	—	—	13.8 (av)	(0–1.5)	−(41–46)	—	—	—
Carboxypepti-dase	Carbobenzoxy-L-tryptophan	16	21	17	19.3	16	−11	−3.4	5	−28
	Carbobenzoxyglycyl-L-phenylalanine	11.8	8.5	−11	14.4	8.9	−18	−2.5	−0.4	−7
	Carbobenzoxyglycyl-L-tryptophan	11.9	9.3	−8.5	15.3	9.3	−20	−3.4	0	−11.5
Pepsin	Carbobenzoxy-L-glutamyl-L-tyrosine ethyl ester	18.3	22.5	14.1	22.1	20.1	−6.5	−4.7	1.4	20.6
	Carbobenzoxy-L-glutamyl-L-tyrosine	18.8	19.6	2.6	23.1	16.6	−21.8	−4.3	3.0	24.4
Urease[b]	Urea	8.2	6.2	−6.8	11.3	9.1	−7.2	−3.2	−2.9	0.9
Adenosine tri-phosphatase	Adenosine triphosphate	7.3	20.4	44.0	14.4	12.4	−8.0	−7.5	8	52

[a] Table II has been reproduced from Lumry (1959) by permission.

[b] These enzymes are thought to form enzyme-substrate compounds in equilibrium reactions, i.e., $\Delta F_{\alpha/\beta}$ etc. are standard thermodynamic changes for the reaction in which the enzyme-substrate compound is formed.

standard enthalpy and standard entropy for the overall reaction and for the half-reactions involved. The enthalpy change is of interest, not only in the context of this reaction but also in reference to other oxidations of NADH.

III. Transferases

The irreversibility of the transmethylation from dimethylacetothetin to homocysteine made calorimetric study of this reaction particularly appropriate.

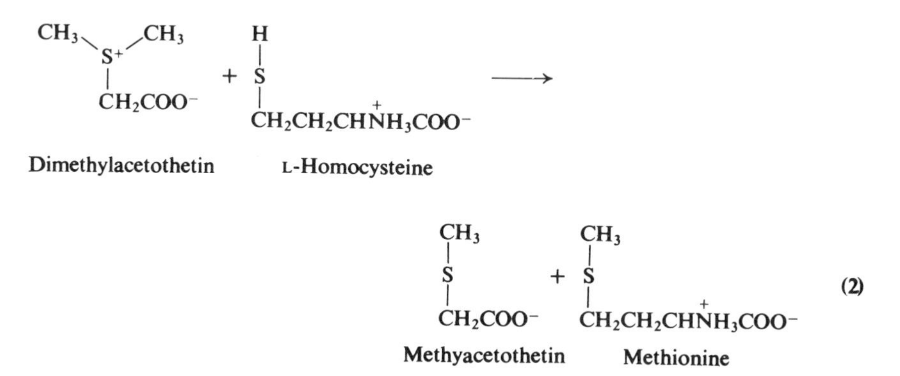

Values obtained for the heats of transmethylation in this reaction and similar reactions in which the methyl donors were betaine and methyl methanesulfonium bromide are given in Table I. Durell *et al.* (1962; Durell and Sturtevant, 1957) have observed that the time course of these apparently very similar reactions differs considerably. Also, the course of heat evolution in transfers from betaine and MMS do not follow a simple kinetic equation. Their data for three transmethylation reactions show that methyl substitution on the nitrogen atom of glycine lowers the heat of ionization approximately 1700 cal/mole. This effect is counterbalanced, however, by a degree in the standard entropy of ionization so that the standard free energy of ionization is relatively unaffected by methyl substitution. The first methyl group has a large effect upon ΔH^o and, especially, upon ΔS^o. Durell and associates, assuming that the entropy differences result from differences in the hydration of the substituted ammonium groups, conclude that the first methyl substitution increases the partial molar entropy of the dipolar ion by 8.5 units but that the second substitution causes a further increase of only 2.5 units. Under the condition of their experiments, pH 7, the three reactions are strongly exergonic. Transfer of metal from DMAT to homocysteine has an unusually large negative standard free energy, and even the least exergonic reaction, the methyl transfer from betaine, appears essentially irreversible.

IV. Hydrolases

Hydrolases, as a class of enzyme, have been better studied than others. Principle emphasis has been upon the hydrolysis of phosphate compounds and of amide and peptide bonds. The great importance of phosphate compounds as energy storage and transfer "agents" in intermediary metabolism has directed the calorimetrists to the measurement of standard free energy of hydrolysis of the phosphate linkages of adenosine triphosphate. Sturtevant (1962) draws attention to the necessity for consideration of ionic states and of buffer suitability in the direct measurement of enthalpy. His comparison of the hydrolysis of ATP, inorganic pyrophosphate, and *p*-nitrophenol phosphate makes the point dramatically apparent. Since the second ionization of phosphoric acid derivatives and the ionization of *p*-nitrophenol lie in the physiological pH range, it is to be expected that the enthalpies of hydrolysis of these compounds are functions of pH. Sturtevant has chosen to study reactions in which there is no net liberation or absorption of hydrogen ions so that the nature of the buffer does not influence the heat data. Thus,

$$XOPO_3H^- + H_2O = XOH + H_2PO_4^- \quad (3)$$

$$YOPO_3^= + H_2O = YOH + HPO_4^= \quad (4)$$

$$ZOPO_3^= + H_2O = ZO^- + H_2PO_4^- \quad (5)$$

The enthalpy changes for reactions of these types are summarized in Table III. Enzymatically catalyzed reactions, to a great extent, have made possible

TABLE III

HEATS OF CONSTANT pH, UNBUFFERED HYDROLYSES OF PHOSPHATE COMPOUNDS AT 25°[a]

Compound	*X*	$-\Delta H$[c]	*Y*	$-\Delta H$	*Z*	$-\Delta H$
p-Nitrophenyl phosphate	$O_2NC_6H_4$	6280	$O_2NC_6H_4$	2760	$O_2NC_6H_4$	−1900
ATP[b]	$^-A{-}O{-}P(=O){-}O^-$	5800	$^-A{-}O{-}P(=O){-}O^-$	5000	$^-A{-}O{-}P(=O){-}O^-$	5800
Pyrophosphate	$^-O{-}P(=O){-}O^-$	5800	$^-HO{-}P(=O){-}O^-$	5800	$^-HO{-}P(=O){-}O^-$	5800

[a] After Sturtevant, (1955a).
[b] A represents the adenylate residue.
[c] Calories per mole.

the direct calorimetry of phosphate compounds. Sturtevant, in his classic paper (1955a), described the hydrolysis of fully charged *p*-nitrophenol phosphate ion to form uncharged *p*-nitrophenol and dihydrogen phosphate. The enzyme orthophosphoric monoester phosphohydrolase derived from prostate gland was used in an acetate-buffered environment. This was a convenient reaction because of the practicality of comparing heat data with results obtained spectrophotometrically. This work finally served to draw attention to the necessity for considering the reaction type and condition in attempting to compare various phosphate hydrolyses. (The point has been further emphasized by Evans in Chapter V.)

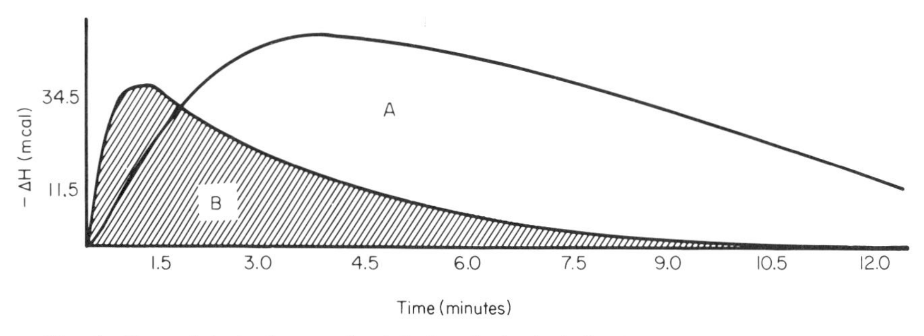

FIG. 1. Curve A is the heat evolved during the hydrolysis of acetylcholine by the enzyme acetylcholine esterase. Curve B represents the heat evolved during a reaction in which the same concentrations of components were present, together with a phosphorous-containing inhibitor (dimethyl dichlorovinyl phosphate).

Studies similar to those of Sturtevant had been undertaken by Ohlmeyer (1945) and Ohlmeyer and Shatas (1952), using a Dewar flask-type isothermal calorimeter.

Sturtevant's measurement of the heat of hydrolysis of acetylcholine has been reported by Wadsö in Chapter III. Brown *et al.* (1968) used the acetylcholine esterase reaction as an assay for the phosphorus-containing inhibitor dimethyl dichlorovinyl phosphate (Fig. 1). Analysis for deoxyribonucleic acid has been described by Brown *et al.* (1968), based upon the DNA hydrolysis catalyzed by deoxyribonucleate oligonucleotide-hydrolase, using commercial reagents. In their impure system, the heat values, though linear, were not proportionate to substrate concentration. This they interpreted to indicate the occurrence of a secondary or interfering phenomenon that contributed to the measured heat values that were thus a function of the enthalpy of hydrolysis.

Possibly because the linkages in most carbohydrates are not easily hydrolyzed by conventional chemical methods, the calorimetric measurement of saccharide heats of hydrolyses only rarely have been underaken. An enzymologic approach allowed Ono and associates (1965) and Takahashi *et al.* (1965a,b) to measure the heats of hydrolysis of a number of biologically important carbohydrates. These workers used gluc-amylase, a catalyst that has been designated an α-1,4, 1,6-glucan 4:6-gluco hydrolase. The enzyme, reported to be electrophoretically pure, has the capability of catalyzing the hydrolysis of both α-1,4 and α-1,6 glucosidic linkages in starch, dextran, and other polysaccharides. This series of papers holds a number of interests, among them the display of methodology for the discrimination of mutarotational heat from the heat of hydrolysis that was primarily sought. In the hydrolysis of the α-1,4 glucosidic linkages it was noted that only the linkage next to the reducing end exhibits exceptional properties. Other linkages in the polymer are little affected by the chain length. Thus in the carbohydrates studied by these workers it appears that the heats of hydrolysis of the α-1,4 glucosidic linkages are essentially additive and that no corrective factor must be taken into consideration, as Sturtevant (1959) had demonstrated necessary in interpreting measurements of heats of peptide bond hydrolyses. This observation of Ono *et al.* (1965) seems to hold promise for the fruitfulness of further calorimetric study of carbohydrates in that structural inference may be made without the necessity for unduly complex analyses.

Another intriguing observation, discussed in detail by Ono and Takahashi in Chapter IV, is that the ΔH of hydrolysis of α-1,4 and α-1,6 linkages are opposite in sign. The ΔH_{av} of α-1,4 glucoside hydrolysis in amylose is -1030 cal mole^{-1}, while the ΔH_{av} of α-1,6 glucoside, though approximately equal to this value in magnitude, is positive in sign. The biological implications of this finding are imposing. Presumably much that is observed in nature about the "preference" of organisms for the 1-4 bond can be readily interpreted in terms of energetics. Certainly study of this system and of the little known gluc-amylase itself seems a promising undertaking. Renewed interest in the classic 1,6 hydrolase, a debranching enzyme, also seems to be suggested by this present data.

Sturtevant and associates (Sturtevant, 1953; Dobry *et al.*, 1952; Dobry and Sturtevant, 1952; Forrest *et al.*, 1956; Rawitscher *et al.*, 1961), using peptydl-amino acid hydrolases and peptydl-hydrolases, studied the energetics of amide and peptide bonds because of the relationship of these to the synthesis and breakdown of proteins intracellularly. From 1952 onward their series of reports alone have provided in essence the microcalorimetric literature for this class of reaction. Huffman (1942) earlier had estimated hydrolysis heats for several dipeptides from combustion heat data. Rawitscher and her collaborators (1961) used the enzyme leucine aminopeptidase and carboxypeptidase in the measurement of hydrolysis heats of six dipeptides. The hydrolysis of poly-L-lycine also has been studied, as have three amino-acid amide reactions all catalyzed by trypsin or by chymotrypsin. In their study, peptides, polypeptides, and amino-acid amides served as substrates for peptidyl-amino-acid and peptidyl-peptide hydrolases. They ran the reactions at 25° in dilute solutions and at neutral pH. The necessity for correction of buffer heats of ionization and to a lesser extent for the heats of ionization of products has been stressed by these authors. (The point has been further evaluated by Wadsö in Chapter III and by Evans in Chapter V.) The data obtained over the several years (since 1952) by these workers indicates that the hydrolysis of peptide bonds produces less heat than had been estimated from combustion data. They also have reported that the heats of hydrolysis of amide bonds to give charged products are more exergonic than those of peptide bonds. Furthermore, because of the heat of ionization of the ammonium ion (about 2000 cal $mole^{-1}$ more positive than the heat of ionization of amino acids), the difference in heats of hydrolysis between amide and peptide bonds is reduced when those reactions are considered that yield uncharged products. Sturtevant (1953) used another enzyme, cathepsin C, to study the hydrolysis of glycyl-L-phenylalanyl-amide acetate. The reaction was carried out in a buffered solution at 25°, and the heat evolution followed a first-order rate law to 90% completion at pH 4.69. Since the products are nearly completely ionized, no correction for ionization heat was thought to be needed.

Kitzinger and Hems (1959) used the asparaginase and glutaminase reactions to allow a comparison of hydrolysis heats of the amide groups of asparagine and glutamine. They concluded that the free energies of hydrolysis of these amide groups were dissimilar and that one therefore could not make the simple assumption that all amide group energies were approximately the same. This was disappointing, because it had been hoped that inference about glutamine could be made from existing knowledge of asparagine and other amino acids. Thermodynamic data for glutamine hydrolysis held special interest because in the glutamine–synthetase reaction the synthesis of glutamine is coupled to the hydrolysis of ATP. The reaction is reversible, and an equilibrium constant known. Hence, if the free energy of hydrolysis of glutamine also had been available, it would have been possible to evaluate the free energy of hydrolysis of ATP. When the assumption was proved untrue by Kitzinger and Hems (1959) that the hydrolysis of glutamine is thermodynamically analogous to that of the asparagine, known from heat of combustion and heat capacity data (Borsook and Huffman, 1938), it became apparent that another approach had to be sought.

The attempt to evaluate thermodynamically the glutamine synthesis was then made in another way, using the conduction-type (heat-burst) calorimeter of Benzinger and Kitzinger (1963). These experiments of Benzinger and his associates provide an important example of the equilibrium method for the calculation of thermodynamic quantities from reaction calorimeter measurements alone. By obtaining measurements in a series of experiments it was possible to determine the chemical equilibrium and thus to calculate the free energy change of the hydrolysis reaction without resorting to techniques other than the reaction calorimetric measurements. These authors ran six successive experiments in which glutaminase was added to ammonium glutamate. The heats resulting from both the synthesis of glutamine and from its hydrolysis could be measured and were, of course, affected by the starting reagent concentration. The ratio of the heats of the forward and reverse reactions then were used to determine the overall equilibrium constant and ΔG in a manner analogous to solution of ΔG from chemical equilibrium data. Data published by Benzinger *et al.* (1959) and discussed in an excellent review (Benzinger and Kitzinger, 1963) is reproduced as Fig. 2. This method was later applied to biologically important analysis of the thermodynamic values for ATPase catalyzed hydrolysis (see below).

Jack bean urease was used in a series of experiments of Brown and associates (1968) to catalyze the hydrolysis of urea. The microcalorimetric measurement was made as an incidental to the development of an analytic technique for the measurement of urea in biological fluids. Urea in urine served also as a substrate, and heat values obtained in these experiments (correlated with chemical analysis for ammonia) indicated that normal urine contained no components that interfered with or otherwise contributed to the measured reaction heat.

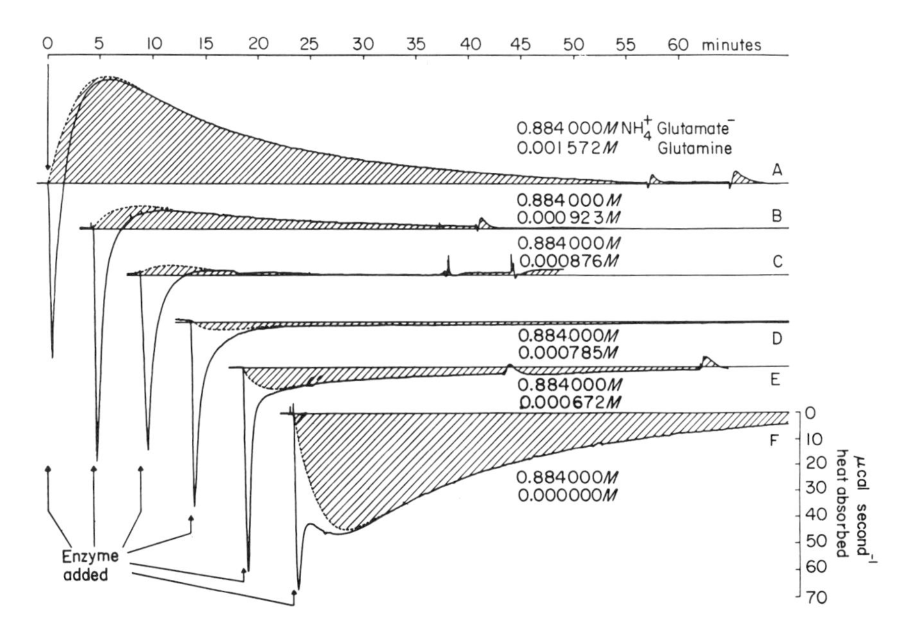

FIG. 2. Curve F: glutamic acid and ammonia mixed with glutaminase, heat absorption due to synthesis of glutamine. Curves E and D: increasing amounts of glutamine added to initial reagents; heat absorption becomes less, and after passing equilibrium concentration, changes sign C, B, and A due to glutamine hydrolysis (from Benzinger and Kitzinger, 1963).

The older experiments of Ohlmeyer and Shatas are of continuing interest. Their study of various phosphate compounds was directed at the pyrophosphate bonds of NAD as well as of ATP and of *p*-nitrophenol phosphate. They concluded that dinucleotide hydrolysis of the P—O—P bonds of NAD yielded energies that were not greatly different from the pyrophosphate bond energies of ATP. Here and in other studies it had been necessary in calculating enthalpy change of adenosine triphosphate hydrolysis to combine thermal data with free energy data obtained in conventional thermodynamic procedures to obtain a value for ATP hydrolysis. All of the calculations were colored by uncertainties present in the several required measurements. In more recent work, Podolsky and Sturtevant (1955) and Podolsky and Morales (1956) used the ATP phosphydrolase activity of myosin as offering a direct method for the measurement of enthalpy change on hydrolysis of ATP. The use of myosin requires the presence of an activating metal ion, and the consequent possibility of a heat contribution from salt formation (excess cation with the inorganic phosphate product of the hydrolysis) existed. Podolsky and Sturtevant (1955) and Podolsky and Morales (1956) used low levels of activating calcium in an attempt to eliminate this possibility. Their value for the ATP hydrolysis enthalpy change, 4700 ± 200, represented a radical departure from values that had been published earlier. The value is of great theoretical importance because it is less than half the 12 kcal/mole value that had been earlier accepted. Direct measurement of ΔH of ATP hydrolysis as approximately 5000 kcal made requisite the reevaluation of the high-energy bond concept. Indeed it may be concluded that the ATP contribution to biological events may not be a consequence of an exceptional enthalpy change upon hydrolysis. Many authors have, since the publications of Podolsky and Sturtevant and Podolsky and Morales, re-formed the concept. They look upon ATP as unique in its ability to participate in energetic events but not as contributing an enthalpy change that is greatly different from that seen upon the hydrolysis of other phosphate esters.

V. Lyases

Kitzinger *et al.* (1956) studied the reaction catalyzed by a carboxylation enzyme isolated from spinach leaves. ΔH of the carboxylation reaction.

Ribulose 1,5 diphosphate + $HCO^{3-} \rightarrow 2$ phosphoglycerate^{3-} + H^+ is -4800 ± 500 cal mole^{-1}. Benzinger and associates (1959; Benzinger, 1956; Benzinger and Hems, 1956) used a malate hydrolase in the development of an equilibrium technique for the calorimetric determination of free energy and entropy changes. The method advanced by these authors permits, in favorable instances, the determination of free energy as well as enthalpy and entropy values using thermal data alone. The method has been described graphically in a review by Benzinger and Kitzinger (1963). The use of the calorimeter to obtain equilibrium constants appears to be a widely applicable procedure in the biochemistry laboratory, especially with the extremely sensitive instruments now available.

REFERENCES

Anonymous (1966). *Sci. Res.* **1**, 32.

Benzinger, T. H. (1956). *Proc. Natl. Acad. Sci. U.S.* **42**, 109.

Benzinger, T. H. (1965). *Fractions* **2**, 2.

Benzinger, T. H., and Hems, R. (1956). *Proc. Natl. Acad. Sci. U.S.* **42**, 896.

Benzinger, T. H., and Kitzinger, C. (1954). *Federation Proc.* **13**, 11.

Benzinger, T. H., and Kitzinger, C. (1963). *In* "Temperature Control and Measurement" (C. M. Herzfeld, ed.), Vol. 3, p. 3. Reinhold, New York.

Benzinger, T. H., Kitzinger, C., Hems, R., and Burton, K. (1959). *Biochem. J.* **71**, 400.

Borsook, H., and Huffman, H. M. (1938). *In* "Chemistry of the Amino Acids and Proteins" (C. Schmidt, ed.). Thomas, Springfield, Illinois.

Brown, H. D., Altschul, A. M., Evans, W. J., and Neucere, N. J. (1964a). *Plant Physiol.* **39**, Suppl., Ixi–Ixii. Abstr.

Brown, H. D., Evans, W. J., and Altschul, A. M. (1964b). *Life Sci.* **3**, 1487–1492.

Brown, H. D., Evans, W. J., and Altschul, A. M. (1964c). *Federation Proc.* **23**, 175. Abstr.

Brown, H. D., Evans, W. J., and Altschul, A. M. (1965a). *Biochim. Biophys. Acta* **94**, 302–304.

Brown, H. D., Neucere, N. J., Altschul, A. M., and Evans, W. J. (1965b). *Life Sci.* **4**, 1439–1447.

Brown, H. D., Chattopadhyay, S. K., Patel, A. B., Shannon, G. R., and Pennington, S. (1968). "Dual Differential Calorimetry, an Analytic Technique for Enzyme-catalyzed Reactions," I.E.E.E. Record (Inst. Elec. Electron. Engr. Publ. 68C 17–Reg–3), 19.4.5.

Brown, H. D., and Pennington, S. N. (1969). Effect of phosphorous inhibition on the kinetics of the adenyl cyclase system. *Proc. 1st Intern. Conf. Calorimetry, Warsaw, Poland.*

Buzzell, A., and Sturtevant, J. M. (1951). *J. Am. Chem. Soc.* **73**, 2454.

Buzzell, A., and Sturtevant, J. M. (1952). *J. Am. Chem. Soc.* **74**, 1983.

Calvet, E. (1963). "Microcalorimetry" (Transl. by H. Skinner). Macmillan, New York.

Calvet, E., and Prat, H. (1956). "Microcalorimetrie." Masson, Paris.

Canady, W. J., and Laidler, K. J. (1958). *Can. J. Chem.* **36**, 1289.

Dobry, A., and Sturtevant, J. M. (1952). *J. Biol. Chem.* **195**, 141.

Dobry, A., Fruton, J. S., and Sturtevant, J. M. (1952). *J. Biol. Chem.* **195**, 148.

Durell, J., and Sturtevant, J. M. (1957). *Biochim. Biophys. Acta* **26**, 282.

Durell, J., Rawitscher, M., and Sturtevant, J. M. (1962). *Biochim. Biophys. Acta* **56**, 552.

Elliott, W. H. (1953). *J. Biochem.* **201**, 661.

Forrest, W. W., Gutfreund, H., and Sturtevant, J. M. (1956). *J. Am. Chem. Soc.* **78**, 1349.

Ging, N. S., and Sturtevant, J. M. (1954). *J. Am. Chem. Soc.* **76**, 2087.

Huffman, H. M. (1942). *J. Phys. Chem.* **46**, 885.

Jacobson, K. P. (1934). *Biochem. Z.* **274**, 167.

Katz, S. (1955). *Biochim. Biophys. Acta* **17**, 226.

Kitzinger, C., and Benzinger, T. H. (1955). *Z. Naturforsch.* **10b**, 375.

Kitzinger, C., and Benzinger, T. H. (1960). *In* "Heatburst Microcalorimetry" (D. Glick, ed.), Vol. 8, p. 309. Wiley (Interscience), New York.

Kitzinger, C., and Hems, R. (1959). *Biochem. J.* **71**, 395–400.

Kitzinger, C., Horecker, B. L., and Weisbach, A. (1956). *Intern. Congr. Physiol. Sci., 20th Brussels*, p. 502.
Lumry, R. (1959). *In* "The Enzymes" (P. D. Boyer, H. Lardy, and K. Myrbäck, eds.), 2nd Ed., Vol. 1, p. 178. Academic Press, New York.
Lumry, R. (1966). Discussion before calorimetry group. Stillwater, Minnesota.
Meyerhof, O., and Lohmann, K. (1932). *Biochem. Z.* **253**, 431–461.
Meyerhof, O., and Schulz, W. (1935). *Biochem. Z.* **231**, 292.
Meyerhof, O., and Schulz, W. (1936). *Biochem. Z.* **289**, 87.
Meyerhof, O., Shatas, R., and Kaplan, A. (1953). *Biochim. Biophys. Acta* **12**, 121.
Ohlmeyer, P. (1945). *Z. Physiol. Chem. Hoppe-Seylers* **282**, 37.
Ohlmeyer, P., and Shatas, R. (1952). *Arch. Biochem. Biophys.* **36**, 411.
Ono, S., Hiromi, K., and Takahashi, K. (1965). *J. Biochem. (Tokyo)* **57**, 799.
Pennington, S. N., Brown, H. D., Patel, A. B., Chattopadhyay, S. K., and Berger, R. L. (1969). Analytical Applications of Microcalorimetry. *Anal. Letter*, 2(8).
Pennington, S. N., Brown, H. D., Berger, R. L., and Evans, W. J. (1969). Stopped flow system for the Calvet-type microcalorimeter. *Proc. 1st Intern. Conf. Calorimetry and Thermodynamics, Warsaw, Poland.*
Podolsky, R. J., and Morales, M. F. (1956). *J. Biol. Chem.* **218**, 945.
Podolsky, R. J., and Sturtevant, J. M. (1955). *J. Biol. Chem.* **217**, 603.
Podolsky, R. J., Kitzinger, C., Benzinger, T. H., Sturtevant, J. M., and Morales, M. F. (1954). *Federation Proc.* **13**, 112. Abstr. 372.
Poe, M., Gutfreund, H., and Estabrook, R. W. (1967). *Arch. Biochem. Biophys.* **122**, 204.
Rawitscher, M., Wadsö, I., and Sturtevant, J. M. (1961). *J. Am. Chem. Soc.* **83**, 3180.
Sturtevant, J. M. (1953). *J. Am. Chem. Soc.* **75**, 2016.
Sturtevant, J. M. (1955a). *J. Am. Chem. Soc.* **77**, 255.
Sturtevant, J. M. (1955b). *J. Am. Chem. Soc.* **77**, 1495.
Sturtevant, J. M. (1959). *Ann. Rev. Phys. Chem.* **10**, 1.
Sturtevant, J. M. (1962). *In* "Experimental Thermochemistry" (H. A. Skinner, ed.), Vol. 2, p. 427. Wiley (Interscience), New York.
Sunner, S., and Wadsö, I. (1962). *In* "Experimental Thermochemistry" (H. A. Skinner, ed.), Vol. 2, p. 239. Wiley (Interscience), New York.
Takashi, K. (1966). *Agr. Biol. Chem. (Tokyo)* **30**, 629.
Takahashi, K., Hiromi, K., and Ono, S. (1965a). *J. Biochem. (Tokyo)* **58**, 255.
Takahashi, K., Yoshikawa, Y., Hiromi, K., and Ono, S. (1965b). *J. Biochem. (Tokyo)* **58**, 251.
Wilson, R., Huffman, L., Brown, H., and Pennington, S. (1969). Effect of metal cations on the glucose oxidase reaction as measured by microcalorimetry. *Trans. Missouri Acad. Sci.* **3**.

CHAPTER VIII

Bacterial Calorimetry

W. W. FORREST

I. Introduction—History and Present Status

One of the most conspicuous properties exhibited by many natural systems undergoing microbial degradation is the evolution of heat; in some cases, with materials of poor thermal conductivity, this may occur so rapidly as to lead eventually to spontaneous combustion. Many efforts have been made to study this gross bacterial thermogenesis; the subject has considerable economic and engineering importance, and the behavior of these systems is now well documented (Carlyle and Norman, 1941; Dye and Rothbaum, 1964). Obviously, however, such measurements cannot be considered microcalorimetry.

Microcalorimetry of bacterial systems has been used mainly in the study of bacterial energetics. The review by Peterson and Wilson (1931) summarizes most of the early work in this field. Unfortunately, much of this earlier work was handicapped by inadequate definition of conditions and unsatisfactory instrumentation. Thus, Bayne-Jones and Rhees (1929) showed that during the early stages of growth more heat was produced by each bacterial cell than during the later stages. In general, limitation of growth by any cause depresses the catabolic activity of bacterial cells, and such limitation appears to have occurred during their experiments. However, more complex explanations have been proposed, and the finding has created a good deal of controversy (Rahn, 1932; Stoward, 1962a).

Modern calorimetric techniques were used in bacterial studies by Prat (1963); while his work was essentially qualitative and descriptive, he showed that the behavior of microbial systems was systematically reproducible. More recent studies have shown that, both in growing and nongrowing bacteria, the processes involved can be studied quantitatively by calorimetry (Battley, 1960; Forrest *et al.*, 1961; Belaich, 1963; Boivinet, 1964). A more complete bibliography is given by Boivinet (1964).

The factors affecting the calorimetric behavior of microorganisms now can be clearly defined; as these differ at different phases of the growth cycle, it is necessary to consider the phases individually.

II. Growing Cells in Pure Culture

A voluminous literature exists on the behavior of bacteria during the growth cycle; detailed consideration may be found in the standard texts (Lamanna and Mallette, 1959).

A. Exponential Growth Limited By Energy Source

In the usual experiments on bacterial growth, a small number of cells is used to inoculate a sterilized growth medium. If this is done in the calorimeter, there may be initially no detectable heat evolution. There are two reasons for this behavior: first, the comparatively low catabolic activity at the beginning of the lag phase and, second, the small number of cells. At the end of the lag phase there is a period of transition when the catabolic activity of unit dry weight of cells and rate of cell division increase until they reach the high rates characteristic of cells undergoing exponential growth; these rates then remain constant at this higher level during growth. The ratio of catabolic activities between growing and nongrowing cells may be as high as 8 to 1 for *Streptococcus faecalis* (Forrest and Walker, 1964). After the lag phase, heat evolution is observed. As the cells begin to divide, the rate of heat evolution increases exponentially, concurrently with the synthesis of new cellular material, so that during exponential growth the rate of heat production per unit of dry weight also remains constant at the highest level reached by the cells (Fig. 1).

During exponential growth a steady state is set up; the rate of synthesis of new çellular material, degradation of energy source, appearance of products of catabolism, and heat production are all accurately described by the same exponential function (Forrest and Walker, 1964). These observations disagree with those of Bayne-Jones and Rhees (1929) (Section I), but more important are the consequences from the point of view of the change in entropy in the system (Prigogine and Wiame, 1946; Stoward, 1962a; Forrest and Walker, 1964).

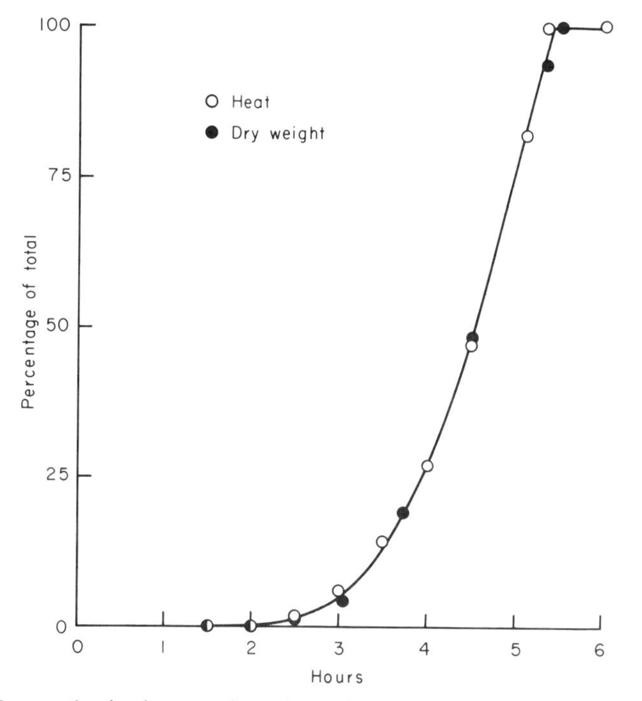

FIG. 1. Heat production by a growing culture of *Streptococcus faecalis*, with glucose as the energy source and growth limited by the energy source (from Forrest *et al.*, 1961). Reprinted by courtesy of the *Journal of Bacteriology*.

In some organisms there is much less difference between the catabolic activities during growth and in nongrowing cells. Belaich and Senez (1967) found that *Zymomonas mobilis* doubled its catabolic activity during exponential growth. With this organism the catabolic activity during the lag phase is comparatively high, and measurable heat production can be observed during the lag phase from the time of inoculation (Fig. 2).

If all other nutrients are present in adequate amounts so that the energy source becomes exhausted first, growth ceases completely. In the case of *S. faecalis* (Fig. 1) and *Streptococcus lactis* (Fig. 3) the saturation "Michaelis" constant for the degradation of glucose is very low, so that the enzyme systems responsible for glycolysis remain saturated until the substrate is almost completely exhausted. The ensuing transition from this exponentially increasing

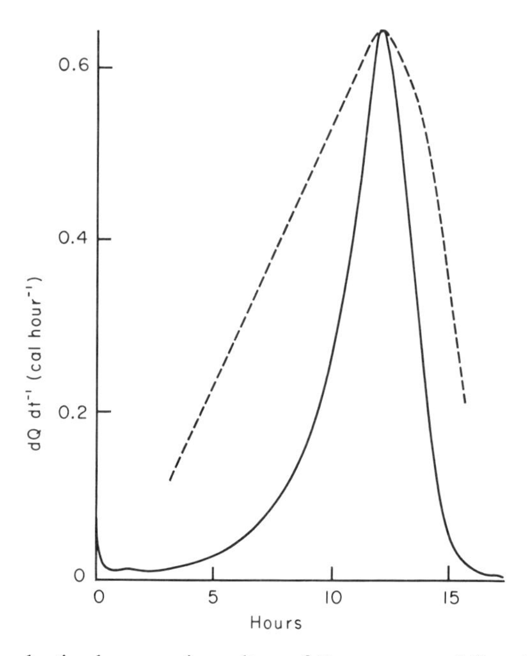

FIG. 2. Heat production by a growing culture of *Zymomanas mobilis*, with glucose as the energy source and growth limited by the energy source (from Senez and Belaich, 1965). Reprinted by courtesy of *les Colloques Internationaux du Centre National de la Recherche Scientifique*.

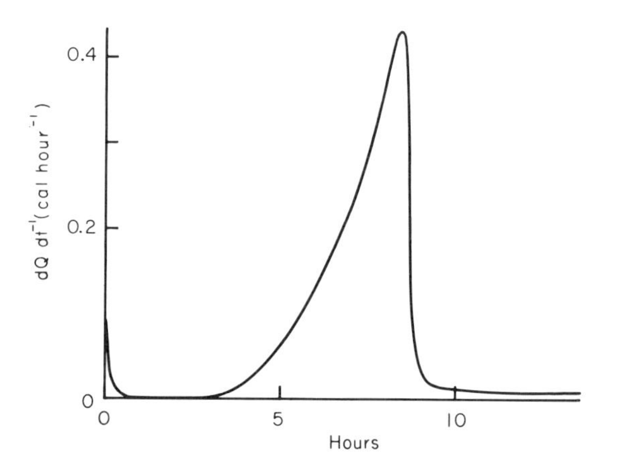

FIG. 3. Heat production by a growing culture of *Streptococcus lactis*, with glucose as the energy source and growth limited by the energy source (from Boivinet, 1964). Reprinted by courtesy of the author.

rate of heat evolution to no detectable rate of heat production is very rapid. In contrast, with *Z. mobilis* grown with glucose as substrate (Fig. 2), the heat evolution at the end of growth decreases quite gradually over a period as the substrate concentration decreases. The saturation constant for glucose in this case is comparatively high, and it appears possible to determine its value experimentally from the calorimetric data (Belaich, private communication).

It is of interest that a fundamental difference exists in metabolic pathways between these organisms: The streptococci catabolize glucose by the Embden–Meyerhof pathway, whereas in *Z. mobilis*, catabolism is by the Entner–Douderoff pathway.

When only exponential growth occurs, with the amount of growth being limited solely by the availability of energy source, heat evolution ceases when the exogenous energy source is exhausted. Such cells exhibit no calorimetrically detectable endogenous metabolism (Figs. 1, 2, and 3). It has been shown (Belaich, 1963; Grangetto, 1963; Boivinet, 1964) that the observed heat evolution, after due allowance for second-order effects such as heats of neutralization, correlates very well with the value of enthalpy change calculated for the degradation of substrate to the observed products. In principle, calorimetry should detect coupled anabolic reactions (Wilkie, 1960; Prigogine, 1961), but it is not practical in such experiments to determine the enthalpy of growth (Senez and Belaich, 1965), which would be expected (Morowitz, 1960) to be less than the experimental error of the measurements.

In this case of exponential growth in a nutritionally adequate medium, all energy available from catabolism of the added energy source, except perhaps a very small amount required for energy of maintenance (Pirt, 1965), is used for synthesis of cellular material. The catabolism of sufficient energy source to produce 1 mole of adenosine triphosphate (ATP) allows the organisms to synthesise 10 gm of new cellular material (Bauchop and Elsden, 1960). The enthalpy change associated with the production of this quantity of ATP in the given system is then defined by the metabolic pathway so that the enthalpy change for biosynthesis of unit mass of organisms, ΔH_b (Boivinet, 1964), is also defined. Boivinet's tables for these quantities in anaerobic growth of *S. lactis* show that in fact constant values are obtained for ΔH_b. Similar measurements have been made under aerobic conditions with *Aerobacter aerogenes* (Grangetto, 1963).

B. Growth Limited by Nutritional Factors

If limitation of other constituents of the medium occurs while an energy source is still present, substrate continues to be degraded, but the catabolic activity of the organisms is depressed, so that heat evolution is now observed at a lower level. This is shown by the experiments of Senez and Belaich (1965) on the effect of phosphate limitation on heat production in *Z. mobilis* and *Escherichia coli*, where calorimetry was employed to give an elegant demonstration of the change in catabolic activity. In this case the effect of phosphate limitation is immediately and completely reversible, and Senez and Belaich give a discussion of the effect in terms of control mechanisms.

In general the decrease in the rate of cell division at the end of the phase of exponential growth is caused either by limitation of some constituent of the growth medium or by accumulation of toxic products. There is a corresponding

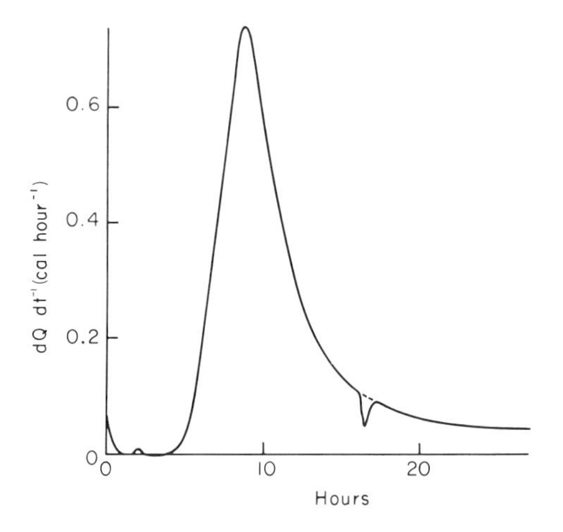

FIG. 4. Heat production by a growing culture of *Streptococcus lactis*, grown with excess energy source (from Boivinet, 1964). Reprinted by courtesy of the author.

decrease in catabolic activity (Forrest and Walker, 1964) and a decrease in the rate of heat production per unit dry weight of bacteria. In aerobic organisms, oxygen limitation may be responsible, so that there can be a change in the pattern of catabolism from oxidative to fermentative, with a corresponding very large change in the products of catabolism and rates of heat production (Stoward, 1962a, b). However, in this part of the growth curve the amount of cellular material produced is still influenced by the amount of energy source added to the medium, though the efficiency of conversion may be lower than during exponential growth (Boivinet, 1964; Forrest and Walker, 1965a).

With a large excess of energy source, the shape of the thermogram is quite characteristic. Growth initially occurs exponentially and then decreases and gradually ceases as the organisms enter the stationary phase. Concurrently, catabolic activity and heat production gradually decline. In the case of *S. lactis* (Fig. 4), the lactic acid produced by degradation of substrate depresses the pH of the medium, so that the growth rate is decreased.

More complex thermograms than those previously discussed also may be encountered (Prat, 1963); secondary growth may occur, or endogenous metabolism may become evident.

C. Endogenous Metabolism

The decrease in growth rate at the end of exponential growth means that ATP is no longer used at a high rate to drive growth processes. Growth may become partly energetically uncoupled (Senez, 1962). Under these circumstances the pool of ATP in the organisms can be shown to increase (Forrest and

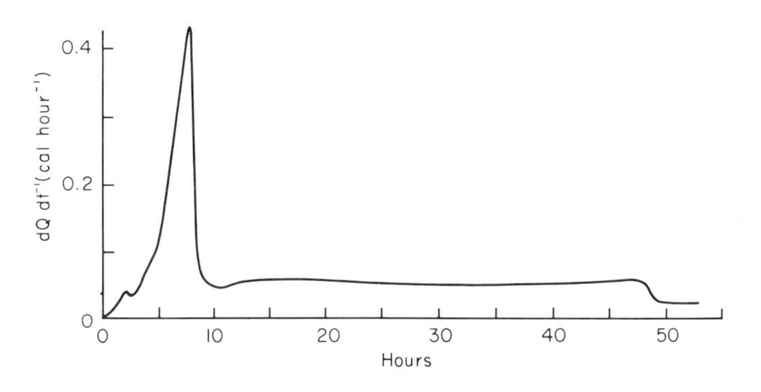

Fig. 5. Heat production by *Escherichia coli* during growth with subsequent endogenous metabolism (from Senez and Belaich, 1965). Reprinted by courtesy of *les Colloques Internationaux du Centre National de la Recherche Scientifique.*

Walker, 1965a). Thus ATP is available for syntheses other than those directly connected with the essential processes of growth. It is found in many organisms that, during the period succeeding exponential growth (Dawes and Ribbons, 1964; Hungate, 1963), reserve materials are accumulated. When exogenous substrates are finally exhausted after this period of accumulation of reserves, endogenous heat production subsequently may occur from the degradation of these reserves. *E. coli* (Sigal *et al.*, 1964) may accumulate large quantities of glycogen, and here endogenous metabolism may continue for long periods after growth has ceased (Fig. 5).

Since these storage materials very often are polysaccharides formed by polymerization of some of the substrates added primarily as energy sources, it is essential in arriving at any thermodynamic balance that accurate assays be carried out for products. Calculations based purely on postulated reaction pathways may be in serious disagreement with the observed data. Thus, Grangetto (1963) reported that *A. aerogenes*, which also accumulates reserve polysaccharides, gave excellent correspondence between the observed heat production and calculated enthalpy change for the degradation of succinate, but with glucose as energy source only about half the calculated heat in fact was observed.

D. Growth in Continuous Culture

The preceding discussion of the heat production during growth has been concerned only with conventional calorimetric measurements in batch culture experiments. In continuous culture systems, where a very high concentration of organisms may be maintained in a steady state at a high rate of growth for long periods (Herbert *et al.*, 1965), large heat outputs may be observed. Such heat evolution is usually considered only as an engineering problem; a large fermenter may dissipate several kilowatts, so that temperature control is necessary. There seems to have been no systematic calorimetric study in such systems, though calorimetric measurements under these conditions would appear to offer advantages in growth studies. The system sets up a steady state that can be closely defined, so that a precise energetic balance could be readily obtained. Under these conditions determinations of enthalpy of growth (Section II, A) might well become practical.

III. Nongrowing Cells in Pure Culture

A. Endogenous Metabolism and Energy of Maintenance

A suspension of microorganisms deprived of exogenous substrate still may exhibit substantial metabolic activity. In many cases, with aerobic organisms, measurable respiration occurs and the course of the metabolism may be followed manometrically, but with many anaerobic organisms there is no readily measurable change. Consequently the endogenous metabolism of anaerobes has been little studied (Dawes and Ribbons, 1962).

S. faecalis in starved suspensions exhibits no detectable respiration and no pH change occurs in the suspending medium, yet calorimetric measurements show that a considerable evolution of heat occurs (Forrest and Walker, 1963; Fig. 6.) This heat evolution is markedly affected by environmental factors (Fig. 7), and its presence is correlated with the ability of the organisms to degrade glucose. The organisms also possess a constant pool of ATP only during this process (Forrest and Walker, 1965b), indicating that the energy made available during this endogenous metabolism is coupled to biologically useful functions by ATP.

The large amounts of heat involved indicate that the process supplying energy is unusual. There is no evidence for the type of reserve material usually found, polysaccharide or poly-hydroxy butyric acid, being accumulated by the

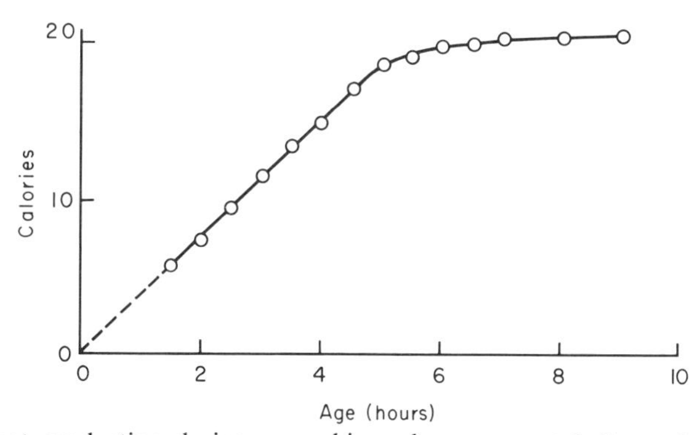

FIG. 6. Heat production during anaerobic endogenous metabolism of *Streptococcus faecalis* (from Forrest and Walker, 1963). Reprinted from *Biochemical and Biophysical Research Communications* by courtesy of the publishers, Academic Press, New York.

organisms. The only materials excreted into the suspending medium during this metabolism are amino acids, yet the process occurring cannot be the hydrolysis of peptide bonds, which could not account for either the large amounts of heat evolved or the production of ATP. Thus, using the calorimetric technique, it has been possible to detect and characterize the endogenous metabolism to the point where normal biochemical studies can proceed.

The existence of endogenous metabolism at a level detectable by microcalorimetry is almost certainly a general property of microorganisms. Many aerobic organisms exhibit endogenous respiratory activity, and Senez and Belaich (1965) have detected substantial endogenous heat production in the anaerobic suspension of *E. coli* (Fig. 5).

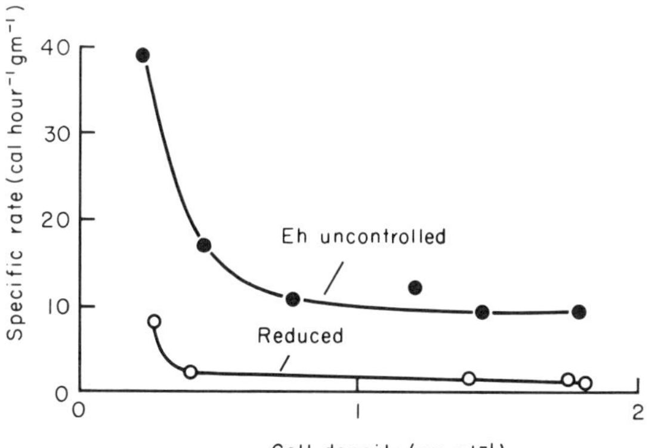

FIG. 7. The environmental factors affecting the rate of endogenous metabolism in *Streptococcus faecalis* (from Forrest and Walker, 1963). Reprinted from *Biochemical and Biophysical Research Communications* by courtesy of the publishers, Academic Press, New York.

B. "Heat of Dilution"

An analogous property to endogenous metabolism is the "heat of dilution" of bacteria. Little is yet known about this. The effect was first reported by Boivinet and Grangetto (1963) as a rapid, large heat evolution when a concentrated suspension of *A. aerogenes* was added to a solution containing an energy source. At that time the effect was thought to be due to the initial interaction between bacteria and energy source, but more recent work (Forrest and Berger, unpublished) shows that the effect also occurs with concentrated suspensions of *S. faecalis* and other bacteria added to an excess of the suspending medium without substrates. The effect is possibly the initial response of the organisms to adjusting their rate of endogenous metabolism to the changed conditions.

C. Catabolism by Resting Cells

For many studies it is most convenient to use resting cells, as metabolic processes continue without the complicating factors of growth and fission. It is easy to remove nutrients present in the culture medium by washing the cells, so that a washed suspension of organisms can at most maintain the status quo. Resting cells are physiologically different from growing cells: Their catabolic activity is lower (Section II, A), and certain enzyme systems may be differentially repressed. However, calorimetric experiments under these conditions lend themselves to simple interpretation. Two processes may occur: endogenous metabolism and, concurrently, the catabolism of added exogenous substrate. The saturation constants for most substrates are very low, so that the kinetics of catabolism are accurately zero order.

Figure 8 shows the course of heat production from glycolysis in a washed suspension of *S. faecalis*. The constant contribution due to endogenous metabolism was determined before adding substrate, so that the net heat of glycolysis could be found by subtraction. Heat production and rate of glycolysis both accurately follow zero-order kinetics, and the rate of heat production is linearly related to the concentration of bacteria in the suspension (Fig. 9).

This accurate relationship holds, in the case of *S. faecalis*, only with cells that have been grown with an excess energy source so that they have been able to accumulate reserve materials and possess endogenous metabolism in the resting state. In strong contrast, cells grown on a medium limited by energy source have no detectable endogenous heat production and have a variable ability to catabolize glucose in washed suspension (Forrest and Walker, 1965b).

In washed suspensions the energy available from glycolysis is completely uncoupled from synthetic reactions (Senez, 1962). In *S. faecalis* energy of

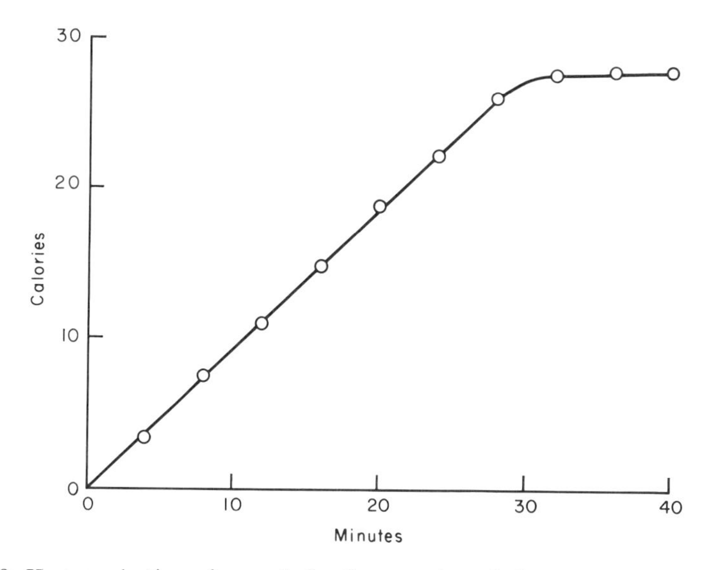

FIG. 8. Heat production of a washed cell suspension of *Streptococcus faecalis*, with glucose as the substrate (from Forrest *et al.*, 1961). Reprinted by courtesy of the *Journal of Bacteriology*.

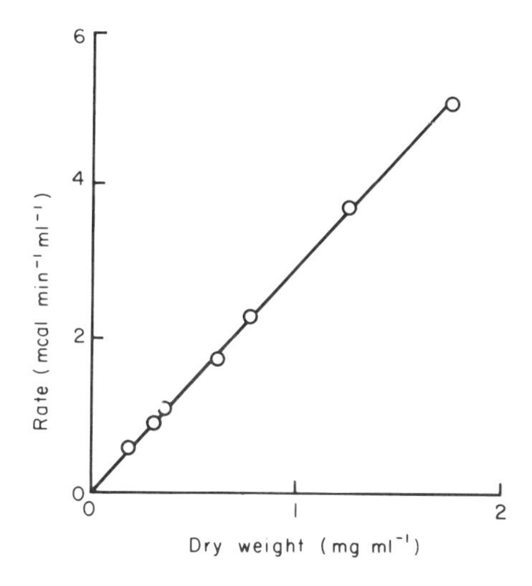

FIG. 9. Rate of fermentation of glucose by washed cell suspension of *Streptococcus faecalis*, as a function of concentration of bacteria (from Forrest *et al.*, 1961). Reprinted by courtesy of the *Journal of Bacteriology*.

maintenance in such circumstances seems to be supplied by the endogenous metabolism proceeding concurrently (Forrest and Walker, 1965b), so that this situation is the extreme case where no coupled reactions need to be considered in arriving at an energetic balance for the catabolism of exogenous substrates. The calorimetric records obtained in such experiments are simple and easy to interpret: This process of resting cells catabolizing a substrate at a constant rate corresponds to a steady state of heat production at an intermediate level of catabolic activity (Forrest and Walker, 1964).

It would seem possible, then, to obtain very good correspondence in such a simple system between the observed and calculated values of the enthalpy change. However, even in this case, some caution is necessary. In glycolysis by *S. faecalis* the observed products are 95% lactic acid and the remainder are fatty acids. For normal biochemical purposes this is a "homolactic" fermentation, but the correlation between observed and calculated values is seriously in error if the assumption is made that the fermentation is 100% homolactic without the actual assay of products. However, if these other products are taken into account, very good correspondence is in fact obtained (Forrest *et al.*, 1961).

IV. Mixed Cultures—Natural Systems

In many natural systems such as occur in the soil and sewage, biochemical degradations are carried out on a complex mixture of natural substrates by an ill-defined mixed population of microorganisms. The end products of degradation are heterogenous, the reactions usually go on slowly for long periods, and there is considerable accumulation of the end products, so that chemical assays in such systems are of limited value.

The calorimetric method has been applied to the study of such ill-defined systems. Thus, van Suchtelen (1931) carried out a series of calorimetric investigations on the microbial energetics of the soil and Walker and Forrest (1964) studied the fermentation occurring in the rumen of sheep.

The ruminal fermentation is a good example of these systems. Here the natural substrates, mainly carbohydrates, are degraded to give the lower fatty acids (volatile fatty acids, VFA). The peculiar physical properties of the rumen contents make it necessary to design specialized apparatus for the study (Forrest *et al.*, 1964), but the methods applicable to the investigation of pure cultures can be extended to this complex system.

A great amount of study has been devoted to nutritional and energetic requirements of higher animals (Blaxter, 1962) by "indirect calorimetry"—analysis of respiratory gases. The ruminant animal is in a special category in these studies since the methods of indirect calorimetry do not give very accurate estimates of the contribution of the ruminal fermentation to the overall metabolism of the animal. Thus the study of the ruminal fermentation is important, both since it gives energetic information that can be correlated with indirect calorimetric data for the whole animal and for the information obtained about a natural mixed culture.

The calorimetric data turn out to be remarkably simple. Despite the great number of concurrent reactions that occur, the kinetics of heat production observed from samples of rumen contents are accurately of first order for long periods (Fig. 10), and the rate of heat production of samples taken from

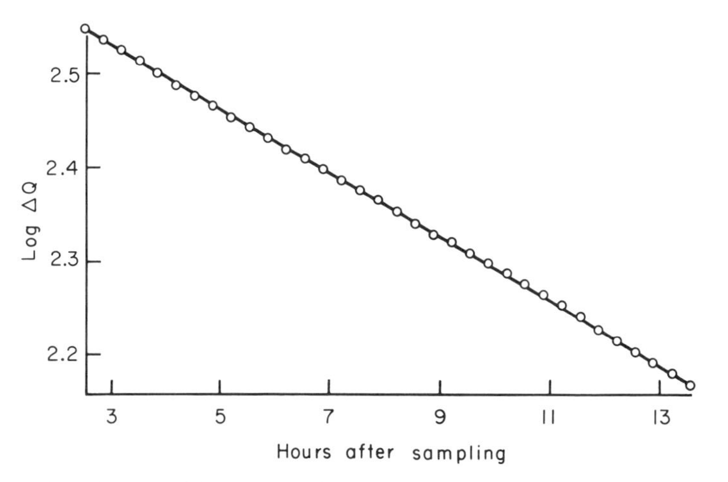

FIG. 10. Kinetic plot of heat production by rumen contents. Data were fitted to an exponential function by the method of Guggenheim (1926). Δt, 11 hours 20 minutes; k_1, 0.076 hour^{-1} (from Forrest, 1967). Reprinted by courtesy of *les Colloques Internationaux du Centre National de la Recherche Scientifique.*

different animals is closely the same. A variation in the rate of heat production does occur between different samples, but this is correlated only with the amount of solid material in the sample. The system in fact is strikingly regular in its behavior, being strongly buffered and poised to maintain anaerobic conditions. Because of this predictable kinetic behavior of the system, the effect of perturbations can be readily detected. Thus additions of sodium ion—or lactate, which is not a normal intermediate in the breakdown of carbohydrates in this system—inhibit the fermentation strongly, whereas the VFA, which are the normal end products of the fermentation, have no effect on the heat production.

It is also possible, within the limitations imposed by the ill-defined nature of the substrates, to correlate the observed heat production with thermodynamic calculations for ΔH in the same way as has been done with pure cultures

(Forrest, 1967), so that it seems quite permissible to make the assumption that in this system heat production is an accurate index of catabolic activity, no matter how complex the processes taking place.

The exponential decay in the rate of heat production is the type of kinetic behavior often found in enzyme-catalyzed degradations, where the concentration of substrate is insufficient to saturate the enzyme. Apparently such behavior is not often found in microbial degradations in pure culture experiments. Certainly the normal pathway for cellulose degradation in this system

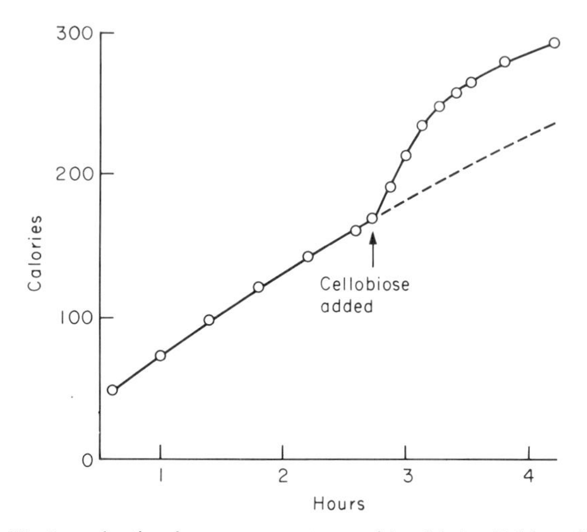

FIG. 11. Heat production by rumen contents with added cellobiose (2 mmoles).

does not show saturation behavior, but if cellobiose, which is a normal intermediate for cellulose breakdown, is added to a sample of rumen contents in active fermentation, there is an immediate increase in the rate of heat production. In this case it is easily possible to saturate that part of the pathway responsible for cellobiose degradation. Zero-order kinetics of the additional heat production due to cellobiose degradation is then observed. This additional heat production is superimposed on the normal exponential rate of heat production, which occurs concurrently (Fig. 11).

The observed heat produced is, however, surprisingly low: for 2 mmoles of cellobiose, it is only −56 cal. This corresponds closely to the enthalpy change (−54 cal; Forrest, 1967) when 2 mmoles of VFA are produced from 1 mmole of glucose residue. Assay of the reaction mixture confirms that in fact only 2 mmoles of VFA are produced. It appears that the process occurring is that, for 2 mmoles of cellobiose, only one half of a mmole is degraded to give 2mmoles of VFA and 3mmoles of ATP (Walker, 1965), and the ATP is used to polymerize the remaining $1\frac{1}{2}$ mmoles of cellobiose to polysaccharide. It is well known (Hungate, 1963) that many rumen organisms store reserve polysaccharides. The correspondence between the observed and calculated heat production for the process is good enough to be reasonably certain that no other reaction could be overlooked, and the analogy with the behavior of organisms in pure culture (Section II, C) in using ATP not required for growth to store reserve materials lends further support to this explanation. Thus, even in such a complex system, calorimetry can give a clear indication of the biochemical processes going on in the system.

V. Conclusions

Heat production in bacteria may occur in steady states at widely different levels of catabolic activity. The definition of the system is simplest at the lower levels of activity, when the concentration of bacteria is constant and the kinetics of the processes occurring are simply described by time-independent parameters. The steady state of exponential growth at the highest level of activity is more complex in that a continuous measure of the increasing bacterial concentration also is necessary in addition to the record of heat production to describe the system fully, though here too the kinetic expressions are simple and the catabolic activity of unit mass of cells remains constant. Thus in any of these steady states the behavior is predictable and regular, so that changes in the behavior of the system can be easily detected and comparative metabolic studies are readily carried out.

In transitions between the steady states, as in the transition from exponential growth to the stationary phase, the catabolic activity of the organisms may change radically, so that accurate description of the system is difficult and more information is required to interpret the results.

There seems little general possibility of using calorimetric experiments of the type presently possible as a tool in the study of anabolic processes within the organism; in general, the catabolic processes occurring are much more obvious. For example, in growth (Section II, A) what is essentially measured is the enthalpy change associated with the process of catabolism, and it is not possible to determine the enthalpy of growth.

The technique is particularly valuable, however, in determining whether or not a postulated reaction occurs, to detect unknown reactions (Section III, B), or to interpret reaction mechanisms (Section IV). Particularly in complex natural systems where the details of the processes going on are too complex to be followed by conventional methods, calorimetry can give an overall view of the process that is obtainable in no other way. Not least of its advantages in such systems is the comparative rapidity with which the determinations can be made. Even in these complex systems quite detailed analyses of mechanisms appear possible using calorimetry as a guide to the biochemical reactions most likely to occur.

REFERENCES

Battley, E. H. (1960). *Physiol. Plantarum* **13**, 192, 628, 674.
Bauchop, T., and Elsden, S. R. (1960). *J. Gen. Microbiol.* **23**, 457.
Bayne-Jones, S., and Rhees, H. S. (1929). *J. Bacteriol.* **17**, 123.
Belaich, J.-P. (1963). *Compt. Rend. Soc. Biol.* **157**, 316.
Belaich, J.-P., and Senez, J. C. (1967). *Colloq. Intern. Centre Natl. Rech. Sci. (Paris)* **156**, 381.
Blaxter, K. L. (1962). "The Energy Metabolism of Ruminants." Hutchinson, London.
Boivinet, P. (1964). Ph.D. Thesis, Univ. of Marseilles.
Boivinet, P., and Grangetto, A. (1963). *Compt. Rend.* **256**, 2052.
Carlyle, R. E., and Norman, A. G. (1941). *J. Bacteriol.* **41**, 699.
Dawes, E. A., and Ribbons, D. W. (1962). *Ann. Rev. Microbiol.* **16**, 241.
Dawes, E. A., and Ribbons, D. W. (1964). *Bacteriol. Rev.* **28**, 126.
Dye, M. H., and Rothbaum, H. P. (1965). *New Zealand J. Sci.* **7**, 97.
Forrest, W. W. (1967). *Colloq. Intern. Centre Natl. Rech. Sci. (Paris)* **156**, 405.
Forrest, W. W., and Walker, D. J. (1963). *Biochem. Biophys. Res. Commun.* **13**, 217.
Forrest, W. W., and Walker, D. J. (1964). *Nature* **201**, 49.
Forrest, W. W., and Walker, D. J. (1965a). *J. Bacteriol.* **89**, 1448.
Forrest, W. W., and Walker, D. J. (1965b). *Nature* **207**, 46.
Forrest, W. W., Walker, D. J., and Hopgood, M. F. (1961). *J. Bacteriol.* **82**, 685.
Forrest, W. W., Stephen, V. A., and Walker, D. J. (1964). *Australian J. Agr. Res.* **15**, 313.
Grangetto, A. (1963). Ph.D. Thesis, Univ. of Marseilles.
Guggenheim, E. A. (1926). *Phil. Mag.* **2**, 538.
Herbert, D., Phipps, P. J., and Tempest, D. W. (1965). *Lab. Pract.* **14**, 1150.
Hungate, R. E. (1963). *J. Bacteriol.* **86**, 848.
Lamanna, C., and Mallette, M. F. (1959). "Basic Bacteriology." Williams and Wilkins, Baltimore.
Morowitz, H. (1960). *Biochim. Biophys. Acta* **40**, 340.
Peterson, W. H., and Wilson, P. W. (1931). *Chem. Rev.* **8**, 427.
Pirt, S. J. (1965). *Proc. Roy. Soc. (London)* **B163**, 224.
Prat, H. (1963). *In* "Recent Progress in Microcalorimetry" (H. A. Skinner, ed.), pp. 111–174. Pergamon Press, London.
Prigogine, I. (1961). "Thermodynamics of Irreversible Processes," p. 27. Wiley (Interscience), New York.
Prigogine, I., and Wiame, J. M. (1946). *Experientia* **2**, 451.
Rahn, O. (1932). "Physiology of Bacteria" McGraw-Hill (Blakiston), New York.
Senez, J. C. (1962). *Bacteriol. Rev.* **26**, 95.
Senez, J. C., and Belaich, J.-P. (1965). *Colloq. Intern. Centre Natl. Rech. Sci. (Paris)* **124**, 357.
Sigal, N., Cattaneo, J., and Segal, I. H. (1964). *Arch. Biochem. Biophys.* **108**, 440.
Stoward, P. J. (1962a). *Nature* **194**, 977.
Stoward, P. J. (1962b). *Nature* **196**, 991.
van Suchtelen, F. H. H. (1931). *Arch. Pflanzenbau (Abt. A. Wiss. Arch. Landwirtsch.)* **7**, 519.
Walker, D. J. (1965). *In* "Physiology of Digestion in the Ruminant" (R. W. Dougherty, ed.), p. 269. Butterworth, London and Washington, D.C.
Walker, D. J., and Forrest, W. W. (1964). *Australian J. Agr. Res.* **15**, 299.
Wilkie, D. R. (1960). *Progr. Biophys. Biophys. Chem.* **10**, 259.

CHAPTER IX

Calorimetry of Higher Organisms

HENRI PRAT

I. Introduction

The features of thermogenesis in multicellular organisms are widely different from those of microorganisms (described by Forrest in Chapter VIII). Heat production per unit of weight of living matter is much smaller. In addition two cases must be distinguished: (1) In plants the thermic flux is related mainly to the rate of cell multiplication as it occurs in bacterial cultures. (2) In animals this source of heat becomes negligible (except in embryonic stages) and the heat production is due chiefly to muscular contractions, secretory activity and cell metabolism, all of which are regulated by nervous and hormonal influences. Great differences result between thermograms of plants and thermograms of animals as we shall have many occasions to point out.

These differences being set aside, as in microorganisms we shall find some common properties to be noted in the analysis of heat production in higher organisms: (1) The shape of the thermograms is specific, i.e., characteristics of a given species of plant or animal under given conditions (as for a bacterial strain). (2) Curve shape can be modified by external agents, i.e., chemical, physical, or biotic, as we have noticed also where microorganisms were studied. (3) The shape undergoes modifications related to internal conditions, namely to aging (senescence).

In order to analyze the heat production it is necessary to use a continuous-recording microcalorimeter, working in isothermal conditions, the best of all for biological purposes being the Tian-Calvet type (see Evans, Chapter XIV; Calvet and Prat, 1956, Calvet *et al.*, 1963). This kind of apparatuses inscribes directly the thermogram, i.e., the recording, minute by minute, of the thermic flux, expressed in calories per hour. In this way, nothing can occur in the physiological functions of the organism under investigation that is not immediately recorded as a change in the outline of the thermogram. The smallest incidents happening in the metabolism are readily registered. Thus, variations in the heat production constitute the most sensitive response of the subject to any external action and the most faithful and delicate indicator of its conditions of activity.

First, we shall consider higher plants, which are the more closely allied to microorganisms. It is of special interest to analyze the thermogenesis of germinating seeds and the gradients of thermogenesis in growing stems and roots and in flowering inflorescences.

II. Thermograms of Germinating Seeds

A. Phases of Physicochemical and Biological Thermogenesis

Let us put into a Tian-Calvet microcalorimeter a small number of dry wheat grains (e.g., 1 gm: about 32 grains), included in the gadget represented on the left of Fig. 1. When we press the bulb *k* an equal weight of water (i.e., 1 gm) is dropped onto the seeds, and the thermogram (right of Fig. 1) undergoes the following transformations:

1. A rapid increase OA of the thermogenesis (up to 1 cal hour^{-1} at 24°C) when the water reaches the grains (in 0).

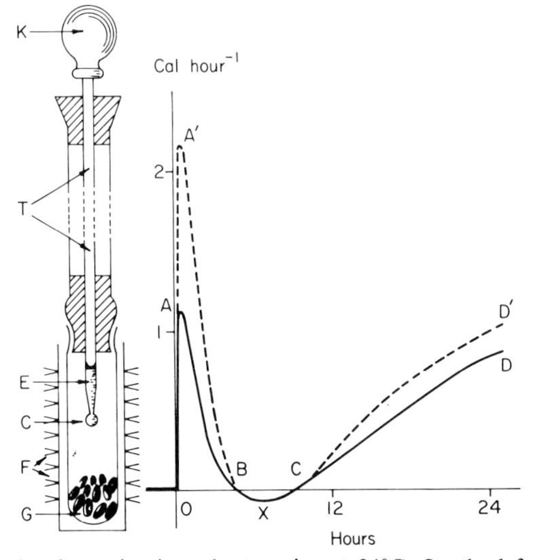

FIG. 1. Thermogenesis of germinating wheat grains at 24°C. On the left: Longitudinal section of the device used for the experiment: T is a glass tube containing 1 cm^3 of distillated water E; K, rubber bulb; C is a calorimetric cell containing 32 dry grains of wheat (Black Hawk Winter), total weight 1 gm. F are thermocouples of the Tian–Calvet microcalorimeter (the apparatus is not represented; for it, see Chapters I and XIV. On the right: The thermogram. At the instant O, the bulb K is pressed, throwing the water onto the seeds. OAB is physicochemical thermogenesis; BXC is dead time; and CD is biological thermogenesis. The dotted curve OA′ BCD′ has been obtained with grains which were dehydrated prior to the experiment (Prat, 1952).

2. A sharp fall AB of heat production after a maximum A is reached in less than half an hour.
3. A dead time BC lasting some hours, when thermogenesis becomes low, null, or even heat absorbing.
4. A slow rise CD of the thermic flux, persisting as long as the process of germination continues.

We have named OAB the physicochemical thermogenesis. It is independent of life, as proved by the fact that it occurs as well with dead seeds. On the contrary CD is the biological thermogenesis, occurring only with living grains. This indicates that the succession of physiological changes constituting the process of germination (cell multiplication in the embryo, digestion of stored reserves by enzymes, etc.) is in progress.

B. INFLUENCE OF PHYSICAL, CHEMICAL, AND BIOTIC FACTORS

The thermogram is reproducible exactly, with seeds of the same lot, every time as long as the conditions of the experiment remain the same. But the slightest change in those conditions provokes modifications in outline of the thermogram; the greater the change, the deeper is the modification. For instance, when the temperature is higher, the duration of the physico-chemical thermogenesis OAB is shortened and its maximum raised, the total calories produced remaining constant (Fig. 2). When the grains are submitted to a prior dehydration, the maximum of the physico-chemical thermogenesis is elevated, without modification of its duration, with the consequence that the total

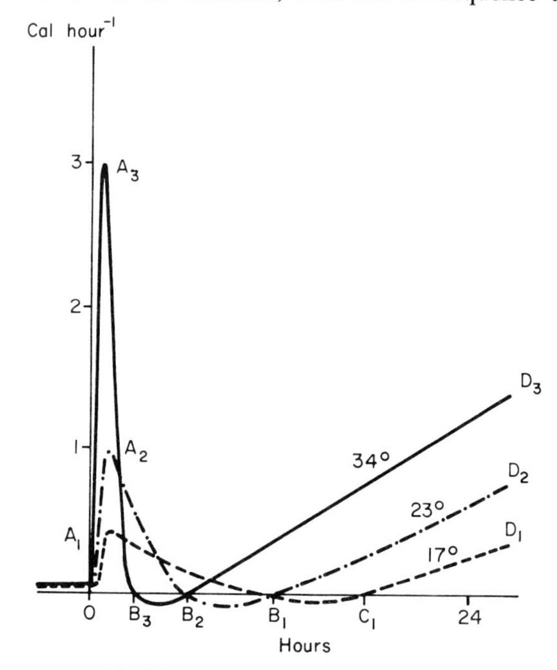

FIG. 2. Thermograms of the Black Hawk Winter wheat at three temperatures: 17°, 23° and 34°C (Prat, 1952).

calories produced is increased (OA′B, Fig. 1). Biological thermogenesis is increased also by ultrasonic treatment. Concerning the influence of chemical factors, some substances reduce the rate of biological thermogenesis (e.g., alcohol, chlorpromazin); others increase it (indolacetic acid, gibberillin, calcium sulfate, potassium nitrate, etc.). Aging of seeds also modifies their thermogram, first by increasing the heat production, then by lowering it, this phenomenon being related to the facts of maturation and dormancy well known to agriculturists.

C. SPECIFIC CHARACTERISTICS OF THERMOGRAMS

We have investigated the thermogenesis of germination of numerous species of cultivated and wild plants. In all cases we have observed the same initial phases: physicochemical thermogenesis, dead time, and biological thermogenesis. But the duration of the phases and the thermic flux produced at each stage vary widely from one species (and even variety) to another, even though conditions of the experiment be exactly the same in all cases. As a general rule, small seeds display, per gram, for the physicochemical thermogenesis, a higher maximum and a shorter duration than do the big ones. This fact is obviously related to the comparatively greater surface offered by the small seeds to contact the soaking water. But it would be an error to interpret the comparison in terms of a mere geometrical relationship. Carrot seeds offer a higher maximum A than celery seeds; however they are bigger (880 seeds gm^{-1} instead of 2300). A similar comparison can be made for radish versus cress (104 seeds gm^{-1} against 552). The shape and numerical features of the thermogram constitute definite characters of comparative physiology which may be used as well as those of comparative anatomy, morphology or histology to identify the species and recognize their relationships. Some species present a strong and short physicochemical thermogenesis, e.g., celery, carrots, tomatoes, leeks, cress, cabbage, and radishes. Other species have a long thermogenesis with a low maximum. Examples of this can be drawn from the cereals. Some display a long dead time, such as leeks and celery, i.e., the biological thermogenesis begins late. Such observations could be of a great interest in agricultural research. Within a given species, the varieties display different characteristics. So the microcalorimetric study can give valuable information for the work of selection of cultivated plants.

III. Gradients of Thermogenesis

A. In Growing Stems and Roots

Plants offer an excellent material to observe gradients of thermogenesis. In a growing stem or root and in a flowering inflorescence we can readily observe with the microcalorimeter that pieces of equal weight taken at different levels give widely different heat outputs. This fact relates first of all to unequal activity of cell-division. Active meristems of the apex can produce ten times the calories per hour that a section of the same organ does wherein mitoses are very rare.

B. In Flowering Plants

But cell division is not the only factor to consider. In flowers at the moment of anthesis the stamens and petals give high relative output of calories even after the cell-divisions have terminated. This is related to an increase of their metabolic rate and is comparable to an autointoxication. Those gradients of thermogenesis associated with flowering can be even more accentuated than those associated with cell-division. An inflorescence can produce a relative heat-output thirty times higher than that of its peduncle.

Figure 3 shows a comparison between the gradients of thermogenesis in a growing asparagus stem A and two inflorescences: a grass B (*Dactylis*) and a compositae C (*Tragopogon*), all at 31°C. The heat fluxes are given in calories per

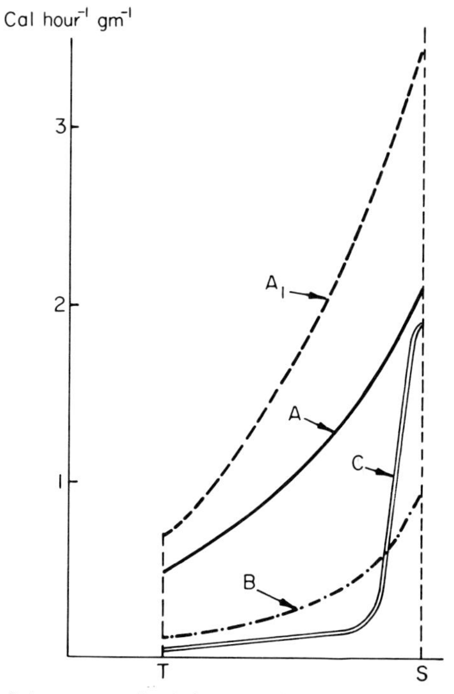

FIG. 3. Gradients of thermogenesis: A is a growing stem of asparagus at 31°C, B is an inflorescence of a grass (*Dactylis*) at 31°C, C is a capitule of a Compositae (*Tragopogon*) at 31°C; A_1 (dotted) is asparagus stem at 37°C; T is basis of the organ; and S is summit (Prat, 1948, 1952).

hour and per gram (cal $hour^{-1}$ gm–^{1}). The letter T indicates the base and S the summit of the organ. The curve A_1 is given also by an asparagus stem, but taken at 37°C. Comparison of curve A_1 with curve A illustrates the fact that heat production is stimulated and the gradients of thermogenesis are increased by raising the temperature.

C. Relations with Cell Division and Sexuality: Integrating Factors

On the whole, we have seen that thermogenesis in plants offers some common features with those of bacteria, namely a general relationship to cell division. However, we notice that, unlike unicellular organisms, there is an influence of integrating factors resulting from hormonal actions (growth hormones, i.e., auxins, and flowering hormones). The most obvious result of this is the establishment of gradients of organogenesis and, hence, gradients of thermogenesis.

Within embryonic stages of the animal kingdom, and especially higher animals, there is a fading of thermogenesis due to cell division and, conversely, an increase and rapidly overwhelming intervention of integration mechanisms (not solely hormonal, as in the plants, but also nervous). In this way the joint neuroendocrine system will regulate the general metabolism as well as the local activity of muscular and glandular tissues, thus controlling the heat output of all parts of the organism.

IV. Thermograms of Insects and Other Invertebrates

A. Economy Periods and Spontaneous Paroxysms

On account of their small size and frequent aptitude for living under confined conditions, insects constitute an excellent material for microcalorimetric investigations. The first example, the cockroach (*Periplaneta*) has an interesting type of metabolism, a fact which is readily apparent at first sight of their thermogram (Fig. 4). This presents an alternation of periods in which the heat output is low and almost uniform or slightly undulated, separated by periods where the thermic flux rises sharply reaching a maximum several times higher than in the previous stage (from two to six times greater in the figured experiment).

The uniform phases correspond to the economy level of activity, characterized by a low rate of basal metabolism, absence of movements, and almost complete suppression of respiratory exchanges. This is related to the fact that the cockroach belongs to the group of "apneic" insects. These possess the ability to temporarily shut the spiracles of their tracheal system and to live in

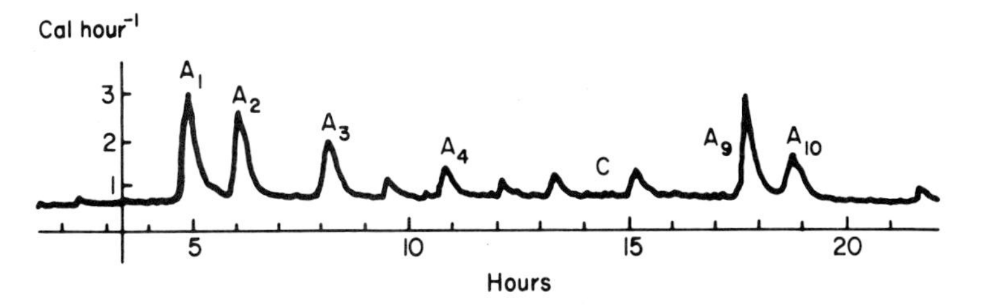

FIG. 4. Thermogram of a cockroach (*Periplaneta americana*) adult male weighing 0.731 gm at 24.9°C. A_1 to A_{10} are the spontaneous paroxysms; and C is the economy level. Notice the diurnal rhythm during the 24 hour experiment (Prat, 1954a).

this way for several hours without breathing, even without connection to the surrounding atmosphere. An important consequence for pest-control results from this observation. The well known resistance of cockroaches to most insecticides during the apneic phase is based upon the fact that toxic substances from outside are not absorbed through the closed tracheal system. Highly valuable in the struggle for life, this ability to live in a state of anoxia is possible because the insect can contract a "debt of oxygen" during the apneic condition. A translocation of oxygen occurs, covering the respiratory needs with accumulation of reduced waste substances, such as lactic acid. During periods of intense thermogenesis, the tracheal system is wide open and admits a strong current of air. Then toxic waste substances previously accumulated are resorbed by oxidation, and the metabolism reaches a high level. This corresponds often, but not invariably, with a period of muscular activity of the insect.

These maxima of thermogenesis are named spontaneous paroxysms, for they occur under perfectly uniform conditions, in the absence of any external stimulation, as an endogeneous rhythm of the animal. They appear at intervals of 1–3 hours, but with different values for maxima, whose distribution reveals a 24-hour (diurnal) rhythm (see Fig. 4). They are under the command of the central nervous system, represented by a double chain of ganglions situated in the head, the thorax, and the abdomen. We have established this fact by performing experiments on sectioned parts of insects, which continue to present separately the spontaneous paroxysms of thermogenesis.

B. Paroxysms Provoked by a Physical or Chemical Agent

Other paroxysms of heat-production can be induced by external stimulation: mechanical (such as a mere contact), optical (a beam of light), or chemical. In this latter case we have noticed several categories of effective agents, for instance, on one hand, the odor emitted by the opposite sex, on another hand, vapors of alcohol.

Figure 5 shows the reactions of a cockroach to repeated olfactory stimulation of whiffs of alcohol vapor. The first dose Z_1 produces a notable response B_1, with a heat production about seven times higher than the economy level. After a rest of 3 hours, a response B_2 to a second stimulation Z_2 is far higher than the first. In Fig. 5 the peak goes off scale, representing at least 20 times the economy level. After that, we observe that the new state of equilibrium reached remains two times higher than the normal economy level, and continues to do so during more than 17 hours. A third whiff of alcohol vapor, in Z_3, initiates an even greater reaction, B_3, with a heat output reaching 30 times the primitive economy level. This fact demonstrates a state of progressive sensitization of the animal. After the third experiment, a stable level is not restored before 15 to 30 hours. In some cases, the perturbation lasts indefinitely, and remains so strong that it finally provokes the death of the insect through exhaustion. A hypothesis could be constructed explaining this observation in terms of alcohol poisoning. The explanation is not acceptable, however, for only males, at a given stage, react in this way. Also, when the olfactory organs, the antennae, are cut, the insect becomes indifferent to alcohol vapors. This observation is sufficient to prove that the induced paroxysms of thermogenesis and the subsequent exhaustion are due to nervous excitation, and not to intoxication. Another fact

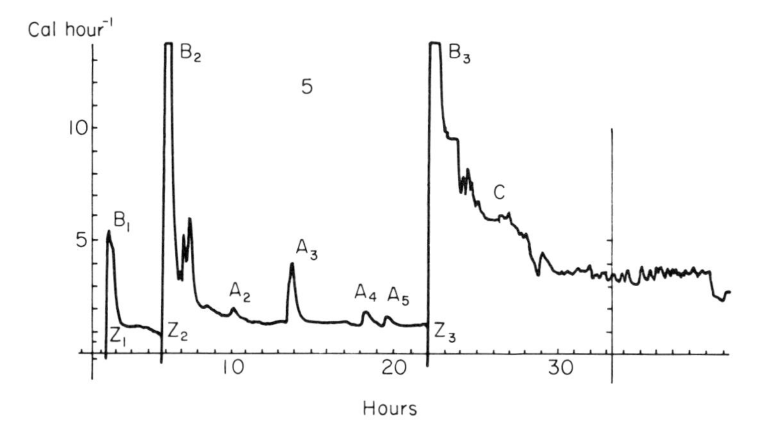

FIG. 5. Paroxysms provoked in an adult male cockroach at 31.5°C by olfactory stimulation. Z_1, Z_2, and Z_3 are points at which the insect received whiffs of alcohol vapor. Note that the induced paroxysms increase in intensity and duration with progressive sensitization of the olfactory system (Prat, 1956).

will corroborate this point of view. It is possible to obtain a thermogram analogous to that of Fig. 5 by using only the sexual odors. We put in the calorimeter an adult male at the "receptive" stage, and submit it to the influence of gushes of air having passed over an adult, sexually active, female. If the male has been, up to this moment, raised separately without communication with the other sex, the first whiff of air containing the sexual odor provokes a paroxysm of the type B_1. The subsequent emissions induce increasing paroxysms of the type B_2 and B_3. There again we observe a progressive sensitization, but it is worth notice that this experiment does not reach the state of death by exhaustion as in the previous case.

These experiments show the value of the microcalorimeter in the investigation of animal behavior, and even in comparative animal psychophysiology. We shall further develop this point below by describing cases of group effect (Section III, E).

C. Influence of Metamorphosis

Microcalorimeters are also valuable tools for analysis of the processes of aging and, especially, a very interesting aspect of entomology, metamorphosis.

Figure 6 shows thermograms of a moth (*Galleria mellonella*) at three stages of its life. In part I, the caterpillar, thermogenesis is of the irregular type, usual

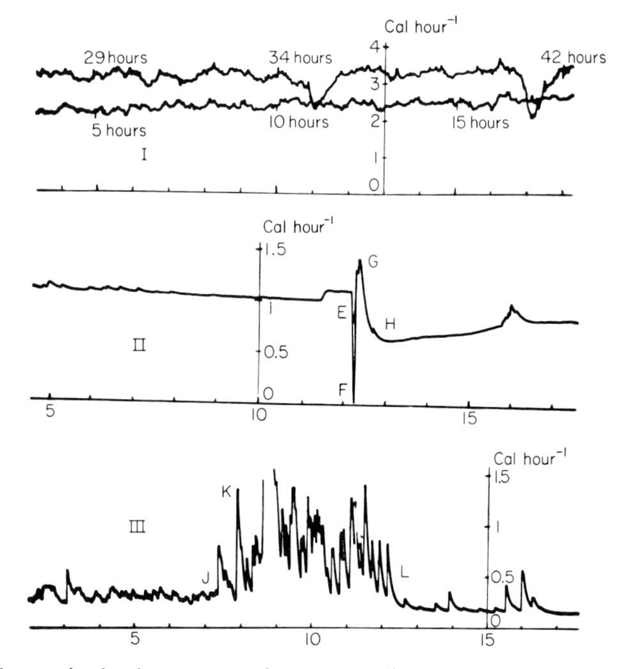

FIG. 6. Changes in the thermogram of a moth, *Galleria mellonella*, during its metamorphoses. Part I is the caterpillar shortly before entering the pupa stage at 24.9°C. In Part II, CD is the chrysalis weighing 0.19 gm at 31.4°C; EFGH is the emergence of the butterfly; HH′ is the rest period (economy); H′ is the first spontaneous paroxysm of the imago stage with faint intensity. III, Male moth weighing 0.06 gm, at 24°C. JKL, a series of strong spontaneous paroxysms representing the period of maximum activity (Prat, 1954a).

among soft skinned invertebrates. In part II, there is first CD, the pupa (chrysalis), with a uniform heat production; then, there is EFG, the emergence of the butterfly, a severe crisis with intense variations of the heat production. This latter stage consists of (1) a sharp fall EF connected with the breaking of the chrysalis skin and the heat absorption due to the expansion of the warm and wet atmosphere of its inside, and (2) a rapid rise FG due to the violent efforts made by the moth to emerge from its prison. Then, in H, a stage of rest begins, lasting about 3 hours, with a low thermogenesis. The butterfly, exhausted by its work to break its chrysalis cocoon, releases and recuperates in a complete immobility, with a low rate of metabolism. In H′ we notice the first spontaneous paroxysm of the imago stage, with a feeble and almost inconspicuous maximum.

In III we have the thermogram of the adult insect: a male moth, which has unfolded and dried its wings and entered first a stable phase of small rapid maxima of thermogenesis (preparation of the flight). Then, in JKL, the insect enters a period of important and almost joint spontaneous paroxysms of thermogenesis, during 5 hours. This period corresponds to an intense agitation. Were the animal free, it should fly and attempt to find a mate. It is the phase of maximum activity and receptivity to external stimulations. Then a period of rest follows, marked only by faint spontaneous paroxysms.

These thermograms illustrate use of the microcalorimeter for the investigation of the process of aging, even when such "catastrophic" events as full metamorphoses are involved. They show also the intervention of integrating factors, i.e., the endocrinal glands (namely *corporae allatae*) regulating metabolism (and thus heat production) at each phase of the life.

D. Specific Characteristics of Thermograms

When we compare the thermograms of several species of insects in order to discover their specific characters, we must always consider homologous stages of the animal life span (larvae, pupae, and imagoes) at comparable states of development. We investigated numerous species of Orthoptera, Coleoptera, Diptera, Lepidoptera, Dictyoptera, etc., and found very different features in their thermograms. These features can be utilized as specific characters of comparative physiology in the same manner that characters of comparative anatomy, morphology, or ecology are used.

We also have investigated other groups of invertebrates: myriapods, crustacea, molluscs, worms, etc. These are noteworthy for the study, in the animal kingdom, of phenomenon that we have described in plants (Section III): gradients of thermogenesis. As a general rule, the head portion of a worm generates per unit weight, a greater heat than the posterior parts. This must be observed in animals deprived of food for several days. Otherwise, the anteroposterior gradient is concealed by fermentations occurring within the intestinal tract.

E. Group Effect

Another very interesting phenomenon can be investigated with microcalorimeters. When we put several insects of the same species into the calorimeter, we notice that the average thermogenesis of each individual becomes lower than it would be if the animal were to be tested alone. A female

Drosophila, weighing one milligram, produces an average thermic flux of 0.03 cal hour^{-1}; however, if we put together five females, their total heat output is only 0.09, instead of $5 \times 0.03 = 0.15$ cal hour^{-1}. The thermogenesis of each insect has been reduced by group effect.

This experiment corroborates observations made on social insects, i.e., honey bees, showing that their metabolism is deeply modified when they are grouped. The result is a collective thermoregulation, realizing a social homeothermy. It is well known that, inside a beehive, a constant temperature is maintained in summer as well as in winter, thanks to this mechanism.

It has been discovered recently that many insects possess glands emitting a gas which reduces fertility; the activity of these glands is proportional to the number of grouped individuals. In the present state of over population, it might, indeed, be useful for mankind to possess such a mechanism of birth regulation!

The group effect we have observed on thermogenesis is probably only one manifestation, among many others, of the complex modifications occurring in the metabolism and behavior of insects every time they are reunited in large numbers. Thus microcalorimeters can be used as faithful and sensitive tools to analyze these important sociological phenomena.

V. Thermograms of Vertebrates

A. Cold-Blooded Vertebrates

For their type of thermogenesis, poikilothermic vertebrates can be compared with some types of invertebrates. Figure 7, part I, shows the thermogram of a snake, *Storeria occipito-maculata*. Note the alternation of economy periods, such as CE when the heat output is low and uniform during several hours, and higher activity periods, such as ABC and EFG which constitute spontaneous paroxysms analogous to those observed in insects. The case is the same for amphibians, such as *Desmognathus fuscus*, whose thermogram is represented in Fig. 7, part II; AB, DE, and GH are periods when the heat production is at the economy level, and BCC′D, EFF′G, and JKLM represent a series of spontaneous paroxysms.

But the difference is not as sharp as it is commonly thought to be between the so-called cold-blooded and warm-blooded vertebrates. The latter frequently possess only an imperfect homeothermy, either at given phases of their life or in response to external agents, physical or chemical. Then they can display temporarily the same behavior as poikilotherms. Modern man himself, needing clothes and heated houses, must be considered as an imperfect "exo" homeotherm.

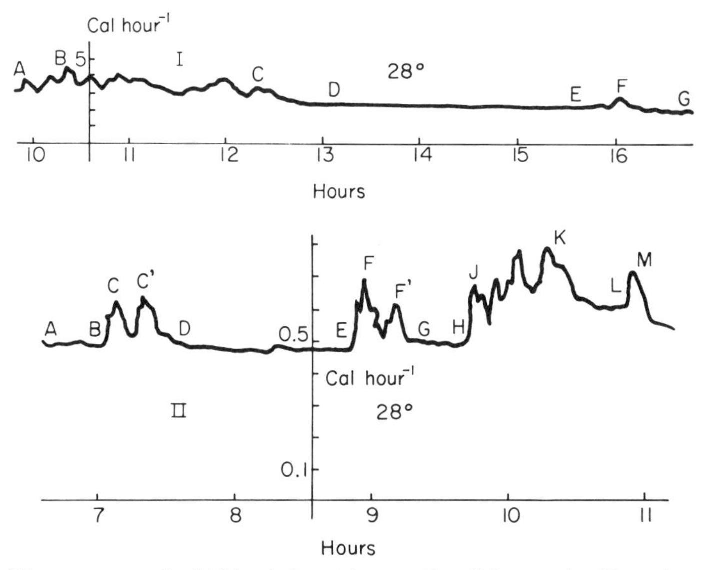

FIG. 7. Thermograms of cold blooded vertebrates. Part I is a snake (*Storeria occipito maculata*), male, weighing 2.85 gm, at 28°C; DE is the economy level of heat output; ABC and EFG are periods of activity. Part II is an amphibian (*Desmognathus fuscus*) weighing 1.05 gm at 28°C; AB, DE, and GH are economy periods; BCC′D, EFF′G, and HJKLM are spontaneous paroxysms (Prat, 1957).

B. WARM-BLOODED VERTEBRATES

Because size of homeothermic vertebrates (mammals and birds), in general, is greater than that of invertebrates and cold-blooded vertebrates, it is necessary to build special types of microcalorimeters in order to study their thermogenesis. Instead of the usual calorimetric cells of 18 mm diameter, surrounded

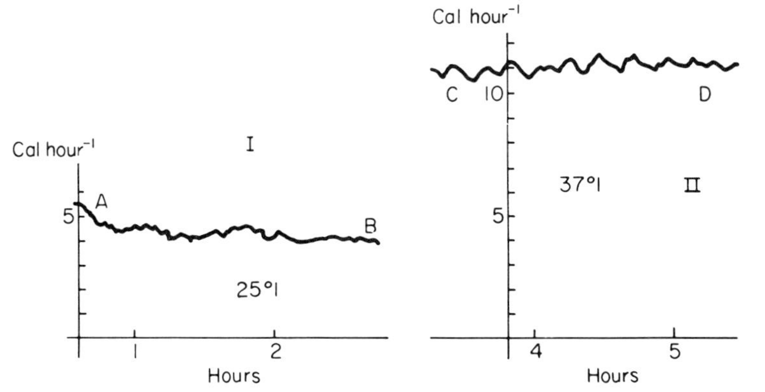

FIG. 8. Thermograms of a young mouse, one day old, weight: 1.45 gm. Part I is at 25.1°C. and Part II is at 37.1°C (Prat 1957).

by 180 thermocouples, we have built large 35 mm cells equipped with 1880 thermocouples. These have enabled us to deal with adult mice, bats, shrews, small birds, etc. The 18 mm cell can, however, contain new born mice, but do not allow continuation of the observation after the first week of life.

Figure 8 shows two thermograms of a one-day-old mouse taken in part I at 25°C, and in part II at 37°C. The first shows an average output of 4 cal hour^{-1} (for a weight of 1.45 gm), the second, 11 cal hour^{-1}. This reaction to external conditions, greater thermogenesis when the temperature is higher, is typically that of a poikilotherm. The reaction of a true homeotherm would be exactly the reverse. In fact, the newborn mouse is not yet homeotherm, and such is the general case among mammals. Even the newborn human baby finds himself in the same state; his temperature is variable according to the external conditions.

C. Influence of Aging

Not only does thermogenesis of young mammals follow passively the variations of the surrounding temperature, but it is, at the beginning, especially low. Figure 9 shows the superimposition, on the same scale, of portions of

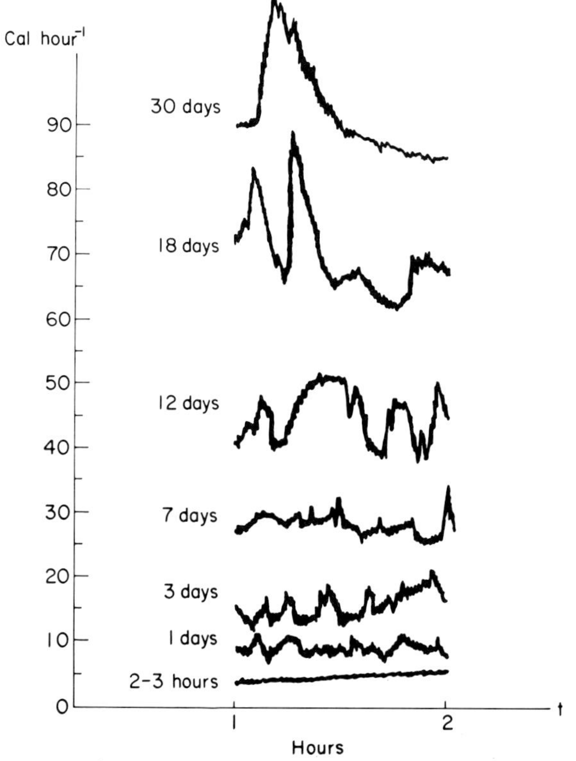

FIG. 9. Thermograms of a mouse taken at various ages from birth to 30 days; the temperature range is 23–27°C (Prat and Roberge, 1960).

thermogram taken on the same animal at successive ages, from 0 to 30 days. The first thermogram, taken 2 hours after the birth, shows a very low average of heat output: 4 cal hour^{-1}. Then, from day to day, the baby mouse increases its weight and its thermogenesis which fluctuates more and more as its spontaneous paroxysms become greater and greater.

Figure 10 shows the variations of the weight P (averaged values for several hundred mice), of the absolute thermogensis A and of the relative thermogenesis

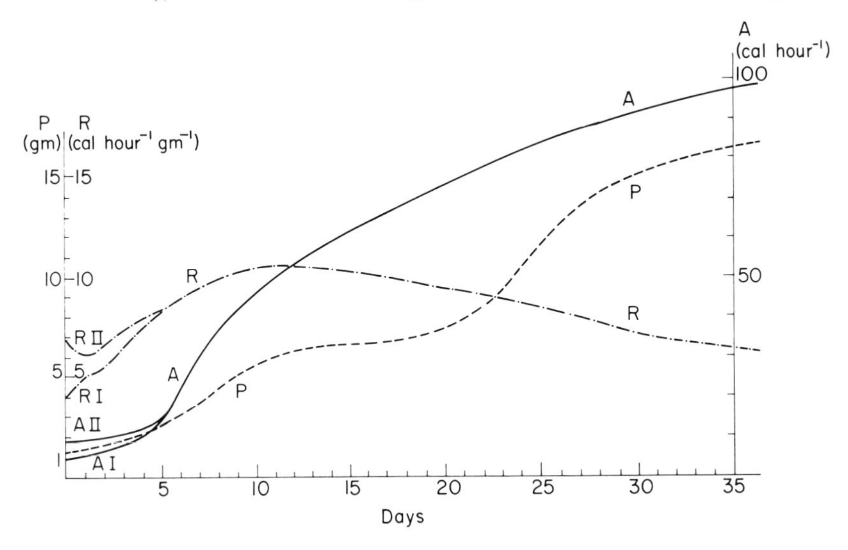

FIG. 10. Variations of the average absolute thermogenesis A, weight P, and relative thermogenesis R as a function of aging of mice from birth to 35 days. The double "roots" of curves R and A are given respectively by two series of experiments at different temperatures: the upper one between 30° and 37°C, the lower between 19° and 26°C. After the fifth day, homeothermy being reached, the two curves merge. These curves represent averages established on 400 mice, of a population as homogeneous as possible (Prat and Roberge, 1960).

R of animals during the course of the first month of life. As the weight and absolute heat output increase steadily, the relative thermogenesis begins to increase, from 5 to 11 cal hour^{-1} gm^{-1}, the maximum being reached around the tenth or eleventh day. Then it decreases regularly.

This fact parallels the observation that, in the human also, basal metabolism begins to increase and, then, decreases after a maximum reached around the tenth year.

D. INFLUENCE OF CHEMICAL FACTORS

What is the cause of these variations, with aging, of the thermogenesis? Obviously the action of the integrating systems. The omnipresent neuroendocrine system regulates thermogenesis as it does other elements of metabolism, growth and development of the young animal. It is a part of the homeostasy of the organism. Overall, this command is chemical, and new pharmacodynamic investigations are able now to give at least a partial answer to questions about the basic mechanisms of control.

Figure 11 shows the influence of a phenothiazine compound: Stematil, on the thermogram of a mouse. In AB we observe the normal thermogram. In Z, Stemetil (40 μg of body weight^{-1}) has been given orally to the mouse and an hour later a deep depression CD of the thermogenesis results. At this time the level is about half of the normal value. In addition, we notice that the heat output becomes perfectly uniform, and this state lasts more than 15 hours.

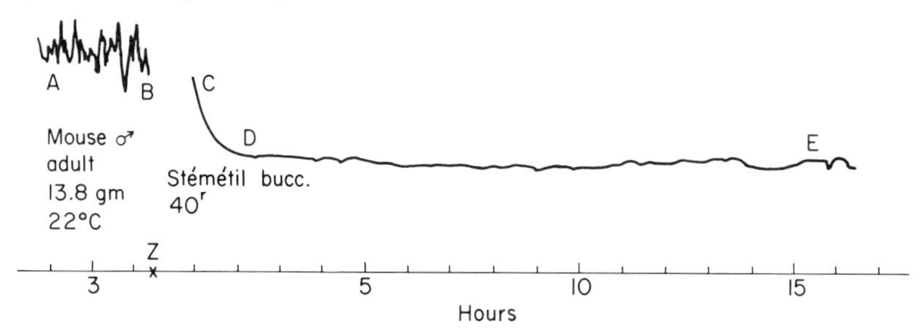

FIG. 11. Influence of Stemetil (derivative of phenothiazin) on the thermogenesis of an adult male mouse weighing 13.8 gm at 22°C. AB is a normal thermogram. In Z a dose of 40 μg of Stemetil gm^{-1} is given orally to the animal; CD is the fall of the thermogenesis; DE is the period of artificial economy level induced by the drug, depression and regularization of the thermogenesis, i.e., artificial hibernation (Prat, 1957).

Thus, by the action of a spasmolytic drug, we have put the thermogenesis of a mammal at the economy level and given to it the general features of the thermogram of a poikilotherm at an inactive period. We have put it artificially in a state of hibernation.

E. Natural and Artificial Hibernation

We shall attempt here only to indicate the wide field that microcalorimetry can open to investigations in hibernation. We have described elsewhere (Prat, 1958, 1959) the influence of sleep inducing and curarizing agents upon thermogenesis. Today it is possible at will, by means of drugs, to put any mammal into a state similar to that naturally reached by hibernating species. This is widely used in medical practice for surgical interventions (chiefly neurological and cardiological) under conditions of artificial hibernation, and also for the psychiatric use of sleep-inducing and relaxing drugs. These interventions have serious implications for the drugs are strongly hypothermizing and reduce metabolism. Microcalorimetry is potentially a key to the analysis of their mode of action.

VI. Conclusion

In this brief discussion we have been able to give only a short account of the innumerable applications of microcalorimetry in the physiological study of higher organisms. This field is large and promising as is also the calorimetric study of microorganisms, which has been examined by Forrest in Chapter VIII.

When we compare multicellular and unicellular beings we can notice wide differences:

1. The relative amount of heat production is far greater in unicellular forms. An important consequence follows: when parasitic bacteria live inside animal or human tissues, they can harm the host cells not only by chemical weapons (toxins) but also by their greater heat production, the bacterial focus acting as a small "thermo-cauter" (Prat, 1962b). Up to now the chemical side of the struggle between the parasite and the host has been the focus of investigation, but it would be interesting also to analyze the physical side of the competition, namely the discrepancies of thermogenesis, eurythermy and stenothermy, between the two antagonists.
2. In multicellular organisms definite integrating systems intervene: only hormonal, trophic and circulatory in plants, determining the establishments of gradients of thermogenesis; nervous and hormonal (neuroendocrine) in animals, controlling metabolism and thus thermogenesis.
3. In plants as in microorganisms heat production is mostly a function of cell multiplication. This effect becomes negligible in animals, except in the first stages of embryonic development, being replaced by muscular and glandular thermogenesis under the control of nervous and hormonal systems. An important thermogenetic influence of the flowering process is also observed in higher plants.

But, in multicellular as in unicellular organisms, there are specific features of the thermograms and modifications of thermogenesis due to aging and to physical, chemical and biotic external factors. The specific features can be used in taxonomy as characters of comparative physiology. The modifications of external origin opens the gate to a great variety of experimentation.

These observations indicate wide application of calorimetry, and, especially, of continuous recording microcalorimeters in the analysis of the behavior of living matter, and the great number of applications of this method, in fields as different as: agriculture (selection of seeds, effects of fertilizers, of herbicides, of growth hormones), pest-control, pharmacodynamy, bacteriology, applied entomology, fermentations industries (breweries, bakeries, wine, cheese, vinegar), dietetics, toxicology, gerontology, etc. We are only at the beginning of the exploration of these fields. The first results enable us to hope for very promising developments.

This is only a short review; we have given more detailed accounts in several publications, enumerated in the following list of references.

REFERENCES

Calvet, E., and Prat, H. (1956). "Microcalorimétrie; applications physico-chimques et biologiques," Masson, Paris.

Calvet, E., Prat, H., and Skinner, H. A. (1963). "Recent Progress in Microcalorimetry. Macmillan (Pergamon), New York.

Prat, H. (1948). *Botan. Rev.* **14**, 603.

Prat, H. (1951). *Botan. Rev.* **17**, 693.

Prat, H. (1952). *Can. J. Botany* **30**, 379.

Prat, H. (1954a). *Can. J. Zool.* **32**, 172.

Prat, H. (1954b). *Rev. Can. Biol.* **13**, 18.

Prat, H. (1956). *Rev. Can. Biol.* **14**, 360.

Prat, H. (1957). *Rev. Can. Biol.* **15**, 336.

Prat, H. (1958). *Therapie* **13**, 833.

Prat, H. (1959). *Therapie* **14**, 594.

Prat, H. (1962a). *In* "Experimental Thermochemistry" (H. A. Skinner, ed.), Vol. II, Chapt. 18, pp. 411–425. Wiley (Interscience), New York.

Prat, H. (1962b). *Compt. Rend. Soc. Biol.* **156**, 1135.

Prat, H., and Roberge, L. (1960). *Rev. Can. Biol.* **19**, 80.

CHAPTER X

Calorimetric Instrumentation: DC Measuring Instruments

JULIUS PRAGLIN

I. Sensitivity Requirements

A. Thermopile Circuits

There will be a wide variation in the sensitivity requirements for individual thermopile microcalorimeters since temperature measurements will be made in some cases with single thermocouple junctions and in other cases with as many as several thousand junctions in series. The voltage output may be as little as $30\ \mu V°C^{-1}$ or as much as 30 to $60\ mV°C^{-1}$. On the other hand, the resolution of changes as small as 10^{-6} degrees may be desired. Thus, even with a system of several thousand junctions, it may be desired to resolve 20 or 30×10^{-9} V through source resistances of as high as 1000 ohms.

B. Thermistor Circuits

While measurement from a thermopile usually consists of connecting a voltage detector across a pair of terminals, the use of a thermistor involves employment of a bridge system such as diagrammed in Fig. 1. The detector is

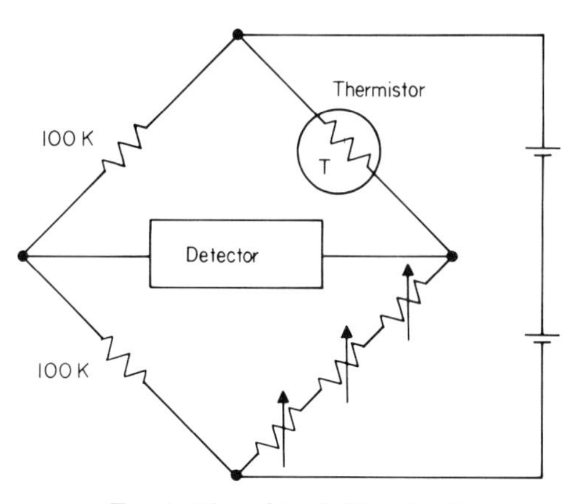

FIG. 1. Thermistor bridge circuit.

either a null balance instrument or, in some cases, the calibrated unbalance from null is used as the output signal. With various thermistors the source resistance of the bridge terminals can vary from about 1K to 100K. Problems will arise from the thermal EMF's caused by the large number of junctions in the bridge and from the high thermal stability necessary in the other bridge arms. Nevertheless, a sensitivity of 10^{-4} °C μV^{-1} is usually attainable with a 100K ohm thermistor, or a theoretical resolution of about 10^{-5} °C if it is assumed that the only system noise would be a thermal agitation noise of about 10^{-7} V peak-to-peak (Hinchey, 1966). An ideal instrument in this case would be able to resolve about 0.1 μV through a 100K ohm source resistance.

II. Limit Imposed on Sensitivity by Thermal Noise

Quite apart from instrument sensitivity, there is a basic limit on resolution imposed by the internal resistance of the sensing circuit. We may represent a thermistor or a thermocouple by a voltage source, S (Fig. 2), which is equal to

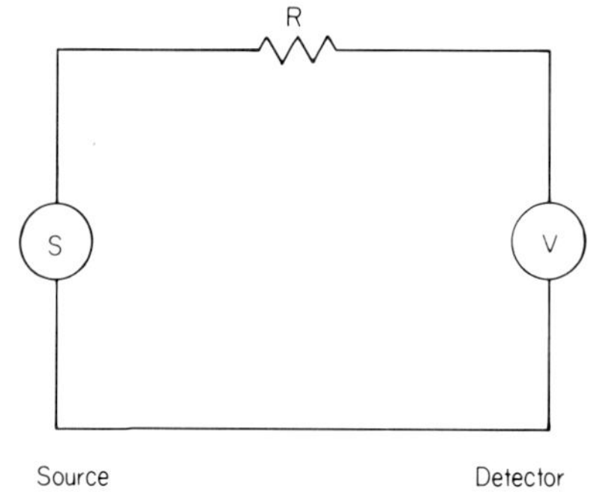

FIG. 2. Circuit for computing thermal noise.

the open-circuit voltage of the sensing device, has zero internal resistance, and is noiseless. In series with S, we place an ideal resistor, R equal to the internal resistance of the device. Equation (1) gives the root mean square noise, E_n, in volts, generated by this resistance or the noise seen by detector, V.

$$E_n = (4KTRF)^{\frac{1}{2}} \tag{1}$$

where K = Boltzmann's constant (1.38×10^{-23} joules/°K), T = Temperature in degrees Kelvin, R = Resistance in ohms, and F = Bandwidth in cycles per second.

If we assume an instrument rise-time of 1 second (about 0.3 cps) at room temperature, Eq. (1) becomes:

$$E_n = 0.76 \times 10^{-10} R^{\frac{1}{2}} \tag{2}$$

The noise, peak-to-peak, E_{pp} which will be observed on a meter or recorder will be about five times the *RMS* value given in (2) or:

$$E_{pp} = 3.8 \times 10^{-10} R^{\frac{1}{2}} \tag{3}$$

Some representative noise values for various source resistances are given in Table I.

TABLE I

NOISE, PEAK-TO-PEAK, 1-SECOND RISE-TIME, FOR VARIOUS RESISTANCES AT ROOM TEMPERATURE[a]

Resistance (ohms)	Noise (V, peak-to-peak)
1	3.9×10^{-10}
10	1.2×10^{-9}
100	3.9×10^{-9}
1000	1.2×10^{-8}
10,000	3.9×10^{-8}
100,000	1.2×10^{-7}

[a] The above values represent the minimum noise possible in a thermopile or thermistor bridge of the tabulated internal resistance using a detector of 1-second rise-time. A detector with a 10-second response would have one third the noise. Conversely ten times faster rise-time will increase the noise by three times as shown in Eq. (1).

It should be realized that Table I is not an argument for minimizing the number of couples or thermistors in a calorimeter since the noise increases only as the square root of the resistance of the added couples or thermistors while the signal increases linearly with each thermoelement. Therefore, while it is desirable to minimize circuit resistance as much as feasible to improve signal to noise ratio with a given number of thermoelements, increasing the number of elements will improve the signal to thermal noise ratio by the square root of the increase of the number of elements. More exactly, the output signal will equal en where e is the emf per couple per °C (or bridge unbalance per thermistor per °C) and n is the number of thermoelements. The thermal agitation noise E_n is proportional to $a\sqrt{nR}$ where a is a constant representing the numerical values in Eq. (2) and R is the resistance of a single element.

Therefore, if we define the signal to noise ratio S as the volts output per °C divided by the RMS noise, then

$$S = \frac{en}{a\sqrt{Rn}} = \frac{en^{\frac{1}{2}}}{aR^{\frac{1}{2}}} = kn^{\frac{1}{2}} \tag{4}$$

since e, a, and R are constant for a particular calorimeter. Thus, increasing the number of sensing elements increases the resolution only by the square root of the number of elements added.

III. The Selection of an Instrument

A. Galvanometers

It should be pointed out that galvanometers are available which will match any available electronic detector in sensitivity. They have the additional advantage of being small enough to be included in the housing of the microcalorimeter, if desired (Calvet and Prat, 1963). However, they are slow, very susceptible to suspension and coil damage on overload, require elaborate precautions to eliminate environmental disturbances, and require cumbersome auxiliary apparatus if recording is desired.

B. Electronic Devices

At this time chopper amplifiers or voltmeters are more usually used to measure the output of thermocouples and thermistor circuits. With source resistances in the range of 1 ohm to about 1 megohm, it is possible to choose a commercially available instrument which will give resolution down to the limits imposed by Eq. (1), the thermal agitation noise. It is, however, not possible at the present state of the art to do this with a single instrument. Therefore, in order that an investigator be able to make a proper choice, a brief discussion of chopper circuitry follows. It is assumed in the following that the bandwidth of measurement is dc to about 0.1 cps (1-second rise-time). Chopper stabilized, data-handling amplifiers are available with faster rise times, but are usually limited to a resolution of a few microvolts.

1. *Basic Chopper Circuitry*

Figure 3 shows, diagrammatically, a typical chopper circuit. $R_1 C_1$ represents an input filter which serves to decrease disturbances due to pick-up of line-frequency voltages. This filter is always desirable to reduce measurement problems but is not practical in measurements where resolution below about 10^{-8} V is desired due to the thermal agitation noise introduced by series resistor R_1. However, where such a filter is not possible, high line-frequency rejection can be obtained by special design of the following amplifier A_1. S_1 is usually a mechanical chopper or switch if voltages below 0.5 mV are to be

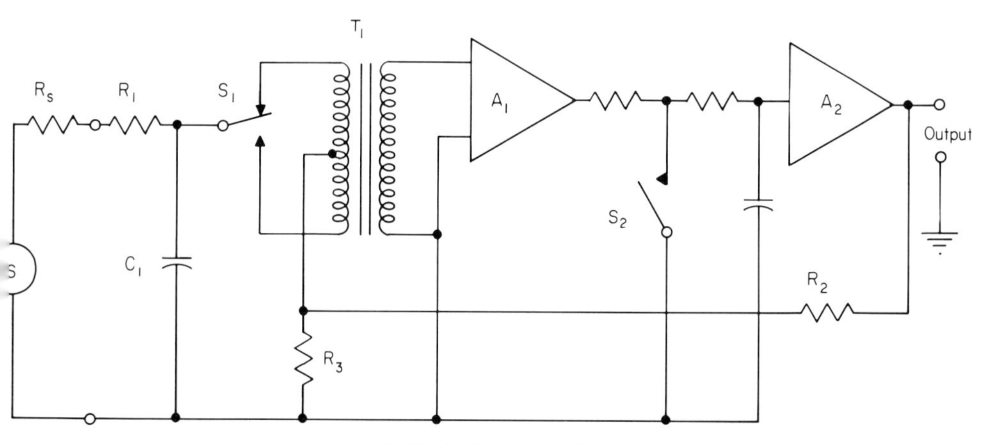

FIG. 3. Typical chopper circuit.

resolved. In less sensitive measurements, photo-choppers (light operated switches) will provide better performance especially with higher source resistances. The chopper is followed by a transformer T, which has the step-up ratio necessary to properly match the input of amplifier A_1 to the source resistance R_s. The chopped dc signal is then amplified as an ac voltage by A_1. It is then converted back to dc by switch S_2, filtered, and further amplified by dc amplifier A_2. The output is used to drive a meter or recorder and is fedback through R_2 and R_3 to stabilize the gain or the metercalibration. A more detailed description is given by Praglin (1966).

2. *Available Instruments*

For the maximum resolution and minimum noise, it is essential that an instrument with the proper transformer ratio in the input circuit be chosen. This ratio may vary from 1:1000 to 1:1 or the transformer may be omitted entirely depending on the circumstances.

The reason that a choice must be made is that while the highest transformer ratio possible will yield an instrument whose short circuit noise will approach the thermal noise in a 0.5 ohm resistor, the input resistance of the amplifier may be too low to give optimum sensitivity. For example, an instrument with a 1:1000 transformer ratio will have a resolution of better than 10^{-9} V. However, at lower sensitivities (perhaps a 1 μV resolution) where the thermal noise often would not be a consideration, the maximum practical source resistance that can be used with such an instrument would not be more than 1000 ohms. By contrast, an instrument which has a resolution of only 0.3 μV can be used with a source resistance of about 10 megohms where thermal noise is not limiting. Thus, it is of the greatest importance to choose a combination of specifications which will best match the particular experimental situation. Table II gives typical performance for several commercially available chopper instruments.

TABLE II

CHARACTERISTICS OF VARIOUS CHOPPER AMPLIFIERS

Resolution zero source resistance[a] (μV)	Approximately equivalent noise resistance (ohms)	Maximum practical source resistance at low sensitivity[b] (ohms)	Stability (μV 24 hours^{-1})	Transformer ratio
0.0003	0.5	1,000	0.01	1:1000
0.003	50	100,000	0.03	1:100
0.03	5,000	10[c]	0.1	1:10
0.3	500,000	100[c]	2.0	1:1

[a] Figures given for a 1 second rise-time. Noise will vary as predicted by Eq. (1) for other response times.

[b] Maximum source resistance cannot exceed the figure in this column if the instrument is to function properly. It will in general be between the figure in this column and the figure in the first column depending on the noise desired as predicted by Eq. (3).

[c] Megohms.

The input resistance of the instruments in the above table will usually be sufficient to cause negligible circuit loading if an instrument is chosen with a realistic sensitivity for the experimental source resistance to be encountered. In general, for the best results the choice should be an instrument which gives just sufficient resolution, but no more than necessary, since it can be seen that a considerable penalty is paid in terms of permissible working source resistance as the short-circuit resolution of the instrument is improved.

The values of zero stability given in Table II will, in general, be sufficient for the particular resolution chosen, but it should be realized that sources of drift will usually be much greater in the experimental setup than in the instrument. Causes of drift are analyzed in more detail by Praglin (1966).

IV. Conclusion

Direct current measuring instruments are available which will give resolution down to the theoretical noise level, however, it is essential that a proper choice be made between resolution and permissible working source resistance if an optimum signal-to-noise ratio is to be obtained.

REFERENCES

Calvet, E., and Pratt, H. (1963). "Recent Progress in Microcalorimetry." Macmillan (Pergamon), New York.

Hinchey, R. J. (1966). Ph.D. Thesis, Purdue Univ., Bloomington, Indiana.

Praglin, J. (1966). "The Measurement of Nanovolts." Keithley Instr., Cleveland, Ohio.

CHAPTER XI

Calorimetric Instrumentation: Indicating Devices and Temperature Measurement and Control

SAM N. PENNINGTON AND HARRY DARROW BROWN

I. Introduction

Our discussion is divided into two topics. Section II includes types of instruments used as indicating devices to allow a visual presentation of reaction progress. The Sections III and IV deal with temperature measuring and control systems suitable for use in connection with microcalorimetric measurement.

II. Indicating Devices

A. Galvanometers

The galvanometer in this application has much to recommend it in terms of sensitivity and signal-to-noise ratio. Because of its simplicity and sensitivity some microcalorimetrists, particularly of the French school, prefer this instrument used in conjunction with a spot-following recorder. Almost all commercially available galvanometers are of the D'Arsonval or moving-coil type, a basic diagram is shown in Fig. 1. The magnetic field of the magnet N-S is concentrated near the coil C by the soft iron bar B; the coil is suspended in the field by a fine conducting wire W which serves as connection point to the coil and as a restoring force. The connection to the opposite end of the coil is made via a spiral S_p which does not exert appreciable force on the coil as it is deflected.

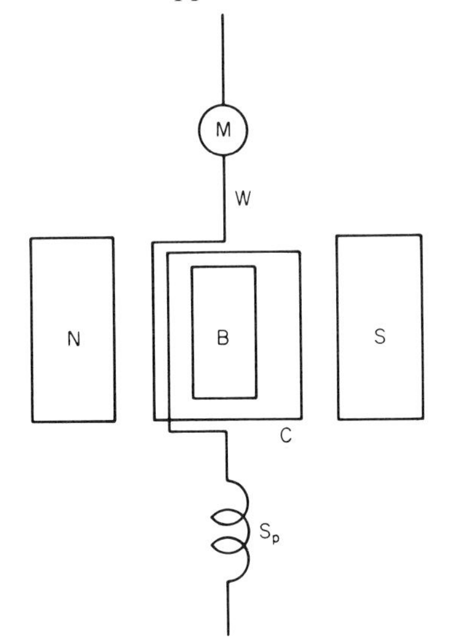

FIG. 1. Diagram of moving coil type galvanometer. M is a mirror, W is a fine conducting wire, C is a wire coil, B is a soft iron bar, N and S are poles of magnet, and S_p is a spiral used for a low torque connection to the coil.

If a current is passed through the coil, the field produced interacts with the magnetic field N-S to produce oppositely directed side-thrusts in a horizontal direction. These thrusts tend to twist the coil out of alignment with the N-S magnetic field. This deflecting force is finally overcome by the restoring force of the connecting wire W. A mirror M attached to W serves to indicate the deflection by reflecting a beam of light focused upon it The measurement of motion of the point of light focused upon a scale distance from the mirror gives a very sensitive indication of current change. Methods for the measurement of the light-spot deflection and for recording it are reviewed below (Section II, C).

B. POTENTIOMETERS

Principally because of the delicate nature of the galvanometer, which often requires elaborate isolation mountings, many workers prefer to use other types of indicating instruments. A potentiometer of the type sometimes used with a microcalorimeter is diagrammed in Fig. 2. This instrument differs slightly from the conventional "text book" type. Instead of the usual voltage divider arrangement, it uses a Lindbeck element or variable current device. This type of potentiometer has the advantage of avoiding sliding contacts within the measuring circuit which might generate thermal voltage. This is, of course, of

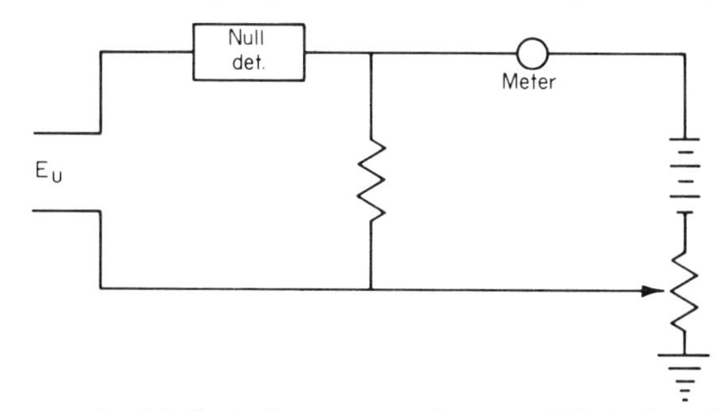

FIG. 2. Diagram of a Lindbeck-element potentiometer. This device eliminates sliding contacts in the measuring circuit which may serve as a source of thermal voltage.

decided advantage when measuring the low-level DC output of thermocouples. The principal drawbacks of the device are the need for a null detector (usually a galvanometer with its inherent faults) and for a sensitive current meter which ultimately limits the sensitivity of the instrument. The fact that a permanent record of the output is not readily obtained also must be considered.

C. SERVO RECORDERS

Servo recorders are probably the most commonly used indicating devices in microcalorimetry. These instruments have the advantage of being rugged, sensitive, and readily applicable to a wide variety of measurements. The basic

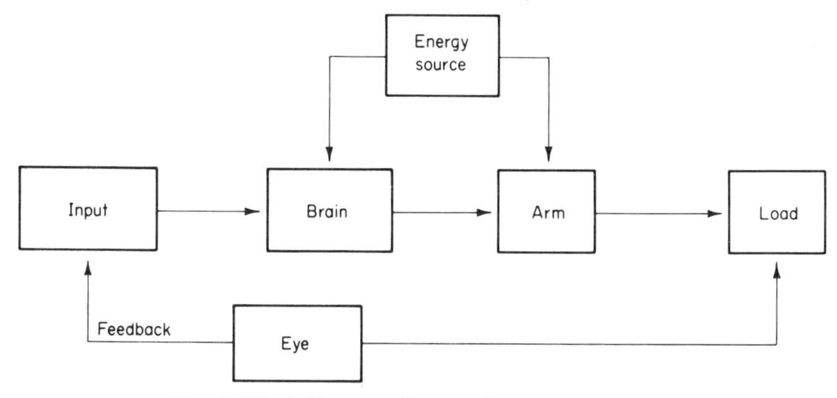

FIG. 3. Block diagram of human "servo-mechanism."

idea of a servo mechanism is simple and a common example is a human performing a manual task such as placing a book in a certain position on a table. The input signal to the motor center of the brain causes a mechanical movement of the arm to place the book into the position on the table (Fig. 3). The important idea is the fact that a sensing device (the eye) measures the difference between the desired and actual positions and feeds this information back to the primary input and enables the device to correct for improper placement of the book initially. This feedback principle is the heart of all servo recorders.

The mechanism of operation of a servo recorder is indicated by the diagram of Fig. 4. The voltage to be measured (Eu) is taken as the input. Eu, after proper

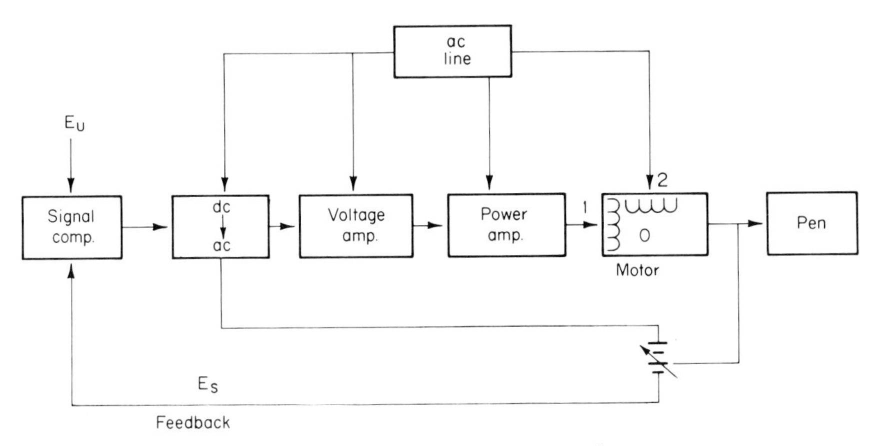

FIG. 4. Block diagram of servo recorder.

amplification and coupling, serves as a source of input to a two-phase induction motor. The induction motor is mechanically linked to a pen-drive system (load) that moves the pen in proportion to Eu. Also connected to the pen-drive is a potentiometer circuit which generates signal Es which is proportional to the amount of pen displacement. The signal Es is continuously compared (feedback) to Eu and the motor drives until $Eu = Es$ or $Eu - Es = 0$. The phase sensitive motor is necessary since the motor must rotate in one direction for $Eu > Es$ and in the opposite direction if $Eu < Es$. This is accomplished by an arrangement such that the motor coil 1 fed by the power amplifier leads (if $Eu > Es$) or lags (if $Eu < Es$) the coil fed by the line voltage 2 by 90 degrees and in turn causes either clockwise or counterclockwise rotation of the motor M. A diagram of the measuring circuit of a commercial servo recorder is given in Fig. 5.

The signal (Eu) to the recorder is generated initially within the thermopile of the calorimeter and its magnitude of change is proportionate to a given heat

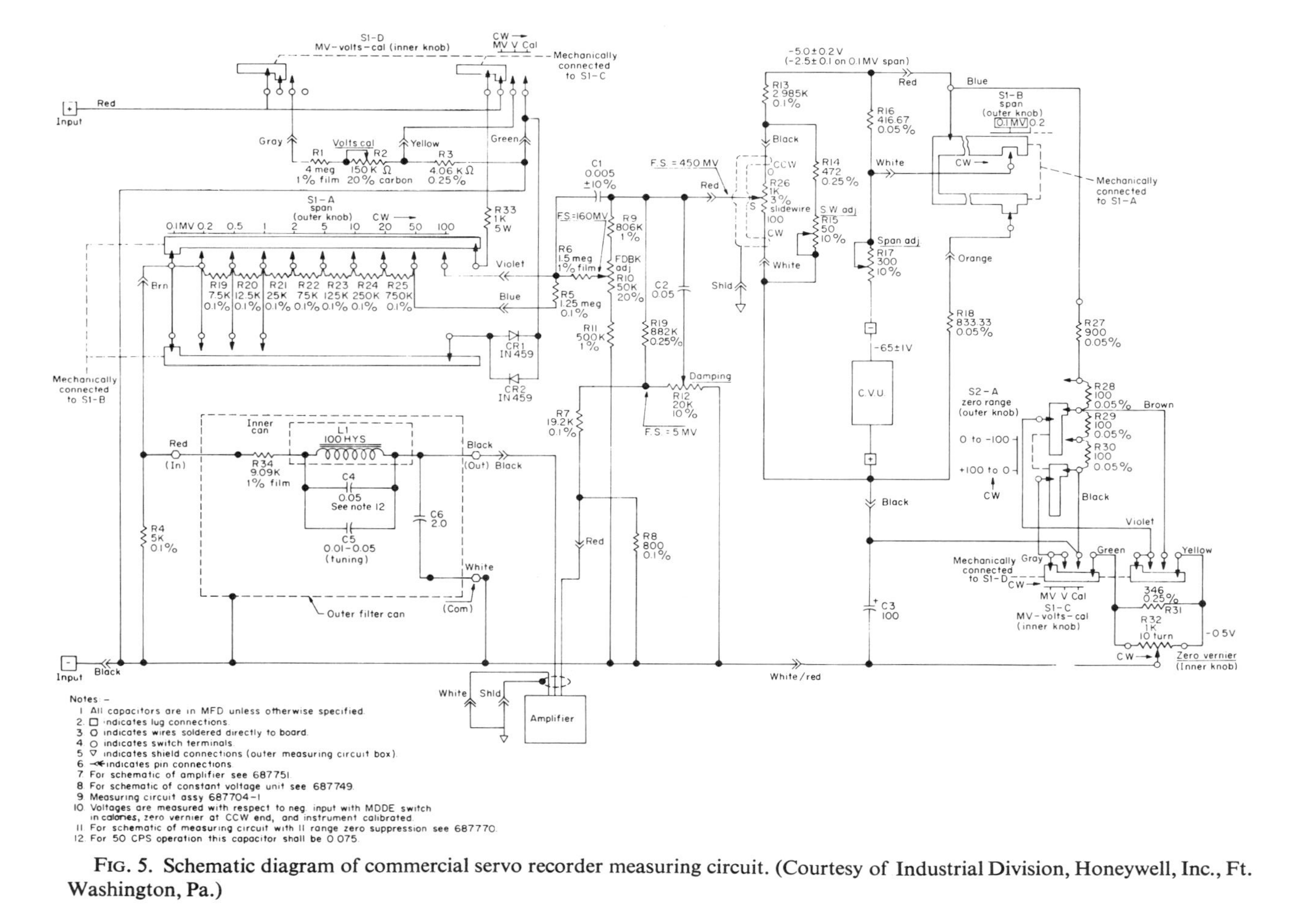

FIG. 5. Schematic diagram of commercial servo recorder measuring circuit. (Courtesy of Industrial Division, Honeywell, Inc., Ft. Washington, Pa.)

change. Generally the signal generated in the thermopile is fed to an appropriate amplifier (discussed by Praglin). To properly match the recorder to the amplifier, the manufacturer's specification should be consulted to determine the magnitude of the output signal and the output impedance of the amplifier.

Because of the sensitivity of modern servo recorder (as low as 0.1 mv full scale), it is possible in some applications to connect the output of the thermopile directly to the recorder. This is particularly true if relatively warm reactions (60 mcal or greater) are to be investigated in a calorimeter that contains a large number of thermocouples within the thermopile.

D. Spot-Following Recorders

The use of a galvanometer as a calorimeter indicator presents some unusual problems from the standpoint of the signal to be fed to the recorder. Since the galvanometer generates an optical signal, a transducer of some type must be employed to convert the optical output to a signal that is compatible with the recorder input. An electro-optical tracking system furnishes such a transduction. Basically the electro-optical tracker functions by focusing a light source or image on a photocathode which in turn generates electrons which are accelerated by an electrical field and focused upon an aperture plate. The aperture plate blocks all of the electron image except that portion that falls off the aperture. The electron multiplier behind the aperture produces a current that is a function of the intensity of the image passing through the aperture. If the image moves, the result is a change in multiplier current. This change in current is detected by a discontinuity detector whose output is fed into a logic circuit which produces an output signal that is proportional to position. The actual device is more complicated than indicated since a servo mechanism is generally included to give precise measurement of movement. In addition, the spot-following device may also track the optical image in which case the multiplier current remains constant and the amount of displacement necessary to follow the image is used as a source of input signal to the recorder. For a more extensive discussion of electron optical tracking techniques, the article by Starer (1967) may be consulted.

III. Temperature Measurement and Control

Temperature measurement is the heart of all calorimetry and thus the subject of primary temperature measurement has been covered elsewhere. However, because of the sensitivity of calorimeters to their environment, it is often necessary to make secondary temperature measurements. The material covered under the present topic is meant to assist the investigator in his choice of temperature measurement and control devices by comparing the sensitivity and other characteristics of the several types of instrumental elements.

A. Thermocouples

The use of thermocouples or thermopiles as the primary heat measuring device in microcalorimetry has been described in Chapter X and this chapter (Section III, A). This section will deal with a brief description of thermocouple theory and secondary application of thermocouples as temperature measuring elements.

Electromotive force is produced in a circuit when the junctions of the two different metals which compose the circuit are maintained at different temperatures. The Seebeck emf [named for its discoverer, Thomas Seebeck] arises because different metals possess different densities of free electrons at a given temperature. The junction serves as a seat of the emf when the two metals are joined to form a circuit such that electrons may diffuse from one metal to the other.

Since the junction is the seat of the emf, the flow of current from one metal to another causes energy, in the form of heat, to flow between the junction and its surrounds. This is known as the Peltier heat after its discoverer, Jean Peltier. It may be shown experimentally that the quantity of Peltier heat that flows is proportional to the current flowing within the circuit and that the direction of the heat flow may be reversed by reversal of the direction of the current flow. Thus, a junction may either be heated or cooled by a current flow and vise versa.

The fundamental equation of thermocouples relates the Seebeck emf ϵ, to the Peltier emf π and the Thomson emf $\sigma\,dt$ which takes into account the fact that an emf may exist along the length of a wire of the same metal if there is a difference in temperature dt along that wire. The total Thomson emf for a wire A where ends are at T_1 and T_2 is

$$\int_{T_1}^{T_2} \sigma_A\,dt \tag{1}$$

Combination of these terms give the Seebeck emf (ϵ_{AB}) for a thermocouple of metal A and B

$$\epsilon_{AB} = (\pi_{AB})_{T_1} - (\pi_{AB})_{T_R} \tag{2}$$

when T_1 is the temperature of the measuring junction and T_R is the temperature of the junction where connection of the measuring device is made to the thermocouple leads, generally 0°C.

By proper manipulation of this Eq. (2) it may be shown that the emf of a thermocouple composed of metals A and B is equal to the difference between the emfs of a thermocouple composed of metals A and C and a thermocouple composed of metals B and C

$$\epsilon_{AB} = \epsilon_{AC} - \epsilon_{BC} \tag{3}$$

at the same junction temperature. Equation (3) allows for the computation of the emfs of a thermocouple of any two metals if a standard metal (generally lead) is chosen since it is found by experiment that the emf of a thermocouple composed of lead and any metal (M) is related to temperature by the equation

$$\epsilon_{Mt} = aT + \tfrac{1}{2}bT^2 \tag{4}$$

where t is the junction temperature and a and b are constant for a given metal M (T_R is taken to be 0°C). Thus by the use of Eqs. (3) and (4) and a table of a and b values, the emf of any thermocouple at a given temperature may be calculated.

Since the output emf of most thermocouples is within the few millivolts range at normal temperatures, a variety of measuring devices may be used. A galvanometer or potentiometer calibrated in temperature or used in connection with a table to relate temperature to emf may be used to measure temperature. The thermocouple output may also be connected to a recorder which has been calibrated and a continuous recording of temperature is thus obtained. A variety of newly devised thermocouple applications have been discussed by Loefller (1967), Toenshoff and Zysk (1967), and Ruehle (1967).

B. Resistive Elements

Included under this heading are both wirewound resistors and thermistors. The wirewound resistor usually consists of a spiral of nickel or platinum wire enclosed in some type of protecting jacket. The nature of the protecting bulb is quite important, particularly in microcalorimetry. The bulbs fabricated of metal offer great mechanical strength, but generally introduce unwanted thermal inertia. Resistance spirals enclosed within glass or ceramic give more rapid response and are to be preferred if their fragility is not a surpassing concern.

The wirewound resistance spiral has a positive temperature coefficient of resistance, i.e., the resistance of the spiral increases with temperature due to the more random movement of the electrons in the wire. The sensitivity of a given resistor is a function of this coefficient and the resistance ratio is on the order of 0.5–1.0 per 100°'s for a given platinum resistor. In order to increase the sensitivity of the resistor, a wirewound spiral may be used without a protecting bulb of any type when the application allows. This is often difficult to accomplish, especially in high temperature applications where oxidation of the unprotected wire changes the electrical and mechanical characteristics of the bulb.

An alternate method of temperature measurement uses thermistors. These devices consist of a sintered mixture of metal oxides such as nickel, copper, iron, cobalt, uranium, and manganese. Thermistors are produced in a variety of shapes and sizes with the two lead wires embedded permanently in the element. Thermistors have a negative coefficient of temperature and much greater sensitivity than wirewound resistance, with a typical resistance range of 5 to 10^2 ohms $50°C^{-1}$ in the $-150°$ to 200°'s range. This increased sensitivity generally means that thermistors are far more applicable to small temperature changes than are wirewound resistors.

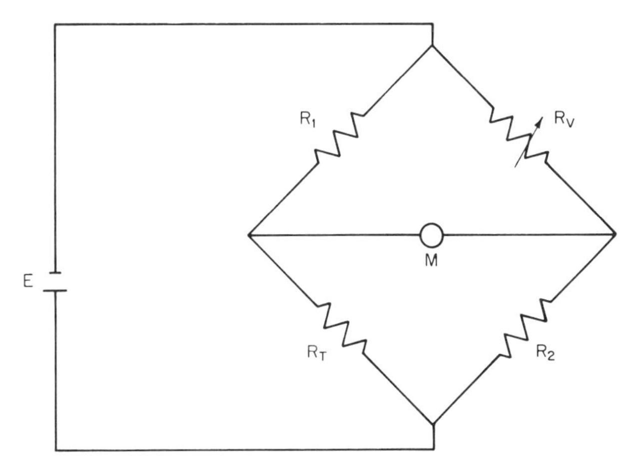

FIG. 6. Wheatstone bridge using a resistive element R_T for temperature measurement, M is the meter, R and R_2 are fixed resistors, R_V is a variable resistor used for balancing circuit, and R_T is a resistive element (thermistor or wire-wound resistant spiral).

The choice of the proper type of thermistor may be approximated from the equation

$$\frac{R_{T_1}}{R_{T_2}} = e^{\beta}\left(\frac{1}{T_1} - \frac{1}{T_2}\right) \tag{5}$$

where R_{T_1} is the resistance of thermistor at absolute temperature T_1
R_{T_2} is the resistance of thermistor at absolute temperature T_2
$e =$ base natural logarithmic (2.718)
β is a constant which is determined by the material composing the thermistor.

Thus, if the temperature range to be measured is known and the sensitivity of the detector is approximated, the proper thermistor may be chosen.

Regardless of which type of element is chosen to measure temperature, a circuit must be used that takes advantage of the characteristics of the resistive element. This device is commonly a Wheatstone bridge, as shown in Fig. 6.

The resistance of the two arms of the bridge may be adjusted using R_V such that

$$\frac{R_T}{R_2} = \frac{R_1}{R_V}$$

the meter (M) will then indicate zero current until the temperature (and resistance) of R_T changes, thus causing an imbalance in the bridge that will be indicated on the meter.

If the meter is replaced by a suitable recording device, a continuous measurement of temperature may be obtained. When using any of the methods described above, temperature calibration may be necessary. The reader is referred to references (Aronson, 1964; Jones, 1967) for a discussion of the variety of calibration systems available.

IV. Temperature Control

To control temperature within the required limits, use is made first of a sensing element such as those described. These elements are in turn used to control a device that controls the temperature.

A. Mercury Switches

Certainly the most common temperature sensing device is the mercury thermometer. A temperature controller using a mercury thermometer is shown in Fig. 7. The rise of the mercury within the thermometer closes the contacts (C)

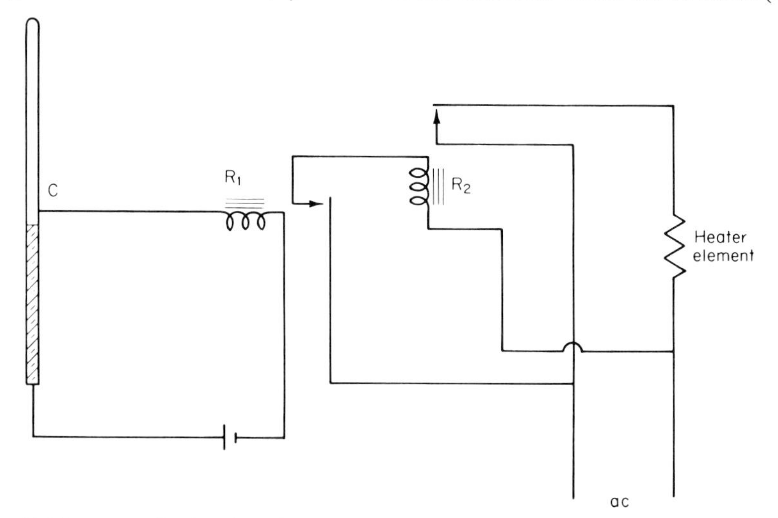

Fig. 7. Diagram of a heating circuit using a mercury thermometer and relays as the controlling device.

thus deenergizing the heater circuit through the normally closed relay R_1. The relays are used to minimize the current flow through the mercury since it is quite susceptible to oxidation. Commercial units based upon this design are available that are capable of temperature control within $\pm 0.1°C$. These devices are capable of carrying a few amps.

B. Tuned Circuits

Various other types of temperature control circuits that utilize relays are available. These generally consist of a triode (or an equivalent solid-state device) whose conduction is controlled by magnetic coupling or through a tuned-grid, tuned-plate circuit such that a relay is operated by the decoupling of the circuit on the movement of a metallic vane or flag into the vicinity of the tuned circuit. The flag is connected to the arm of a temperature-indicating meter. When the proper temperature is indicated, the heater circuit relay is opened. Lowering of the temperature causes movement of the meter arm with resulting movement of the flag such that the grid circuit is either coupled, decoupled, tuned, or untuned, depending upon the type of relay used to effectively energize the heater circuit (Bukstein, 1961).

C. Thyratrons

More sensitive temperature control may be obtained from other devices although more complicated and therefore expensive apparatus is necessary.

The precise control of temperature involves a change in concept as to control of the heating element as compared to the relatively simple on-off system exemplified by the mercury switch. The lag time involved in an on-off system prevents its use in really precise temperature control. This limitation may be overcome by a system which has an output continuously variable with input signal (proportional). A proportioning control system eliminates the lag of an on-off system. The set up of such a system is shown in Fig. 8.

The transducer of Fig. 8 is a thermocouple whose emf is fed directly to a suitable amplifier which in turn supplies a signal to the thyratron whose function will be described.

Another common form of temperature transducer is constructed using thermistors or wirewound resistors within a bridge arrangement. The signal to the amplifier is generated by the amount of imbalance within the bridge caused by the change in resistance of the element with temperature. The amplifier output is fed to a proportioning controller such as a thyratron.

Since the thyratron is in an ac circuit, it conducts only during the positive half of the cycle; however, the portion of the positive cycle during which the thyratron conducts is determined by the grid bias which is in turn determined

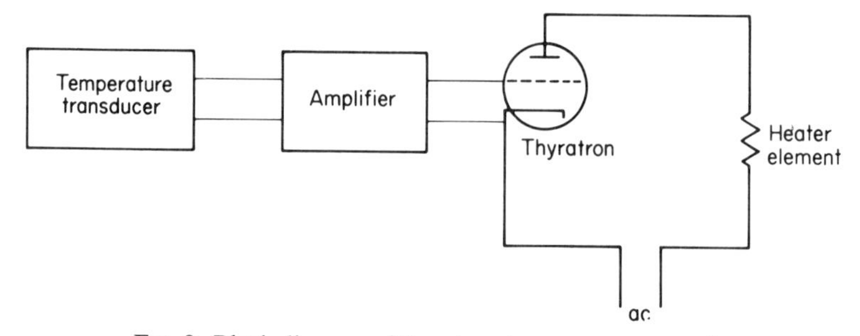

FIG. 8. Block diagram of thyratron temperature controller.

by the output of the temperature transducer. At a low bias, the thyratron conducts early in the positive cycle thus allowing more current to flow through the heating unit as compared to the current that would flow with a high bias causing the thyratron to fire late in the cycle. The actual circuitry (Fig. 8) is complex despite the essential simplicity of the concept. In practice, commercial units are available to combine amplification and thyratron circuitry as well as the circuitry of the temperature transducer into a single unit. Thus, the only burden to fall to the calorimetrist is the need to assemble the heating element and to place the temperature-indicating device within a properly designed oven. The thyratron temperature controller is generally capable of carrying 10 to 15 amps and thus will handle most resistance heating devices used in calorimetry.

D. SATURABLE REACTORS

If higher amperages are to be carried by the temperature controller (such as maintenance of room temperature by resistance heating) another type of control device must be employed. One such device, a saturable reactor, is shown in Fig. 9. The control winding C is composed of many turns of fine wire such that a few milliamps will saturate the iron core I and thereby decrease the reactance of the core to the ac current, thus allowing more current to flow

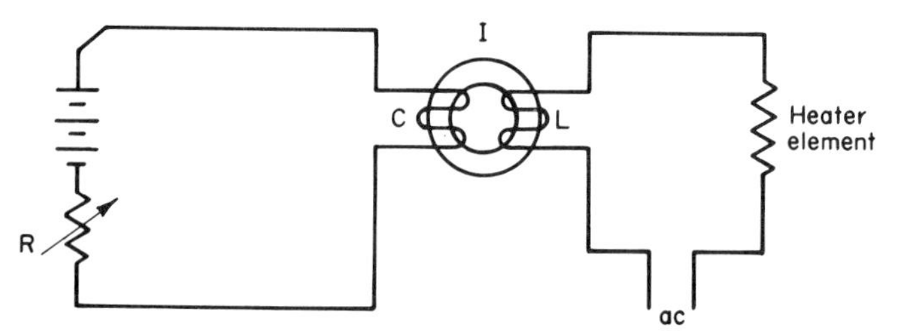

FIG. 9. Diagram of heating circuit using a saturable reactor as a controlling device. C is the control winding, I is the iron core, and L is the load winding.

through the heater element. The variable resistor R can therefore control several amps in the load winding L by varying a milliamp current in the control winding. If the source of current to the control winding is replaced by a temperature transducer whose current output is proportional to temperature the device will serve as a temperature controller. A suitable temperature transducer for this application is a thermistor within a Wheatstone bridge as shown in Fig. 6. The meter connections are taken to the control winding such that a bridge imbalance causes current to follow in the control windings. R_V is used to adjust the circuit so that no current flows at the desired temperature.

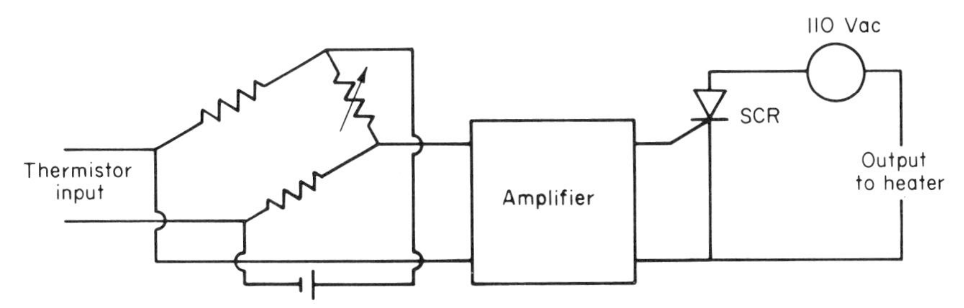

FIG. 10. Simplified diagram of a heating circuit using a thermistor as the sensing device and a silicon controlled rectifier (SCR).

An alternate method of temperature control using a silicon-controlled rectifier (SCR) is shown in Fig. 10. This controller functions in a manner similar to those previously described in that the amplifier output controls the SCR which in turn modulates the line voltage and feeds the heating element. These devices are reasonably inexpensive and have the advantage of carrying high current loads.

REFERENCES

Aronson, M. H., ed. (1964). "Temperature Measurement and Control Handbook." Instruments Publ., Pittsburgh, Pennsylvania.

Bukstein, E. (1961). "Industrial Electronics Measurements and Control." Howard W. Sams, Indianapolis, Indiana.

Jones, E. W. (1967). *Instr. Control Systems* **40**, 135.

Loeffler, R. F. (1967). *Instr. Control Systems* **40**, 89.

Ruehle, R. A. (1967). *Instr. Control Systems* **40**, 113.

Starer, R. L. (1967). *Instr. Control Systems* **40**, 103.

Toenshoff, D. A., and Zysk, E. D. (1967). *Instr. Control Systems* **40**, 109.

CHAPTER XII

Calibration and Test Reactions for Microcalorimetry

ROBERT L. BERGER

I. Introduction

Procedures for the calibration of calorimeters may be placed naturally into four categories. Every physics student is well acquainted with the determination of the mechanical equivalent to heat (White and Manning, 1954) which can, of course, be used as a fundamental calibration. The most classical method in calorimetry, however, has been based upon the use of an electrical heater (Calvet, in Skinner, 1962). A known resistance is used and a carefully measured, timed current produces a known joule heating. More recently a number of chemical methods have been employed (Irving and Wadsö, 1964; Richard and Rowe, 1922; Sturtevant, 1955; Bernhard, 1955; Podolsky and Morales, 1955) with the advantage that heat is generated simultaneously throughout the calorimeter cell rather than at a series of points. Lastly an alpha or beta particle emitting radiation source may be used (Rothchild, 1964). The method chosen will depend upon the response time of the calorimeter, the reaction time to be studied, the nature of the calorimeter, and the sensitivity of the calorimeter. Since this Chapter is meant to be a practical guide rather than a treatise, several simple, reliable methods will be given for laboratory use. Details may be found in the fairly extensive literature on the subject (Sturtevant, 1959; Calvet and Prat, 1963; Kitzinger and Benzinger, 1961).

For many reactions a very simple calorimeter, consisting of two Thermos flasks can be used. Gutfreund (1965) has used such a system to study the reaction of cytochrome C. Heat levels giving temperature rises of several tenths of a degree were expected. In general, however, a much more sensitive instrument is required. These have been discussed at length elsewhere in this book. There are basically two types: heat-conduction and nonisothermal instruments. From a calibration standpoint one may think of them as heat-measuring and temperature-measuring instruments. Thus the calibration for the heat-conduction instrument is a measurement of the instrument constant. That is to say, how many calories does so many microvolts for so many seconds equal. No correction for unmeasured-heat loss is necessary if the half-time of the reaction is not much longer than the decay time of the instrument. The non-isothermal apparatus, however, must be corrected for its heat-conduction losses. The electrical heating equivalent must be known for each solution if calories are to be obtained, and microvolts now give calories directly so that in a lossless instrument a uniform voltage output would represent zero heat production beyond the initial rise. This would represent continuous heat production in the heat-conduction instrument. Nonetheless, the actual calibration of the instruments is done in the same way so that a description of the electrical heater method will be the same for any calorimeter, only the use of the data will differ.

II. Radiation

The electrical heater method has the disadvantage that it produces heat at various fixed points in the solution rather than throughout the entire solution. Thus heat must diffuse through the solution and this puts a lag in the thermal response of the system. In order to overcome this, two techniques have been employed. One is the use of alpha or beta particles whose range in water is less than 60 microns. Plutonium (Rothchild, 1964) and Polonium (Davids and Berger, 1964) have been used. Sources can be obtained from the Monsanto Chemical Corporation, Dayton, Ohio. However, the evolution of hydrogen makes the liquid samples dangerous unless used immediately. We have used P-32 as an energy source, but it has the disadvantage of a short half-life and dangerous X-ray production from bremstrahlung at high activities. A far better system would be Promethium-147, peak β energy of 0.223 MeV, half-life 2.5 years, or Technetium-99, peak β energy of 0.29 MeV, half-life 2.12×10^{15} years (see Handbook: Anonymous, 1960). These could be obtained in various source forms, such as a paint, or a distributed source, such that all of the betas are absorbed inside the calorimeter. The maximum range in water for Promethium-147 would be 0.06 mm and for Technetium-99, 0.98 mm. Calorimetric calibrations could be supplied with the sources by the Monsanto Chemical Co., Dayton, Ohio. The problem with these sources is that they continuously produce heat and it is difficult, therefore, to bring the system to equlibrium, insert the source and watch the heat rise. With a flow calorimeter, this could be overcome by flowing the isotope into the calorimeter encased within microspheres and held in a colloidal suspension. A simpler method is to use known chemical reactions. Unfortunately, only a very few reactions have been worked out well enough for use as standards. These will be discussed and instructions for their use will be given below.

At present, most calibrations are done by electrical means and by chemical reactions. Gunn has discussed the problem of standards in great detail and those interested in this problem should read his paper (Gunn, 1965). In general most workers in biochemical calorimetry feel that three or four reasonably well-known reactions can be used with an error of 1–2%. Work certaintly needs to be done on the development of special reactions and several such projects are under way. These will be mentioned when the reaction is discussed.

III. Electrical Calibration

The circuit shown in Fig. 1 enclosed in the dotted lines is a simple version of the main circuit. More detailed discussions are given by Skinner (1962) and Laws (1938). For many purposes the simple circuit will be adequate, certainly so for the discussion that follows. The heat produced in the heater, R is given by the following equation:

$$Q = \text{Power} \times \text{time} = R \times I^2 \times t$$

where I is the current flowing in the heater and t is the time for which it flows. The units are such that if I is in amps, t in seconds, R in ohms, Q will be in Joules. This may be converted to calories by dividing by 4.181 at 20°C (Weast, 1956, p. 2708).

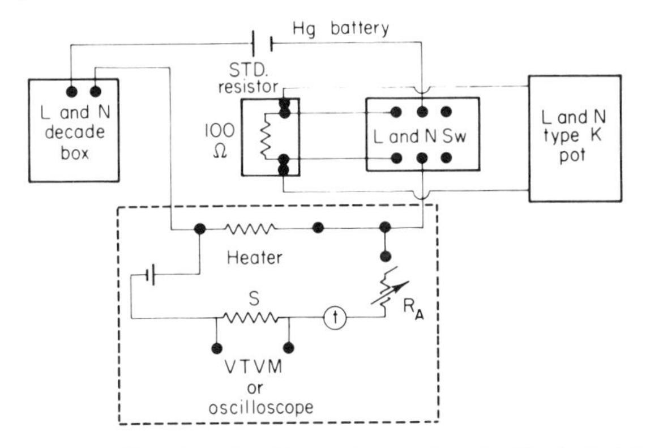

FIG. 1. Calorimeter calibrating circuit; portion enclosed with dotted lines is a simple version of the main circuit.

TABLE I

PHYSICAL CONSTANTS OF THERMOCOUPLE AND HEATER WIRES

Alloy or metal gauge	Resistance		Density (gm cm^{3-1})	Thermal conductivity (100°C cal sec^{-1} cm^{2-1} cm^{-1} $°C^{-1}$)	Heat capacity (C_p) (cal gm^{-1} $°C^{-1}$)	EMF–mV (0°–100°C; relative to platinum)	Diameter (mils)
	Ω mil^{-1} at 68° 100 ft^{-1}						
Copper	10.321		8.90	0.9340	0.092	0.76	
30		103.2					0.030
32		164.1					7.950
34		260.9					6.305
36		414.8					5.000
38		659.6					3.965
40		1049.0					3.145
Constantan	294.0	28.45[a]	8.40	0.0506	0.094	−3.51	
Chromel	425.0	41.00[a]	8.73	0.0460	0.107	+2.84	
Alumel	177.0	17.10[a]	8.60	0.0710	0.125	−1.29	

[a] Resistance value for alloys is obtained by multiplying conversion factor given by resistance of copper.

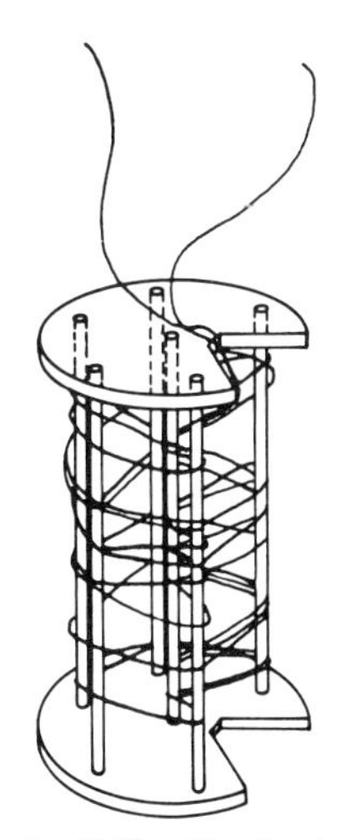

FIG. 2. Calibration heater.

As a practical example, if we use a 14-foot piece of No. 34 Constantan wire with which to wind our heater we can see from Table I that it will be 103.6 ohms. Such a heater, for use in the microcalorimeter described in Chapter XVII is shown in Fig. 2. It is most important to minimize thermal lag in the construction of the heater by applying only a thin coating of insulation. The material marked 3M in Table II has been found satisfactory. The resistance of this heater unit must be carefully measured in place in the solution of the calorimeter, in the case of nonisothermal instruments, or inside the thermopiles and solution cell in heat conduction units. Since precision power supplies are readily available, one may be used in place of a battery. If a battery is used, a dummy load should be inserted until the voltage becomes constant though with

TABLE II

INSULATING MATERIALS

Name	Company	Method of application	Time	Reistance to immersion in NaCl	Coating, thickness available
Glyptol	General Electric	Dip or paint	2 hours at 125°C	Poor	1 mil
3M	Minnesota Mining	Dip or paint	None	Good	1 mil
Parelyene C	Union Carbide	Deposit	None	Excellent	1 μ up
Kel-F	Minnesota Mining	Dip	None	Good to Excellent	1 mil

a mercury battery this would not be necessary. For most work a General Radio standard 100 ohm resistor will be of adequate precision. Any simple potentiometer, vacuum tube voltmeter, or oscilloscope may be used as a voltage indicator.

Having made a careful resistance measurement, and wired and tested the circuit, we are now ready to calibrate. Let us suppose that a total heat rise of 30 mcal is desired. Let us take the heat capacity of the calorimeter to be ($C_p = 4$ ml $\times$ 0.9988 at 20°C) (Weast, 1956, p. 2080) 3.995 gm-cal, with 4 ml of

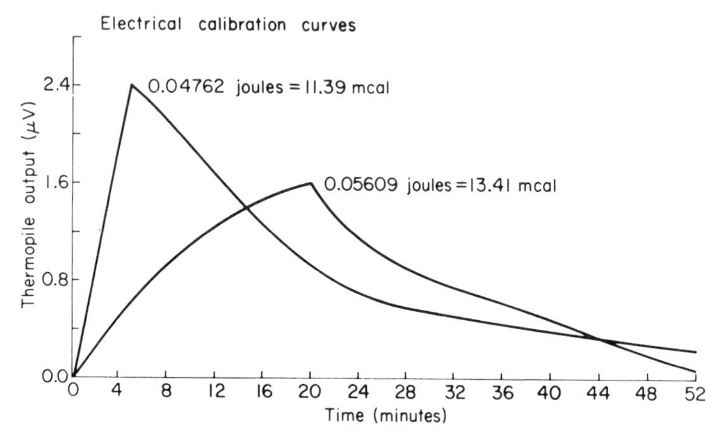

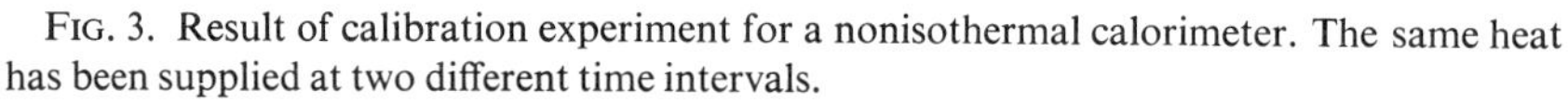

FIG. 3. Result of calibration experiment for a nonisothermal calorimeter. The same heat has been supplied at two different time intervals.

distilled water in the cells. In order to proceed we must first make a conversion to joules. Thus, 0.1254 joules are needed. If this is to be supplied in one minute, then a current of

$$I = \sqrt{0.1254\,(60 \times 103.6)}$$
$$= 0.00447 \text{ amps}$$

If S, the standard, is 100 ohms, then we would expect a voltage drop across it of 0.447 V. Since S is readily known to 0.1 %, I is known to 0.2 % providing the voltage is read to 0.1 %. This will give 0.4 % error in I^2. If R is known to 0.1 % from the earlier measurement and the time, t, is known to 0.1 %, the error in the heat will be 0.6 %. Thus we see the need for care in our measurements. The temperature rise will be given by:

$$Q = V \times \rho \times C_p \times \Delta T, \text{ and in this case}$$

$$0.030 = 4.000 \text{ cm}^3 \times 0.998231 \text{ gm/cm}^3 \times 0.9988 \text{ cal gm}^{-1}\,°\text{C}^{-1} \times \Delta T$$

thus $\Delta T = 0.0075$°C assuming no conduction losses.

If a 25 junction thermopile were used, at 1 mV °C^{-1}, then 7.5 μV would be produced. Results of such an experiment are shown in Fig. 3 for the nonisothermal calorimeter. The same heat has been supplied at two different time intervals. Note the considerable loss due to heat conduction for the longer heating time. If work is to be done under these circumstances, it is necessary that a data correction program be worked out. Such a program is briefly described in Chapter XVII. Details can be found in papers by Berger and Davids (1965; Davids and Berger, 1969). A similar problem arises in the calibration of conduction calorimeters. Details of a correction program have been given by Calvet (in Skinner, 1962) and by Kitzinger and Benzinger (1961), as well as Davids and Berger (1969). While the above procedure should always be done in order to check the instrument, for biochemical work in general, when great precision is not needed, the testing and calibrating with a known chemical reaction is to be preferred.

IV. Chemical Reaction Calibration

The most common reaction in use is the combination of NaOH with HCl. The base should be present in excess so that corrections for carbonate formed by CO_2 absorption do not have to be made. The reaction yields 13.5 kcal mole^{-1} at 25°C. See Richards and Rowe (1922) for details, and Chung (1960) for a good detailed description of the experimental technique.

In general, one simply determines the voltage obtained when a particular number of meq of HCl are neutralized and this becomes the calorimeter constant. As an example, in the case of the heat conduction instrument, using 15 ml cells, 15 ml of 0.01 *M* NaOH is loaded into each cell. Into the tare is put 0.1 ml of NaOH and in the second cell 0.1 ml of HCl of the appropriate strength to give the desired heat. Thus 0.05 *M* acid would give 67.5 mcal. Prior to mixing a good base line is obtained, the reactants are then mixed and the integrator allowed to run until the base line is again reached. A second mixing is then done which should yield almost no heat. The number of integrator counts for 67.6 mcal is now divided into the above number and the instrument constant is determined. That is, in any subsequent reactions the number of counts is multiplied by the instrument constant, expressed as millicalories per count, to give millicalories. One should always do a number of test reactions over a concentration range to cover the heats expected in order to determine the reproducibility of the measurement.

In the case of the nonisothermal reaction calorimeter a similar procedure is employed. However, for instantaneous reactions, one reads the number of microvolts at the peak of the curve which is now equal to the heat, that is, the constant is now millicalories per microvolt. A simple table can then be made up by varying the concentration of HCl so that even if there are losses, reactions faster than one or two seconds can be determined simply from the microvolt reading obtained. Again heating from the mixing process should be checked.

This may need to be subtracted from the reading. If longer reactions are to be run, then a heat data correction program will need to be worked out, if it is not supplied with the instrument.

One final word of caution. In carrying out chemical reactions, the volume of liquid supplied is most important since it will change the heat capacity of the system. Clearly for the most precise work, this should be determined for each solution. For much work in biochemistry, if transfer pipettes and Lang-Levy pipettes are used the heat capacity values will probably not vary beyond the other experimental errors. If there is a question, this should be checked using the electrical heaters first in distilled water and then in the experimental solution. Eq. 1 will then give C_p.

Table III gives values of the Tris-HCl reaction for use in calibration. Since CO_2 does react with the carbo-amino groups, it is best to have Tris in excess. Material from Sigma Chemical Corp. has been found to be very good. The National Bureau of Standards is currently working on a standard Tris, issued

TABLE III

HEATS OF SOLUTION[a b]

		$\epsilon = 8907 \pm 8$	
Substance	mmoles	$10^4 \times \log R_i/R_f$	$-\Delta H$ kcal mole^{-1}
HCl	0.988	13.33	12.02
	1.076	14.22	11.77
	1.047	13.93	11.85
	1.013	13.56	11.92
	1.026	13.67	11.87
			Mean 11.89 ± 0.04
HOAc	1.501	19.44	11.53
	1.579	20.56	11.59
	1.365	17.76	11.59
	1.402	18.31	11.64
	1.817	23.78	11.65
			Mean 11.60 ± 0.02

[a] From the heats of solution of aqueous hydrochloric acid, 1.690 mmole gm^{-1} in aqueous THAM solution ($\Delta H = -11.89 \pm 0.04$ kcal mole^{-1}) and in 0.1 M sodium chloride ($\Delta H = -0.53 \pm 0.01$ kcal mole^{-1}); initial pH = 8.05. The heat of ionization of the protonated form of THAM is calculated to be $+11.36 \pm 0.04$ kcal mole^{-1} (ionic strength 0.1). This value should be compared with the value obtained calorimetrically at ionic strength 0.013 of $+10.93 \pm 0.10$ kcal mole^{-1}. For ionic strength 0.02 (Berger *et al.*, 1968b, $\Delta H = +11.36 \pm 0.2$ kcal mole^{-1}.

[b] From Nelander (1964).

as Standard Reference material 734. The mixing of solid Tris with HCl gives 7.5 K cal/mole, see Irving and Wadsö (1964). At an ionic strength of 0.01 the heat of neutralization is 11.36 K cal/mole and is temperature invariant (Bernhard, 1955). While the neutralization reactions give a reasonable calibration of the energy response of the calorimeter, they do not give a check of the utility of the instrument for the study of reaction kinetics. A simple reaction will be discussed which makes an excellent example. The heat determinations given are probably not better than ±20% but the rate data can be checked optically and are somewhat better.

The Iodine Clock reaction has been studied for nearly 100 years (see Bunau and Eigen, 1962). A particularly useful series of reactions can be set up to demonstrate the method of Initial Rates using the following scheme:

$$IO_3^- + 8I^- + 6H^+ \rightarrow 3I_3^- + 3H_2O$$

The method involves measuring the time required for a definite small amount of iodate to be consumed. This is done by determining the time required for the iodine produced by the reaction (AsI_3^-) to oxidize a definite amount of a reducing agent, in this case, arsenious acid. This does not react directly with iodate but does react with iodine as rapidly as it is formed. When the acid has been completely consumed, free iodine is liberated and produces a blue color by complexing with a small amount of soluble starch present.

TABLE IV

IODINE CLOCK REACTION: 32.2°C AND pH 5

Solution	Initial volumes of solutions (ml)				Time to starch endpoint (seconds)	ΔH to starch end point (kcal mole^{-1})
	I	II	III	IV		
[a] H_3AsO_3	5	5	5	5	935 I	22 ± 1
[b] IO_3^-	5	10	5	5	470 II	
[c] Buffer A	65	60	40		912 III	
[d] Buffer B				65	300 IV	
[e] I^-	25	25	50	25		

[a] H_3AsO_3. 0.03 *M*. Should be made up from $NaAsO_2$ and brought to a pH of about 5 by addition of HAc.

[b] KIO_3. 0.1 *M*.

[c] Buffer A. Pipette 100 ml of 0.75 *M* NaAc solution, 100 ml of 0.22 *M* HAc solution, and about 20 ml of 0.2% soluble starch solution into a 500-ml volumetric flask. Make up to the mark with distilled water [yields (H^+) = 10^{-5} *M*].

[d] Buffer B. Pipette 50 ml of 0.75 *M* NaAc solution, 100 ml of 0.22 *M* HAc solution, and about 10 ml of 0.2% soluble starch solution into a 250-ml volumetric flask. Make up to the mark with distilled water [yields (H^+) = 2×10^{-5} *M*].

[e] KI. 0.2 *M*.

The reaction may be written at about pH 5 as,

$$H_3AsO_3 + I_3^- + H_2O \rightarrow HAsO_4^{=} + 3I^- + 4H^+$$

and the overall reaction up to the starch end point as

$$IO_3^- + 3H_3AsO_3 \rightarrow I^- + 3HAsO_4^{=} + 6H^+$$

Clearly considerable buffer must be present in order to keep the pH constant. Details of the method can be found in Shoemaker and Garland (1962). Table IV shows a typical set of experiments which can be used to test the calorimeter or as an experiment on initial rates.

A second reaction useful in calibration and applicable to use of the calorimeter for the study of reaction kinetics is that of carbon dioxide and water catalyzed by the enzyme carbonic anhydrase, Kernohan (1965). Its use as a standard reaction for testing a flow apparatus has been discussed recently (Berger *et al.*, 1968a). Table V below indicates the experimental design. The reaction, carried out in imidazole buffer containing *p*-nitrophenol indicator, is:

$$H_2O + CO_2 \rightleftharpoons H^+ + HCO_3^-$$

For this reaction to proceed from left to right the hydrogen ion must be removed. This then is followed by the spontaneous fast reaction

$$\begin{aligned} &\text{Buffer} \rightleftharpoons \text{Buffer} \\ H^+ + &\text{base} \rightleftharpoons \text{Acid} \end{aligned}$$

The progress of the second reaction is determined by measuring the absorbance of the p-nitrophenol at 410 mμ. By choosing the enzyme and indicator concentrations, it is possible to vary the rate of the reaction and its associated optical density change over a wide range.

The thermal values at various temperatures are yet to be determined but they most likely will follow those for the bicarbonate reaction plus the buffer hydrogen ion heat. For slower work, a much greater dilution, 0.01 mg/ml, of the enzyme can be used and this brings the reaction to about 180 seconds half-life. The above reaction permits one to do a test reaction for slow kinetics as well as to determine the mixing properties of the reaction calorimeter. It is to be hoped that a more careful investigation of some of the reactions might be carried out in the future to provide us with much better standards.

For rapid thermal studies, where millisecond reaction times are to be measured, the bicarbonate-HCl reaction is useful for calibrations. Table VI gives a summary of the data at different temperatures (Roughton, 1941; Berger and Stoddart, 1965). In general one would carry out a calibration of the rapid reaction apparatus as follows: Starting with the thermocouple as close to the mixer as possible, a 0.02 *M* Tris and 0.02 *M* HCl solution prepared so that the Tris is half neutralized and CO_2 free, is flowed through the mixer and observation tube. Readings on a microvoltmeter, or a galvanometer, are

TABLE V

OUTLINE OF EXPERIMENTAL DESIGN FOR CARBONIC ANHYDRASE AS A CALIBRATING REACTION

Experimental conditions		Rate constant (second^{-1})		Half time (msecond)		
Assigned number	Concentrations of enzyme[a] after mixing (mg ml^{-1})	Expected	Measured	Measured	Total O.D. change during reaction	Number of experiments
1	10	—	37.4 ± 1	18.5 ± 0.5	0.0886 ± 0.0044	11
2	40	1497	149.5 ± 4	4.6 ± 0.1	0.0868 ± 0.0044	11
3	400	1497	1320 ± 120	0.52 ± 0.05	0.0653 ± 0.0044	4
4	400	1497	1370 ± 180	0.51 ± 0.05	0.0609 ± 0.004	6

[a] Lot No. S1020, Mann Research Laboratories, New York, New York.

TABLE VI

CARBONATE DEHYDRATION RATE CONSTANT

Temp.	Conc. $NaHCO_3$[a]	$ksec^{-1}$				Number of experiments		ΔH
(°C ± 0.1)	(moles $liter^{-1}$)	Optical[b]	Std	Thermal	Std	Optical	Thermal	(cal $mole^{-1}$)
3.8	0.08	2.49	0.31	3.13	0.29	7	8	2100
	0.04	3.15	0.32	3.12	0.26	6	8	
	0.02	2.48	0.11	2.79	0.26	4	4	
18.0	0.04	15.11		15.80	1.5	2	3	1413
24.1	0.08	25.89	1.88	21.77	0.59	7	8	970
	0.04	20.60	1.21	22.70	2.96	6	8	
	0.01	19.00	1.98	17.1	3.15	6	6	
36.9	0.04			49.04	9.08		6	460

[a] HCl concentration, after mixing, one half that of $NaHCO_2$.
[b] 590 mμ.

taken at zero, 2, 4, 6, etc., centimeters down the tube. Thus with a single Cu-Constantan junction 9.2 μV would be expected from the reaction. If this is constant up the tube, good mixing has been achieved. For continuous flow work the procedure is now repeated using bicarbonate-HCl (Fig. 4). For stopped-flow calorimetry a position one to two centimeters above the point of mixing is chosen, the flow started and then stopped and the reaction curve traced out. Since the reaction is usually over in 1 second, no corrections are

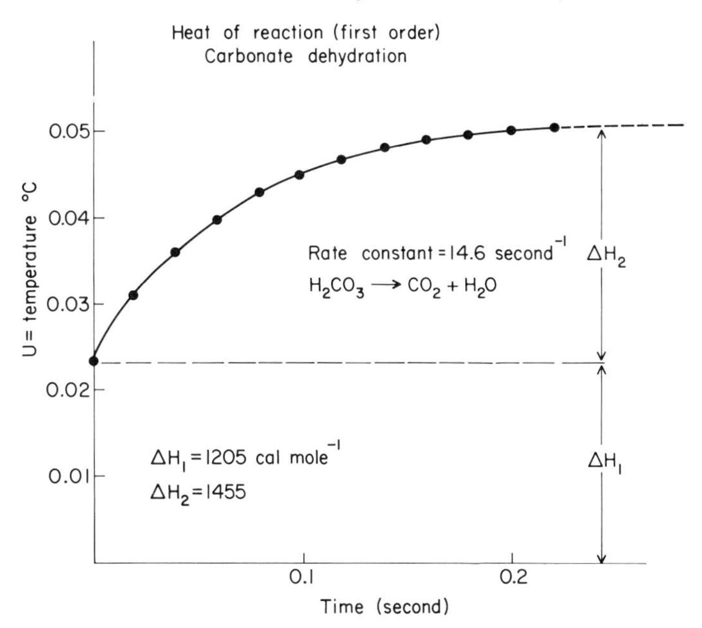

FIG. 4. Heat of reaction (first order) for carbonate dehydration.

needed for a well designed system. However, both heat conduction and instrument response time correction programs have been worked out and should be referred to if these corrections are needed (Berger and Davids, 1965; Stoddart and Berger, 1965).

This has been a very brief outline of calibration methods, but it is hoped it will lead the newcomer to have confidence in his equipment. Much work remains to be done in the development of adequate standards for solution calorimetry in biochemistry and it is to be hoped that with the advent of commercial calorimeters such work will be forthcoming.

REFERENCES

Anonymous (1960). Radiological Health Handbook, PB121784R. Office of Tech. Serv., U.S. Govt. Printing Office, Washington, D.C.

Berger, R. L., and Davids, N. (1965). *Rev. Sci. Instr.* **36**, 88.

Berger, R. L., and Stoddart, L. C. (1965). *Rev. Sci. Instr.* **36**, 78.
Berger, R. L., Borchardt, W., and Balko, B. (1968a). *Rev. Sci. Instr.* **39**, 486.
Berger, R. L., Chick, F. Y., and Davids, N. (1968b). *Rev. Sci. Instr.* **39**, 362.
Bernhard, S. A. (1955). *J. Biol. Chem.* **218**, 961.
Bunau, G. V., and Eigen, M. (1962). *Z. Physik. Chem.* (*Frankfurt*) **32**, 527.
Calvet, E., and Prat, H. (1963). "Recent Progress in Calorimetry." Macmillan (Pergamon), New York.
Chung, D. M. F. (1960). M.S. Thesis, Utah State Univ., Logan, Utah.
Davids, N., and Berger, R. L. (1964). Unpublished report of the National Heart Institute. (Not to be considered part of the Scientific Literature but available from the authors.)
Davids, N., and Berger, R. L. (1969). "Currents in Modern Biology, Supplement." (in press).
Gunn, S. R. (1965). *J. Phys. Chem.* **69**, 2902.
Gutfreund, H. (1965). "An Introduction to the Study of Enzymes," p. 157. Blackwell, Oxford.
Irving, R. J., and Wadsö, I. (1964). *Acta Chem. Scand.* **18**, 195.
Kernohan, J. C. (1965). *Biochim. Biophys. Acta* **96**, 304.
Kitzinger, C., and Benzinger, T. H. (1961). *Methods Biochem. Analy.* **8**, 309.
Laws, F. A. (1938). "Electrical Measurements," 2nd Ed. McGraw-Hill, New York.
Nelander, L. (1964). *Acta Chem. Scand.* **18**, 973.
Podolsky, R. J., and Morales, M. F. (1955). *J. Biol. Chem.* **218**, 945
Richard, T. W., and Rowe, A. W. (1922). *J. Am. Chem. Soc.* **44**, 684.
Rothchild, Lord V. (1964). *Proc. Roy. Soc.* (*London*) **B159**, 291.
Roughton, F. J. W. (1941). *J. Am. Chem. Soc.* **63**, 2930.
Shoemaker, D. P., and Garland, C. W. (1962). "Experiments in Physical Chemistry," Chapt. IX, Experiment 26. McGraw-Hill, New York.
Skinner, H. A., ed. (1962). *Exptl. Thermochem.* **2**, 177.
Stoddart, L. C., and Berger, R. L. (1965). *Rev. Sci. Instr.* **36**, 85.
Sturtevant, J. M. (1955). *J. Am. Chem. Soc.* **77**, 255.
Sturtevant, J. M. (1959). "Techniques of Organic Chemistry," 3rd Ed., Vol. 1, Pt. 1. Wiley (Interscience), New York.
Weast, R. C., ed. (1956). "Handbook of Chemistry and Physics," 38th Ed. Chem. Rubber Publ. Co., Cleveland, Ohio.
White, M. W., and Manning, C. K. V. (1954). "Experimental College Physics," 3rd Ed., p. 142. McGraw-Hill, New York.

CHAPTER XIII

The Calorimeters: Adiabatic Calorimeters

T. ACKERMANN

I. Principles of Design

As pointed out in Chapter VI, the ΔH values accompanying conformational changes of macromolecules in solution can be determined by measuring the heat capacity of the sample as a function of temperature in the temperature region of the transition. In this section we shall deal with a brief description of adiabatic calorimeters which have already been used for measurements of this type. In order to be able to evaluate the excess heat capacity C_c it is necessary to make the primary heat capacity measurements with the greatest possible precision. It is probably true to state that only the adiabatic heating methods are capable of the required accuracy. The two main techniques using adiabatic heating, namely the direct methods (single-vessel calorimeters) and the comparison methods (twin-vessel calorimeters) are discussed in the following sub-sections.

One important consideration in designing calorimeters to work near room temperature is that a number of versatile materials which are not usable at extreme temperatures behave quite satisfactorily between 250°K and 400°K. These include, for example, thermosetting resin coatings and adhesives, and rubber-like gasket materials. We pay particular attention, therefore, to the ways in which such materials may be used, and to the precautions their use necessitates.

In all the direct methods the primary experiment is the determination of the total heat capacity of the calorimeter vessel and its contents by measuring the rise in temperature consequent upon the addition of a measured quantity of electrical energy. In order to obtain the heat capacity of the liquid specimen itself, the heat capacity of the empty calorimeter vessel must be determined independently. In principle, the simplest way of doing this is to carry out a calibration experiment with the calorimeter vessel filled with a reference liquid whose heat capacity is known accurately and whose thermal conduction characteristics are similar to those of the test liquid. This procedure is in fact the basic comparison method, but carried out with the maximum of experimental inconvenience.

It is inherent in the direct methods that all the electrical energy liberated must be strictly accounted for, together with the secondary energy effects such as residual heat leak etc. In terms of an overall precision of 0.05 %, if the residual heat leak exceeds 0.1 % of the average heating power, the former must itself be known to better than 10 %. Since these limits are beyond the capabilities of any except a jacketed adiabatic calorimeter, we are dealing only with adiabatic calorimeters in which the calorimeter vessel is surrounded by an insulating space and by an adiabatic shield whose temperature is adjusted to match that on the outer surface of the calorimeter vessel. In this context the term "adiabatic" means zero net heat transfer between the calorimeter vessel and the shield rather than merely that the temperature of the shield is kept more or less equal to that of the calorimeter vessel. In a recording adiabatic calorimeter the direct method may be realized in terms of continuous heating of the calorimeter. This method consists in setting up a constant rate of heating or, more usually, a constant heating power, and measuring both the power and the rate of heating. The latter is usually done by determining the time taken for the temperature to increase by a predetermined amount. The operating equation is then

$$C = (\dot{Q} - \dot{q})(dt/dT) \tag{1}$$

the residual heat-leak rate $\dot{q}$ is known as a function of the "heat-leak coefficient," governing the transfer between the calorimeter vessel and the shield. This coefficient usually changes only slowly with the temperature, so that $\dot{q}$ may in some cases be kept effectively zero over heating intervals of several degrees. The heat capacity of the calorimeter vessel plus the liquid specimen is then compared to that obtained in the blank experiment under equivalent conditions. Provided only that uniformity of temperature is maintained over the whole extent of the outer surface of the calorimeter vessel during heating, this is in principle the most accurate way of realizing the direct method of calorimetry, since differences in the temperature distribution inside the calorimeter vessel under different conditions of filling become irrelevant in steady-state heating.

The term comparison method is sometimes used, in a broad sense, to cover the whole field of calorimetric techniques using standard reference substances for calibration. In a more limited sense, the meaning of the term comparison method is restricted to the technique pioneered by Joule (1845), that is to say the use of twin calorimeter vessels, one containing a reference substance, and the other one containing the substance under investigation. The early history of the comparison method was reviewed by Richards and Gucker (1925); their paper includes an extensive bibliography. In the high-precision twin-vessel calorimeter described in Section III below, the heating power applied to one or other of the twin calorimeter vessels is adjusted to equalize the heating rates. This method is particularly well suited to continuous heating procedures and hence to recording twin-vessel calorimeters, since the electrical power input to the heaters can be controlled continuously during a long heating period so as to keep the instantaneous temperature difference between the two vessels negligibly small. It is essential to this method to be able to determine precisely the small corrections to the heating power of one of the twin vessels, and this is achieved by separating the additional heating power required for the liquid specimen of the greater heat capacity in a fully balanced twin calorimeter. The main practical advantage of this method is that it is comparatively simple to determine with adequate precision the current intensity in the auxiliary heater. The operating equation in this case, using the same nomenclature as above, is

$$C_s - C_r = I_a^2 R_a(dt/dT) \tag{2}$$

where C_s is the heat capacity of the liquid specimen and C_r is the heat capacity of the reference liquid. The subscript a refers to the auxiliary heater in the specimen vessel and dT/dt is the common heating rate in deg sec^{-1}. It is obvious from Eq. (2) that this is the only comparison method which gives the difference between the total heat capacities of the filled specimen and reference calorimeter vessels directly, without additional information about either C_r or C_s. The detailed theory of one application of this method is set out in Section III, below.

The basic requirements for a calorimeter vessel for liquids are that it shall contain the liquid specimen absolutely without leakage, and that its internal and external geometry shall be compatible with the operation of an automatically controlled adiabatic shield. The second requirement implies, among other things, that the external area of the vessel proper be minimized with respect to the internal volume available to the specimen, and that the temperature be uniform over the whole external surface. These requirements are common to both the direct and the comparison methods. A third requirement, particularly important in the direct method, is that the total heat capacity of the calorimeter vessel must be small in comparison with that of the liquid specimen. Ease of operation is an important requirement for successful calorimetry. It is pointless to develop an elaborate and sophisticated apparatus if repairs and breakdowns extend unreasonably the time lapse between effective measurements. The commonest sorts of breakdown include electrical circuit failures, so that an additional requirement is that all parts of the calorimeter vessel are easily removable for testing and repairing. Another essential requirement of the design of any calorimeter vessel is that the heat liberated in the heating element must be distributed quickly through the specimen and the

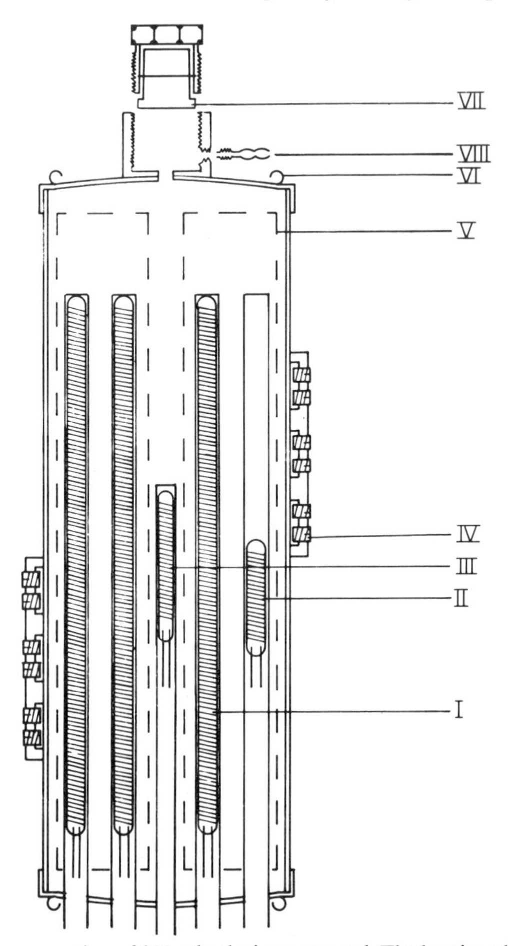

FIG. 1. Vertical cross-section of 300 ml calorimeter vessel. The heating elements (I and II) are isolated from the liquid by tubular casings. The smaller heating elements (II) are used as auxiliary heaters in the twin-vessel calorimeter because of their lower thermal inertia. III is the glass encapsulated platinum thermometer in tubular casing, IV is the screw clamps for thermocouple junctions, and V indicates silver vanes (see Fig. 2). VI indicates little hooks for suspension wires. The valves (VII) and nozzles (VIII) are discussed in Section III below. The entire calorimeter vessel is gold-plated.

vessel proper. Two distinct considerations are important in this respect. Firstly, the temperature difference between the actual resistance element of the heater and the calorimeter vessel proper must be minimized. Secondly the temperature gradients in the liquid sample must be kept small.

In practice a reasonable way of reducing the temperature gradients inside a metal calorimeter vessel is to extend the internal surface in contact with the liquid, so that no part of the liquid is more than a few millimeters away from a metal surface providing a conduction path to the heater. This is achieved in a calorimeter vessel of relatively large diameter by inserting a set of metal vanes whose extremities are in good thermal contact with the vessel proper (see also Section II). Figure 1 shows the construction of a vessel made for the recording adiabatic calorimeters described in Sections II and III below. The vessel is made of silver and has an internal capacity about 300 ml, suitable for liquid samples of volume between 250 and about 275 ml. The heating elements (I and II in Fig. 1) are encased in tubular casings in good thermal contact with silver vanes distributed over the entire volume of the liquid sample. A platinum resistance thermometer (glass encapsulated, 3 cm long by 3 mm wide, 100 ohm) is inserted in a central well in the lower part of the vessel. The thermocouple junctions (IV in Fig. 1) are screw-clamped onto the outer surface of the vessel. The components of the calorimeters are connected by solder, and the upper lid is fitted with a valve (VII in Fig. 1) which seals against a polytetrafluoroethane gasket. The heating elements comprise "constantan" filaments, insulated by spun glass sleeving, wound onto silver rods. The resistance wire terminates well above the ends of the silver rods, so that the temperature difference which builds up between the resistance wire and the rod and casing does not result in any uncompensated heat loss to the exterior, while the low heat capacity of the resistance wire ensures that it comes rapidly to thermal equilibrium with the casing and the silver rod once heating ceases. The arrangement of the heaters and silver vanes in the calorimeter vessel is illustrated by the plane section shown in Fig. 2, which is self-explanatory.

At temperatures between room temperature and 100°C a liquid may be used to transport heat to and from the shield structure as in the calorimeter assembly shown in Fig. 5, Chapter VI, and this is the principle of the "wet" shield calorimeter in which the jacket is a vessel forming the outer boundary of the insulating space surrounding the calorimeter vessel. The alternative method—the "dry" shield, widely used at high temperatures—has also been used in work near room temperature. The dry shield may either be a separate structure inside the insulating jacket, or it may form part of the jacket vessel as in the adiabatic calorimeters explained in the following sections. The dry shield is heated by attached electrical windings in which the current is controlled by an automatic regulator. Again, it is necessary, that the temperature is uniform over the whole internal surface of the shield.

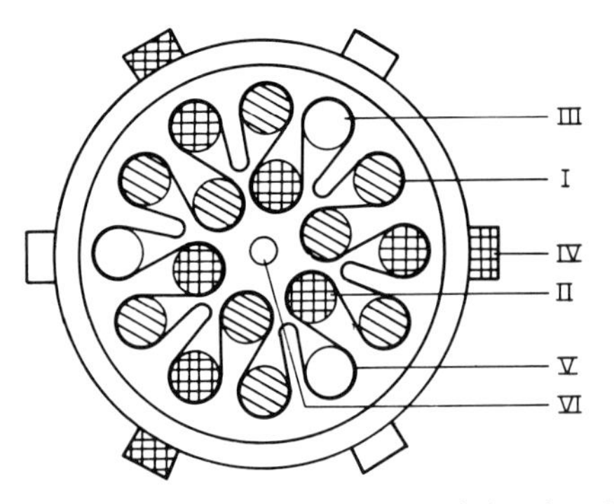

FIG. 2. Schematic plane section of a calorimeter vessel showing the arrangement of the heating elements (I and II), silver vanes (V) and temperature sensing elements (IV and VI). I indicates the main heating elements, II indicates auxiliary heating elements, III indicates tubular casings for additional heating elements, IV indicates clamps for thermocouple junctions and VI indicates central tubular casing for platinum thermometer.

II. The Adiabatic Calorimeter for Heat Capacity Measurements on Liquids and Solutions (continuously heated adiabatic single-vessel calorimeter with platinum wire resistance thermometer operating an automatic timer and print-out unit)

Space does not permit an extensive review of the many calorimeters reported in the literature, so the reader is referred to the bibliography. It will be useful, however, to describe briefly three calorimeters which have been used for measurements of the heat capacities of solutions in the temperature region of the helix-random coil transition.

As an example for the direct method, an adiabatic single-vessel calorimeter assembly is shown schematically in Fig. 3. The calorimeter vessel which has been described in Section I, (Fig. 1) is supported by the massive copper lid of the dry shield. The lid of the adiabatic shield is fixed to the mounting bar by two tubular supports which are made of argentan (german silver) to minimize the heat conduction between the shield and the mounting bar. The vessel is suspended from the underside of the shield lid by thin steel wires. The stepped inner ring of the shield lid is threaded to screw into the top of the shield body. This arrangement facilitates the assembling and dismantling of the apparatus while ensuring good thermal contact between the lid and the body. The body of the shield is made from a cylindrical copper tube of 1 cm wall thickness. This shield, controlled by a two-valued response regulator, has been shown to give a precision better than 0.1 % on the heat capacity results throughout the quoted temperature range (Ackermann and Rüterjans, 1964). The rate of heat leak from the calorimeter vessel into the steel wires is smaller than that through the electrical leads. The shield heating element is a Teflon-insulated resistance wire wound in grooves on the outer surface of the heavy copper tube. The heater windings are covered by a closely fitting sheath of 1 mm thick copper sheet. Additional heating tapes (not shown in Fig. 3) are attached to the base and to the lid of the shield. The three heating elements are normally connected in series, and so the resistances of the lid and bottom heaters are chosen so as to

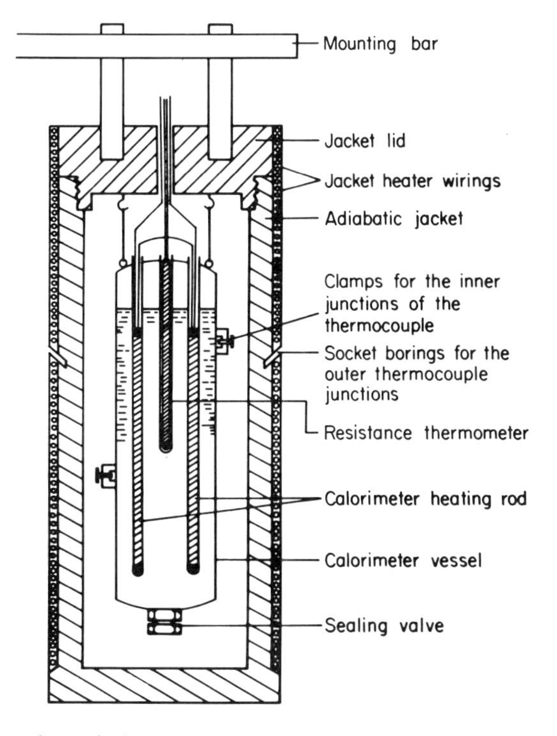

FIG. 3. Schematic vertical cross-section of the single-vessel calorimeter assembly.

partially compensate the extra heat loss through the supporting tubes. Since the copper body of the shield may be assumed to be at effectively uniform temperature, the shield junctions of the control thermocouple are located in the mica-insulated socket borings shown in Fig. 3. A central hole through the lid provides passage for the leads to the calorimeter vessel heater and thermometer and to the calorimeter vessel thermocouple junctions. The entire shield assembly is surrounded by glass wool insulation in an outer container (not shown in Fig. 3). Despite its large ratio of total heat capacity to heated surface area the thermal lag of the shield does not exceed about 15 seconds. The degree of temperature uniformity over the inner surface of the shield is adequate for the precision quoted above provided that the shield heating rate does not exceed 0.01 deg sec^{-1}. The two-valued response regulator is similar to that for the twin-vessel shield (see Section III, below). A detailed description is given elsewhere (Wittig and Schilling, 1961).

To obtain the heating rate of the calorimeter vessel it is necessary to measure accurately the time interval between successive temperature measurements. If a platinum resistance thermometer is used, it may be connected to a d.c. bridge in which the measuring current is passed continuously. The timing device is then included in the bridge circuit. The balancing resistance is increased by equal small increments corresponding to temperature increments of the order of 1 deg, at roughly equal time intervals, depending on the heating rate; this can be done either manually or automatically, see Section III, for example. The timer then records the times at which the thermometer bridge is in balance with successive resistance increments. The main heating rate is then observed directly as a function of temperature. In practice it is convenient to use an automatic timer and print-out unit, and a typical diagram of circuit connections is shown in Fig. 4.

The balancing arm of the Wheatstone bridge thermometer circuit is a decade-dial resistance box, operated by a telephone-type step relay. If the initial resistance of the decade arm is greater than that of the resistance thermometer, then as the temperature rises the light beam from the mirror galvanometer in the Wheatstone bridge circuit moves towards its zero position, where it illuminates a photocell. This triggers a series of operations whose sequence is controlled by a servomotor-driven camshaft operating mechanical switches. First the count from a double electronic timer is transferred to a tape print-out unit, and the camshaft commences to turn. Next the photocell is locked to prevent retriggering on the return sweep of the galvanometer image, the bridge voltage is transferred to a dummy resistance, and the step relay is operated, increasing the bridge balancing resistance by one unit. The bridge power supply is then reconnected (swinging the galvanometer image to maximum deflection) and the trigger photocell is resensitized, ready for the next balance point of the thermometer bridge. The pulses which are fed continuously into the counter unit are supplied by a quartz crystal oscillator, so that the precision of the time measurements is potentially better than that of all the other measured properties.

Probably the most important source of error in this type of timing unit is due to drifting of the galvanometer zero. The image position corresponding to zero coil current in the galvanometer is always subject to changes due to variation in ambient temperature and humidity, and with most galvanometers a continued large deflection (such as occurs immediately after the bridge resistance is increased) shifts the zero towards the deflection. An improved adiabatic calorimeter assembly in which the platinum wire resistance thermometer is replaced by a digital quartz thermometer is described in Section IV, below.

The print-out unit is designed to work as an adding machine, and prints three rows of digits for each equilibrium passage of the bridge circuit. The first row gives the programmed resistance value (i.e., the resistance of the platinum thermometer when the bridge passes through its balance point). The second row gives the number of time-pulses counted between successive balance points, and the third row gives the total time, equal to the sum of all the heating periods from the commencement of the experiment. Thus the printed tape represents a numerically recorded time vs. resistance function from which the corresponding temperature-enthalpy variation of the calorimeter vessel and its contents can be calculated. The thermometer resistance value printed in the

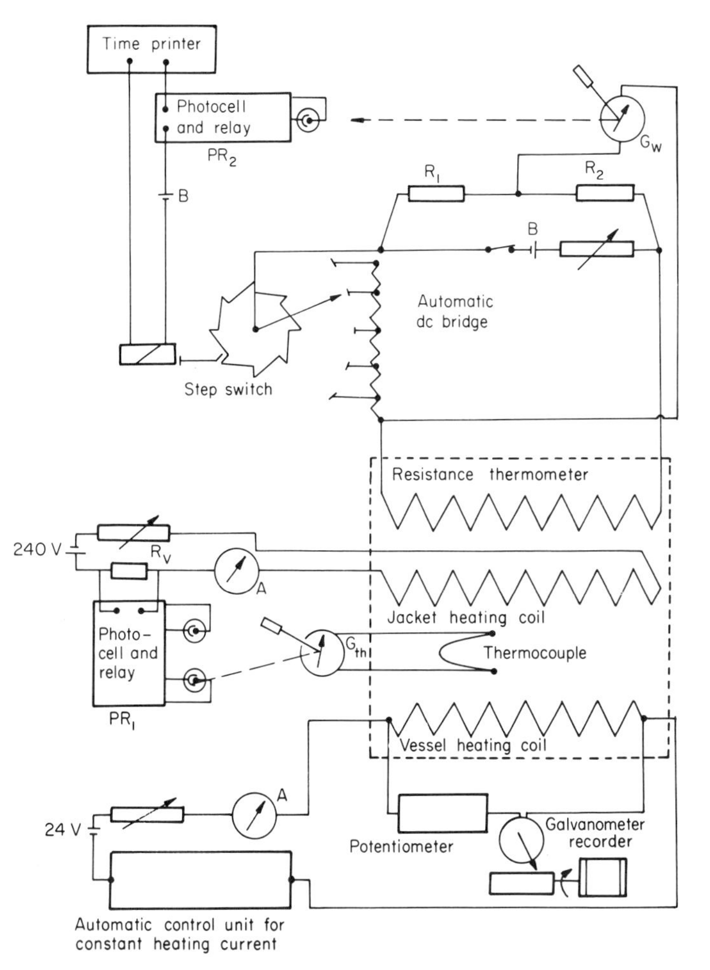

FIG. 4. Block diagram of circuit connections for the recording adiabatic calorimeter operating an automatic timer and print-out unit.

first row gives the temperature of the system. The second row value is immediately related to the heat capacity of the system, while the corresponding enthalpy increase can be obtained from the third row.

If the measured time interval between two successive balance points of the thermometer bridge is denoted by Δt and the corresponding constant increment in the thermometer resistance is denoted by ΔR_T, then the heat capacity of the system (calorimeter vessel and contents) is given by

$$C = I \cdot E \left(\frac{dR_T}{dT}\right) \cdot \left(\frac{\Delta T}{\Delta R_T}\right) \tag{3}$$

where (dR_T/dT) is the temperature coefficient of the thermometer resistance at that temperature, I is the current through the heater and E is the recorded potential drop across the heater. Equation 3 shows that for a particular heat capacity the time interval, Δt, varies inversely with the heating power $I \cdot E$. The decreasing of $I \cdot E$ increases the accuracy of the timing but decreases the accuracy with which $I \cdot E$ can be measured and at the same time reduces the rate at which experiments can be completed. Increasing $I \cdot E$, on the other hand, increases the difficulty of adiabatic control and may cause errors due to variations in the temperature distributions inside the calorimeter vessel as the temperature increases. For heat capacity measurements on aqueous solutions in calorimeter vessels of the type shown in Fig. 1, a heating power about 1 watt has proven satisfactory. The complexity of the numerical calculations to be carried out on the primary data depends upon the nature of the problem being studied. For example, if high precision is unnecessary, Eq. 3 may be used to evaluate the heat capacity (vessel plus contents) over a single temperature increment directly from the measured value of Δt. If the total heat capacity changes only very slowly with increasing temperature, it may suffice to use the arithmetic mean of a series of measured Δt values (this can be done automatically by the adding machine of the print-out unit described above). In general, however, a series of values of $\Delta t/\Delta R_T$ will be fitted to a polynomial in T, using a "least squares", or other equivalent procedure. The experimental procedures and methods of calculation just described depend on the nature of the investigation rather than upon the use of the various automatic devices. In particular, the experimental procedure to be followed with a manually controlled calorimeter is implicit in the foregoing discussion, while the extent to which the data are processed automatically will be determined more by economic factors than by the methods to be employed.

III. The Recording Adiabatic Twin Calorimeter (continuously heated adiabatic twin calorimeter for direct determinations of differences between heat capacities of solutions and a reference heat capacity of an equivalent amount of pure solvent)

As an example for the comparison method, a recording adiabatic twin-vessel calorimeter is described in this section. A schematic diagram of the calorimeter assembly is shown in Fig. 5. Since the calorimeter includes multiple adiabatic shields with a relatively large number of thermocouple junctions fixed to the calorimeter vessel, removing the calorimeter vessel for filling would disturb the whole arrangement and the necessary reproducibility of experimental conditions would be very difficult to achieve. Consequently special arrangements must be made for filling the twin vessels. Several arrangements involving permanently-connected filling pipes have been reported. The main disadvantage of such arrangements is that the filling pipes inevitably increase the quantity of heat exchanged between the calorimeter vessel and its surroundings. Another disadvantage is that the masses of the specimen and reference liquids must be calculated from indirect weighings. In order to avoid these difficulties, the two calorimeter vessels are mounted in a metal frame, together with the two sets of thermocouples. The frame is designed to be a good sliding fit into the metal cylinder forming the inner part of the adiabatic shield. The top ring of the frame, carrying all the electrical terminals in insulated clamps, can be screwed onto the underside of the lid of the adiabatic shield which is, in turn fixed to the mounting-plate of the apparatus. The assembling of the apparatus is completed by drawing the cylindrical part of the adiabatic shield up over the frame and fixing it to the shield lid. Thus, by first removing the shield cylinder, the entire inner part of the calorimeter may be removed for filling, and then replaced, without disturbing either of the thermocouple circuits. The filling connection is made through flexible Teflon hoses, each of which has a threaded metal nozzle which screws into the top of the filling nipple. The other end of the Teflon hose terminates in a three-way valve enabling the calorimeter vessel to be connected either to a vacuum pump or to a calibrated filling reservoir.

To fill the twin calorimeter vessels, the frame is first weighed with the calorimeter vessels empty and dry. A measured quantity of the solution to be studied is then transferred as liquid from the calibrated reservoir to the previously evacuated first calorimeter vessel. The vessel is then sealed and the frame is weighed again. The amount of pure solvent to be put into the second calorimeter vessel is then calculated from the weight of solution in the first vessel and known weight composition of the solution. Approximately the correct amount of solvent is then transferred from the second calibrated reservoir to the previously evacuated second calorimeter vessel. After the second vessel is sealed the frame is again weighed. If the two charges are not exactly matched, the amount of solvent to be added to the second vessel is calculated. This vessel is then reconnected to the filling apparatus and the additional solvent put in. Again the vessel is sealed and the frame is weighed once more.

This procedure is repeated until the previously calculated mass of the filled

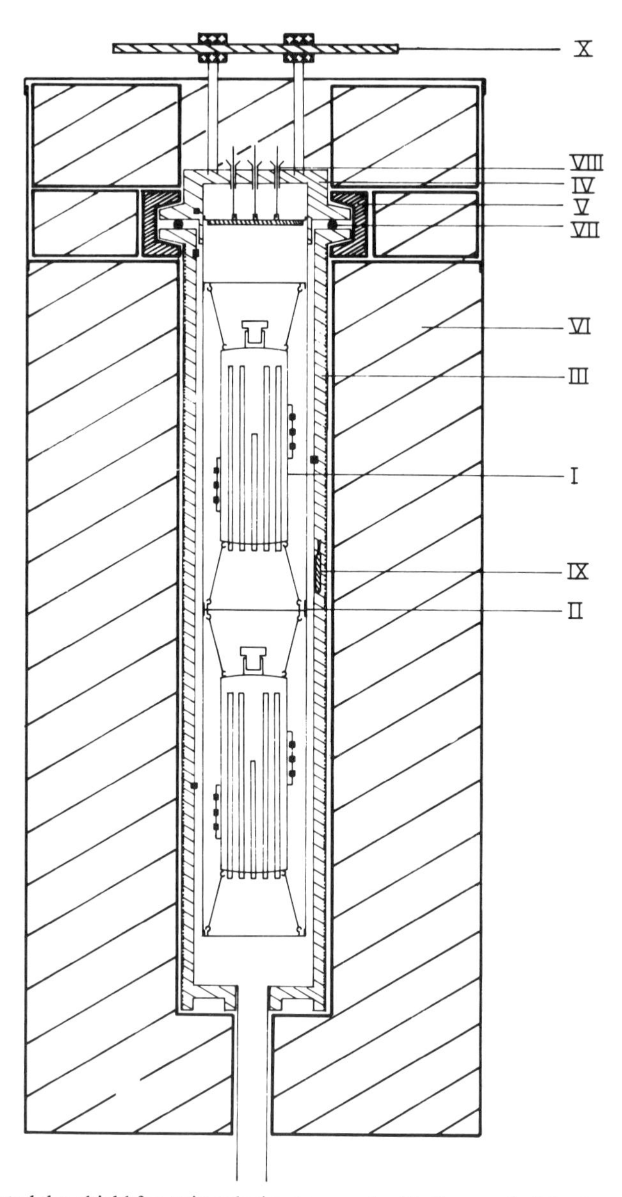

FIG. 5. Evacuated dry shield for twin calorimeter vessels. I is the vessel of the type shown in Fig. 1, II is the calorimeter vessel supporting frame, III is the shield body, IV is the shield lid, V is the lid securing clamps, VI is the Teflon wool insulation in an outer container, VII is the Teflon gasket, VIII is the electrical lead exit seals, IX is the auxiliary resistance thermometer for calibration measurements and X is the mounting plate.

system is confirmed by the final weighing. If too much solvent has been added, the filling nipple is unscrewed completely and removed and solvent is sucked out through a Teflon siphon. If the filled system is assembled, the interior of the shield vessel is evacuated. The general design of the shield body is similar to the single-vessel shield described above, so that the main differences arise from the provision of vacuum-tight seals and the need to accommodate two calorimeter vessels. The shield is fixed to the shield lid by clamping their flanges together, vacuum sealing being effected by the Teflon gasket (VII in Fig. 5). Thermal contact between the lid and the body is improved by the clamps (V in Fig. 5), of which the inner part is 6 mm thick copper contained in a steel former to give the requisite strength. The electrical leads pass to the outside of the shield lid through metal-to-glass seals. The shield vessel is connected to the vacuum line by the argentan pipe at the bottom. The stationary pressure in the insulating space inside the shield vessel does not exceed 10^{-4} torr at 400°K. The inner frame which supports the calorimeter vessels fits into grooves in the body of the shield. The arrangement of the shield heater is essentially the same as that in the single-vessel shield (see Section II).

The signal source of the shield regulator is the multijunction thermocouple mounted between the outer surface of the upper calorimeter vessel and the shield frame. The main parts of the circuit are shown in the block diagram of circuit connections (Fig. 6). The control thermocouple is connected to a mirror galvanometer, the reflected image from which actuates the photocell relay. The photocell relay is arranged as a quasi-linear response regulator; it is essentially a two-valued response regulator incorporating an electronic circuit which effects a Gouy-type modulation (Gouy, 1897). The "high" and "low" heating powers are defined by the series resistances, R in Fig. 6, a constant dc voltage being applied to the whole heater circuit. The positive response of the regulator (corresponding to a negative signal, i.e., shield colder than upper calorimeter vessel) is to short-circuit one of the series resistances thereby increasing the shield heating power. A second quasi-linear response regulator (not shown in Fig. 6) is used to control the lid of the adiabatic shield, because of the limited thermal contact between the lid and the body. This is not necessary in the single-vessel shield shown in Fig. 3 since the thermal contact is much better in this case.

The method of incorporating the Gouy modulation in the relay circuit in this apparatus is especially suited to calorimeter regulators by virtue of the low frequency of the modulating signal, so it is worth while to describe it in detail. The circuit is shown in Fig. 7; it was first developed by Wittig and Schilling (1961). Instead of the two photocells being connected in a simple bridge whose out-of-balance is amplified to operate the electromagnetic relay, each is connected to the grid of the controlling pentode of a sweep circuit. The two sweep circuits are connected to a polarized double-coil relay. In Fig. 7 the

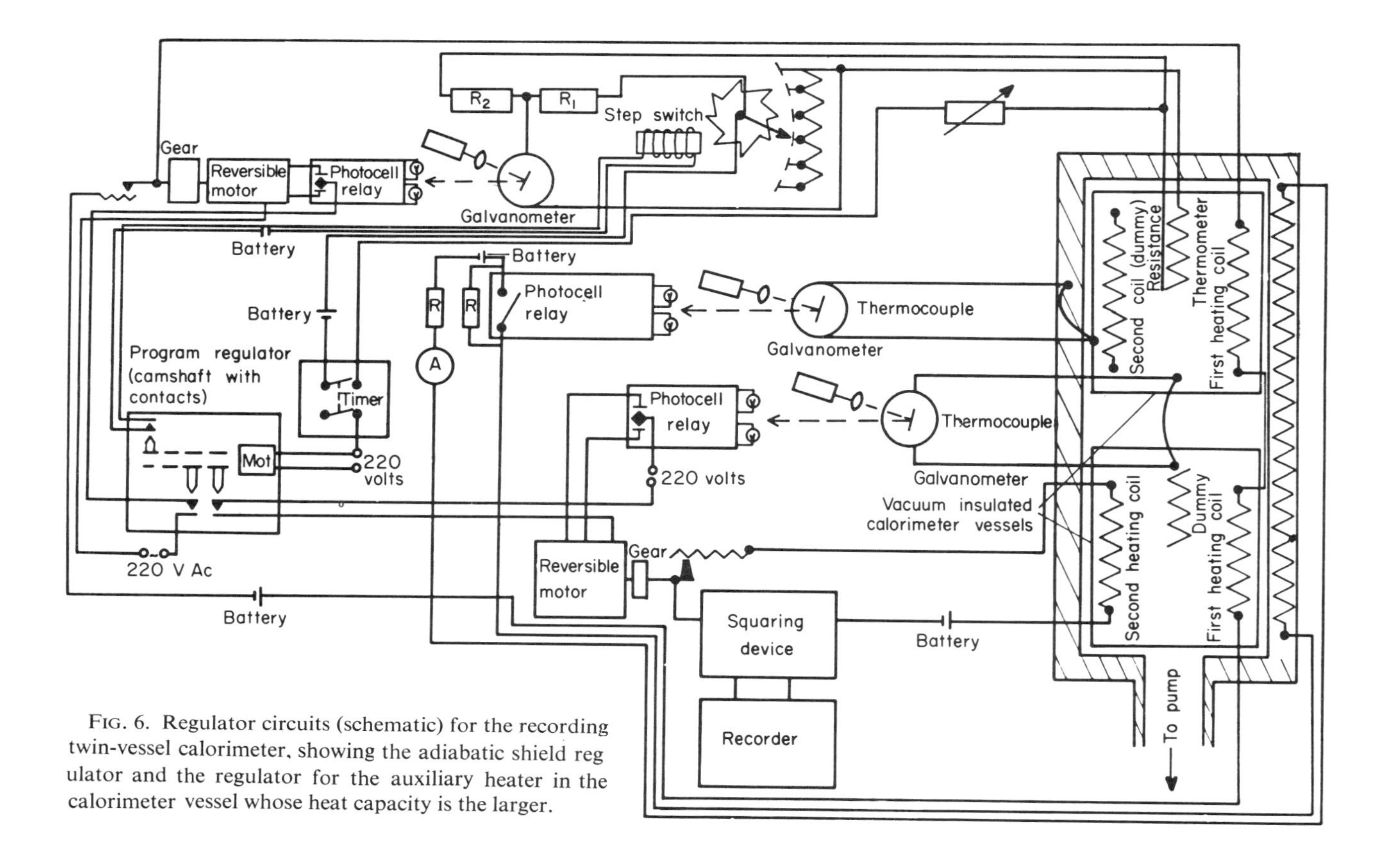

FIG. 6. Regulator circuits (schematic) for the recording twin-vessel calorimeter, showing the adiabatic shield regulator and the regulator for the auxiliary heater in the calorimeter vessel whose heat capacity is the larger.

photocells are denoted by P_1 and P_2. When the two-way contact S of the polarized double-coil relay A is switched to position 1, as shown in Fig. 7, the capacitor C_1 is shunted by the 40 ohm discharging resistor. The charging rate of the capacitor C_2 is then controlled by the suppressor-grid current in the pentode R_2, and the cathode current, i.e., by the intensity of the light on the photocell P_2. When the potential across the capacitor C_2 attains the ignition potential of the glow tube G_2, the capacitor is discharged and the discharge

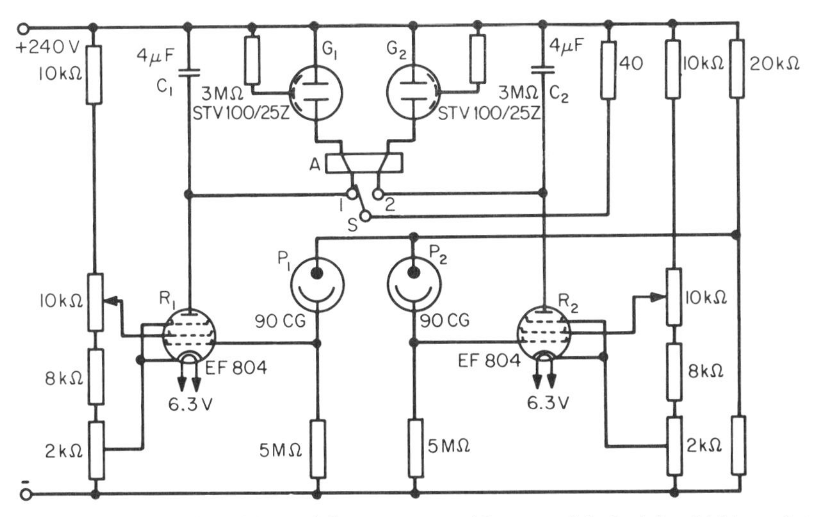

FIG. 7. Photocell relay with quasi-linear response (Gouy modulation) for shield regulator of recording adiabatic calorimeters. A is the polarized double-coil relay, P indicates 90 CG photocells, R indicates EF 804 pentodes, G indicates STV 100/25 discharging tubes, and C indicates matched capacitors. The electronic tubes are all German; replacement by equivalent British or American tubes necessitates the adjustment of other components.

current switches the relay contact S over to position 2. The capacitor C_2 is now shunted by the 40 ohm resistor. C_1 commences to charge at a rate depending upon the illumination of the photocell P_1. This continues until the potential across C_1 attains the ignition potential of G_1, when the contact S is switched back to position 1, and the whole sequence is repeated. The function of the photocell relay contacts shown in Fig. 6 is effected by a second two-way contact linked to S (not shown in Fig. 7). In position 1, this second contact short-circuits the series resistance in the shield heater circuit, while in position 2 the heater series resistance remains in circuit.

The zero position of the galvanometer image is adjusted to give equal illumination of the photocells P_1 and P_2, and the 10 kilo-ohm potentiometers are then set so as to make the ratio of the periods of the two sweep circuits effectively unity. Under these conditions, the mean heating rate of the shield at zero signal is midway between the "high" and "low" heating rates. The effect of a positive signal (shield hotter than calorimeter vessel) is to increase the illumination on P_2 and to decrease that on P_1 by about the same amount. This decreases the charging period of C_2 (during which the "high" heating power is applied to the shield) and increases the charging periods of C_1 (during which the "low" heating power is applied). Thus the total effect of the signal is to reduce the main heating rate of the shield by an amount roughly proportional to the magnitude of the signal. In other words the circuit behaves as a linear response regulator for galvanometer deflections less than half the distance between the centers of the photocells. The actual performance of this regulator of course depends on how accurately the "high" and "low" heating rates of the shield are initially adjusted relative to the heating rate of the calorimeter vessel, and on how much either or both the calorimeter vessel heating rate and the mean shield heating rate changes with increasing temperature. It is comparatively simple to ensure that the mean signal never exceeds 0.001 deg, even with a control thermocouple of only six or eight junctions. As indicated in Fig. 5, the entire shield assembly is surrounded by Teflon wool insulation in an outer container.

The principle of the comparison method has already been explained in Section I. It is obvious from Eq. 2 that, if the current intensity, I_a, in the auxiliary heater is regulated so as to keep the heating rate, dT/dt of the specimen vessel equal to that of the reference vessel, then provided that the resistance, R_a, of the auxiliary heater and dT/dt are known and remain constant, I_a^2 gives a measure of C_s–C_r. It follows that a continuous record of I_a^2 over an extended heating period gives the value of the difference between the heat capacity of the solution and the heat capacity of the pure solvent at any point in the temperature range covered. This procedure has the great advantage that the effects of small fluctuations of temperature gradient in the two vessels can be neglected. Since the total change of the resistance of a properly selected constantan element between 300°K and 400°K is only a few tenths of one percent, the resistance of the heaters may be regarded as being independent of temperature over this sort of temperature range. Thus the main factor limiting the reliability of the value of C_s–C_r determined from a record of I_a^2 is the extent to which dT/dt changes as the temperature increases. For a given current intensity in the main heaters of the two vessels (connected in series), dT/dt depends on the heat capacity of the calorimeter vessel filled with the liquid having the lower total heat capacity. Up to about 400°K the temperature coefficient of the specific heat of water and of dilute aqueous solutions is very small. The total change in the specific heat of water between 300°K and 400°K is less than 2%. Thus the influence of the temperature-variation of the specific heat of water on the results obtained from the simpler form of the recording twin calorimeter, using a constant current in the main heaters, can be neglected.

In the case of nonaqueous solvents, however, where the temperature dependence of the specific heat is greater than in the case of water, it is necessary to keep the heating rate constant, or to monitor it continuously throughout the experiment. Since this requirement is almost peculiar to this type of investigation we describe here the principle of an effective temperature-program regulating device. The electronic circuit of the device is illustrated in the upper part of Fig. 6. The temperature of the apparatus is controlled by a resistance thermometer (a platinum thermometer of 100 ohm nominal resistance), and the variable reference rheostat of the circuit of the Wheatstone bridge can be operated by means of a step-switch relay such as that commonly used in telephone dialing systems. The current passing through the coil of the mirror galvanometer depends upon the deviation of the instantaneous resistance of the platinum thermometer from the reference value corresponding to the setting of the dial resistance box. When dc voltage from a battery is applied to the bridge circuit, an out-of-balance current, corresponding to a deviation of the instantaneous temperature of the platinum thermometer from the corresponding programmed reference value, will result in a deflection of the galvanometer light beam from its zero position between the photocells. The photocells are connected to an amplifier and relay circuit which actuates a reversible servo motor; the sense of rotation of the servo motor is determined by the sign of the galvanometer deflection, but its speed is fixed. The servo motor operates a 10-turn precision-wound helical potentiometer which controls the current in the main heater circuit, reducing the power input to the main heaters whenever the instantaneous temperature indicated by the platinum thermometer is higher than the programmed reference value, and vice versa. The whole process operates intermittently, the dc supply being connected to the thermometer bridge for a short interval every 30 seconds, say, while the servo motor relay circuit includes a cam switch which cuts off the ac supply after a fixed interval, so that the servo motor rotates through a fixed small angle at each cycle of operation, and correspondingly the power input to the main calorimeter heaters changes by one small (positive or negative) step every 30 seconds. On completion of each cycle of operation the resistance indicated by the dial resistance box is automatically increased by one step (0.1 ohm, say). Within a few minutes of the regulator commencing to operate the heating rate of the calorimeter vessels attains a constant value, which is determined by the pre-selected operating cycle period of the temperature-program regulator and the step value of the dial resistor box in the thermometer bridge circuit. Since this resistance box consists of decadic units (not shown in Fig. 6) connected in series, continuous operation is possible, so that the pre-selected constant heating rate can be maintained throughout the entire temperature range up to 400°K.

The temperature-program regulator controls the heating rate in the upper calorimeter vessel, which is always filled with the liquid having the smaller total heat capacity. Thus the platinum resistance thermometer in the upper calorimeter vessel is connected into the bridge circuit of the temperature-program regulator; the corresponding platinum thermometer in the lower vessel is a dummy, included to ensure that the total heat capacities of the twin vessels are as nearly equal as possible. The auxiliary heating is then always applied to the lower calorimeter vessel, while the second heating coil (or assembly of heater rods) in the upper calorimeter vessel is also a dummy and is not connected to the power circuit. The auxiliary heater regulator keeps the current intensity in the auxiliary heater in the lower calorimeter vessel at the value which is necessary to maintain the twin-vessel system in thermal balance. As may be seen from Fig. 6, this regulator is essentially similar to the temperature-program regulator, except that the sensing element of the auxiliary heater regulator is a multijunction thermocouple mounted between the two calorimeter vessels. This is connected to the galvanometer through a circuit arranged to give maximum sensitivity. The light beam from this galvanometer actuates the photocell relay which in turn switches a reversible servo motor resistor unit similar to that in the automatic temperature-program regulator. This unit is also operated intermittently by the timing device which controls the operation of the temperature-program regulator, so that the auxiliary heating power in the second heater element in the lower vessel is also varied in definite steps. In other words, each heating rate check on the upper vessel is followed by a check on the temperature-balance between the two vessels, and the current intensity I_a in the auxiliary heater system in the lower vessel is automatically adjusted to minimize the temperature difference between the two vessels. To be effective, the thermocouple-galvanometer unit must be capable of indicating temperature differences of the order of 10^{-5} deg.

Finally, the square of the auxiliary heating current, I_a^2, is recorded so that a plot of the excess heat capacity of the solution is obtained. If an X-Y recorder (not shown in Fig. 6) is available, the X-scale can be operated by an additional thermocouple measuring the actual temperature of the calorimetric assembly, so that the recorder gives directly a plot of C_s–C_r against temperature.

The final operating equation for this apparatus then takes the form:

$$C_s - C_r = I_a^2 \cdot R_a \, dR_T/dT \cdot \Delta t/\Delta R_T, \tag{4}$$

where dR_T/dT is the temperature coefficient of the platinum thermometer and $\Delta t/\Delta R_T$ is the time programmed for a 1 ohm resistance increment in the platinum resistance thermometer. The temperature derivative of R_T does of course vary with temperature, so that the recorded value of C_s–C_r has to be corrected to allow for this whenever the maximum accuracy is required.

IV. An Improved Adiabatic Calorimeter (application of the digital quartz thermometer in an adiabatic single-vessel calorimeter)

As stated in Section I, ease of operation is an important requirement for successful calorimetry. Since the image position corresponding to zero coil current in a mirror galvanometer is always subject to changes due to variation in ambient temperature and humidity, the precision of the timing procedure described in Section II will depend upon the reproducibility of this zero mirror image position. Recently, a high-precision digital quartz thermometer has been developed*. The temperature readings obtained from this instrument are independent of variations in ambient humidity. The precision of the measured

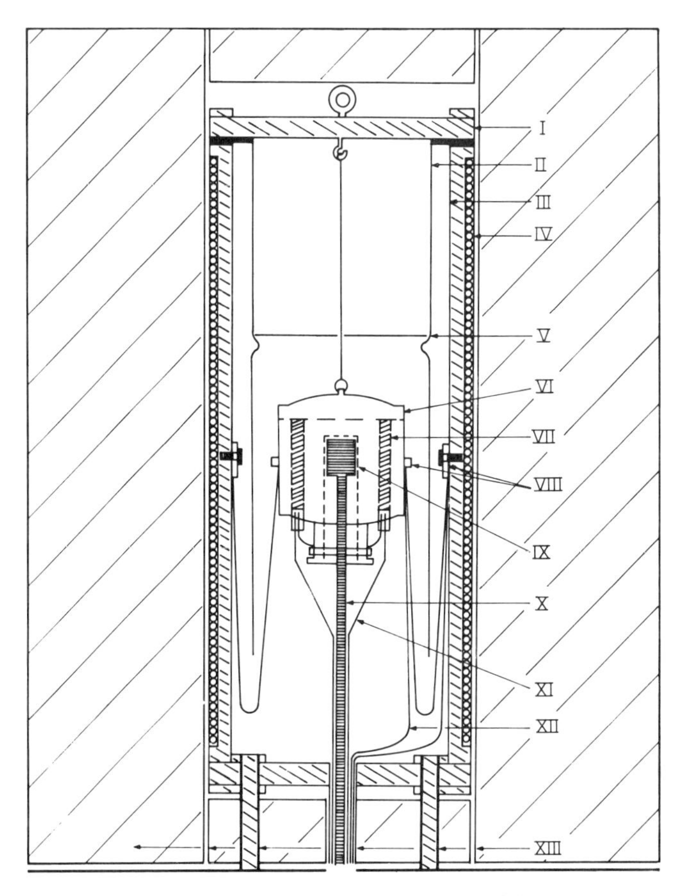

FIG. 8. Schematic vertical cross-section of an improved adiabatic calorimeter. I is the shield lid, II is the radiation shield, III is the shield body, IV is the shield heater wirings. V is the lid of the radiation shield, keeping the suspending wire in its central position, VI is the calorimeter vessel, VII indicates heater elements, VIII indicates thermocouple junctions, IX is the temperature sensor probe of the digital quartz thermometer, X is the coaxial cable, XI indicates leads to the heating elements, XII indicates leads to the control thermocouple junctions, and XIII is the Teflon wool insulation in outer containers.

* The digital quartz thermometer is manufactured by the DYMEC division of Hewlett-Packard Co., Palo Alto, Calif.

temperature of the probe is not affected by parasitic electromotive forces as in galvanometer circuits. In this Section we shall describe an adiabatic single-vessel calorimeter which incorporates the quartz sensor of the digital quartz thermometer instead of the platinum wire resistance thermometer used in the apparatus for continuous recording of the total heat capacity.

The principle is essentially the same as that of the adiabatic calorimeter described in Section II. A schematic vertical cross section of the apparatus is shown in Fig. 8. The adiabatic jacket is made of a massive copper tube as in the calorimeter shown in Fig. 3, and the entire calorimetric assembly is surrounded by Teflon wool insulation in an outer container. In order to reduce the amount of sample required for the measurements on solutions of poly-nucleotides, the volume of the calorimeter vessel proper is reduced to 100 ml, so the vessel is capable of about 80 ml of liquid specimen. The electric heaters are arranged in thermal contact with silver vanes as in the calorimeter vessel shown in Figs. 1 and 2. The temperature sensing element of the digital quartz thermometer is incorporated in a central well in the inner part of the sealing valve (IX in Fig. 8). A 12-foot length of flexible coaxial cable is permanently attached to the sensor probe. The cable is sealed to the probe body, and is terminated at the other end with a watertight connector mating with the associated sensor oscillator. For a description of this instrument and its operational principle the reader is referred to the instruction manuals (which are distributed by the manufacturer of the dy 2801 A thermometer). The arrangement of the jacket heater wirings and of the control thermocouple is similar to that explained in Section II. A central hole through the bottom of the jacket vessel provides passage for the coaxial cable and for the leads to the calorimeter vessel heater and to the control thermocouple. The calorimeter vessel is suspended from the underside of the shield lid by thin steel wire. An additional radiation shield (silver) is inserted in the insulating space between the outer surface of the calorimeter vessel and the inner surface of the adiabatic jacket. The axial location of the steel wire is adjusted by means of thin-walled lid of the silver-shield (V in Fig. 8). A block diagram of the circuit connections is presented in Fig. 9, which is self-explanatory.

The experimental procedure is the same as for the adiabatic calorimeter described in Section II, except that the temperature increment ΔT for a given time interval Δt is obtained from the quartz thermometer readings. The "reset" of the quartz thermometer read-out unit is triggered by an electronic timer of the type described in Section II. A card or tape punching unit attached to the BCD output may be used to transfer the primary temperature data to the digital computer. Various arrangements of this kind have been described in recent years (Sturtevant, 1937; see also Osborne *et al.*, 1960). The quotient $\Delta t/\Delta T$ is then calculated, and the total heat capacity of the calorimeter (vessel and content) is obtained from Eq. (1) in which $\dot{Q}$ is calculated from the

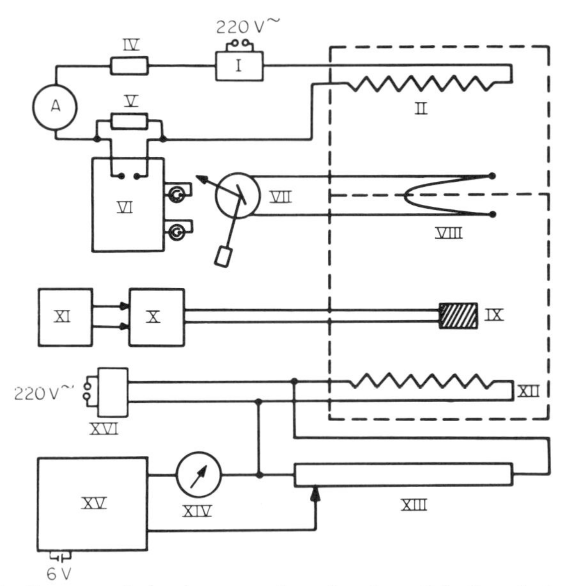

FIG. 9. Block diagram of circuit connections for the adiabatic calorimeter shown in Fig. 8. I and XVI are the controlled constant voltage supply, V and IV are the resistors, II is the jacket heating element, VI is the photocell relay, VII is the galvanometer, VIII is the jacket control thermocouple, XI is the electronic timer, X is the sensor oscillator of the digital quartz thermometer, IX is the temperature sensor probe of the digital quartz thermometer, XII is the calorimeter vessel heating element, XIII is the resistor (decadic voltage divider), XIV is the galvanometer and, XV is the voltage control unit.

heating power in the heater rods of the calorimeter vessel. In other words, the operating equation in this case, using the same nomenclature as above, is

$$C = I^2 R \cdot \Delta t / \Delta T \quad (5)$$

This calorimetric apparatus has been shown to give a precision better than 0.05% on the heat capacity results throughout the quoted temperature range (Ackermann *et al.*, 1967).

Recently, high-precision calorimeters for heat capacity measurements on small amounts of liquid solutions have been described (Privalov and Monaselidze, 1964; Sturtevant, 1966). The reader is referred to the original papers for a detailed description of these recording adiabatic twin calorimeters which are similar to that described in Section III.

REFERENCES

Ackermann, T., and Rüterjans, H. (1964). *Ber. Bunsenges. Physik. Chem.* **68**, 850.
Ackermann, T., Klump, H., and Neumann, E. (1967). Unpublished observations.
Gouy, M. (1897). *J. Phys.*, *3me Ser.* **6**, 479.
Joule, J. P. (1845). *Mem. Proc. Manchester Lit. Phil. Soc.* **2**, 559.

Osborne, D. W., Flotow, H., and Otto, K. (1960). Personal communication, Chem. Div., Argonne Natl. Lab.
Privalov, P. L., and Monaselidze, D. R. (1964). *Dokl. Akad. Nauk SSSR* **156**, 951.
Richards, T. W., and Gucker, F. T. (1925). *J. Am. Chem. Soc.* **47**, 1876.
Sturtevant, J. M. (1937). *J. Am. Chem. Soc.* **59**, 1528.
Sturtevant, J. M. (1966). Personal communication.
Wittig, F. E., and Schilling, W. (1961). *Chem. Ingr.-Tech.* **33**, 554.

CHAPTER XIV

The Conduction-Type Microcalorimeter

WILLIAM J. EVANS

I. Introduction

Biochemical calorimetry has in recent years become a notably important phase of reaction calorimetry. This trend would seem to indicate that it will assume an increasingly important role in the biochemical and allied sciences. That the trend is only recent may appear somewhat puzzling in view of the fact that practically all reactions proceed with a measurable change in heat content. Thus aside from the general applicability of the calorimetric method, it is made the more puzzling when one considers that the quantity i.e., change in enthalpy directly derived from calorimetric measurement is usually a prerequisite to the estimation of free energy.

The fact that calorimetry has yet to become a routine tool, such as ultra-centrifugation and electrophoresis, in the biochemical laboratory can undoubtedly be related to the complexities associated with the measurement of heat. Heat, unlike quantity of matter, for example, cannot be confined. To the present, no heat insulator corresponding in perfection to electrical insulators has been developed. This and other difficulties has necessarily led to considerable elaboration of calorimetric equipment. There is a further particular difficulty with which the worker in biochemical calorimetry must contend, and that is, he must in general, deal in microcalorimetry. This term is reserved, in the strictest sense of the word, for those calorimeters capable of detecting a change in temperature of one one millionth degree centrigrade. Microcalorimetry presents rather formidable difficulties over and above those encountered in ordinary calorimetric studies, even with simple materials. Further, not all microcalorimeters are suitable for biochemical work. However, in spite of these difficulties, significant advances in design of microcalorimeters especially suited for biochemical work (Calvet, 1948; Buzzell and Sturtevant, 1951; Kitzinger and Benzinger, 1960), has led to commerical production of some. This coupled with the relative ease of fabrication of certain of these designs now makes this equipment readily available to the biochemical laboratory.

Skinner has reviewed instrument development (Chapter I) so that we need only briefly consider classification. Calorimeters are either isothermal or adiabatic, though some calorimeters do not belong strictly to either category. In the isothermal calorimeter, quantity of heat is measured by the amount of isothermal phase change it produces in the calorimeter contents. The Bunsen ice calorimeter (Bunsen, 1870) provides one of the best examples of an isothermal calorimeter. With this instrument the heat evolved melts a certain quantity of ice with a concomitant change in volume which is proportional to the heat liberated. Aside from the disadvantages of having to work only at 0° (other substances can be employed in place of ice, however) a serious objection to this instrument type is the relatively large thermal inertia which renders them unsuitable for thermokinetic work (Calvet and Prat, 1956). Swietoslawski discusses other disadvantages, in particular where slow reactions are to be studied (Swietoslawski, 1946). In the adiabatic method of calorimetry, heat exchange between the calorimeter and its environment is eliminated if zero thermal head is maintained, a condition which can only be approximated in actual practice.

This method of calorimetry was first used and brought to a high degree of development by Richards (1909) and his co-workers. The classic example of a microcalorimeter of the adiabatic type is that developed by Lange and associates (Lange and Messner, 1927; Lange and Robinson, 1931). Although many modifications have been made calorimeters of this same basic design are in use today. Gucker *et al.* (1939) described an improved version which had a wider range of applicability. Both the Lange type and Gucker's microcalorimeter require approximately a liter of solution for their operation. Obviously in a great many instances neither of these instruments would be suitable for biochemical work. Moreover, they are adapted for only processes of short duration. The greatest advantage of the adiabatic procedure lies in its application to the study of prolonged experiments (Sturtevant, 1959), and it has, in fact, been the most widely recommended for the study of slow reactions. However, the method was not applied to these processes until the development of accurate devices for integrating the varying electrical power required for compensation. Sturtevant (Buzzell and Sturtevant, 1951) described a highly sensitive twin adiabatic calorimeter which employed electromechanical computers for measuring the electrical compensation energy. This calorimeter was especially suited to thermokinetic studies. Perhaps its chief disadvantage was the length of time required for equilibration.

On the basis of constructing microcalorimeters and testing their suitability for biochemical work during the course of some years, it is the author's opinion that the following criteria should be realized before a calorimeter can be considered adequate for general work in this field:

1. The calorimeter should require small amounts of material for its operation. This stems, of course, from the fact that often as not biochemical materials are available only in small amounts.
2. It should be suitable for the study of both fast and slow reactions as both types are encountered in biochemical work.
3. No stirring should be required, as often the heat developed by stirring may be larger than that of the process under investigation. This is especially true where small amounts of heat are evolved over long periods of time.
4. The calorimeter should be very sensitive, with a reasonable degree of accuracy. It might be well to mention here that temperature sensitivity of a calorimeter can sometimes be misleading. Though minimum detectable temperature describes one phase of calorimeter sensitivity, the quantity of ultimate interest is the minimum, total change in heat content which can be reliably measured. For example, a calorimeter sensitive to one microdegree with a heat capacity of 1000 cal may have a maximum heat sensitivity of 0.001 calorie per mm galvanometer deflection. However, there is no implication that this quantity of heat could be measured with a reasonable degree of reliability. It may well be in this hypothetical case that 0.1 cal is the minimum change in heat which this calorimeter can measure with any reasonable degree of reliability. Quantities of heat of this order of magnitude and far less are more the usual than unusual in a considerable number of instances in biochemical work.
5. Due to the lability of many biological materials, among other considerations, the calorimeter should have a relatively short equilibration time.
6. The calorimeter should have a low thermal inertia so as to be suitable for kinetic studies. In this connection some means of continuously recording the thermal output is highly desirable.

The above criteria exactly describe some of the salient features of the Calvet heat conduction microcalorimeter which is based on a prototype described over 40 years ago. Various aspects of this calorimeter will be described in some detail.

II. Design Considerations

A. Tian (1923) introduced a technique for microcalorimetry which, though similar in principle to that used by Callendar earlier (1910) has provided an important base for the development of the field. One of the chief conclusions which came from Tian's investigations in microcalorimetry was the fact that stirring was not required if one dealt with surface temperature. It has also been pointed out by Whipp (1934) that it is advantageous to deal with surface- rather than internal-temperature, since it is surface temperature which determines heat losses. Another very important consequence of Tian's work was the development of multi-jacket thermostats which served to avoid temperature fluctuations in the main calorimetric container. This type of thermostat was used by Ward (1930) and by A. V. Hill in his classic investigations on the heat of production of nerve (Hill, 1932, 1933a, b).

Calvet (1948) made important modifications, and, over a number of years with a considerable number of refinements to this calorimeter, has markedly increased its scope of application. The Calvet calorimeter, as it has become more commonly known, has been described in great detail in numerous publications (Calvet, 1948, 1956; Calvet and Prat, 1956; Calvet and Prat, 1963; Atree *et al.*, 1958; Benjamin and Benson, 1962). In essence it consists of a thermostated metal block of good thermal conductivity in which is contained twin microcalorimetric elements. Each of these elements is covered by an array of closely fitting thermocouples which are connected in opposition. One of the elements serves as a tare, the other containing the process under investigation. If the cell containers are covered by a sufficient number of regularly arranged thermocouples, and, provided these are affixed normal to the surface of the cell wall, then each individual junction may be considered to be at a uniform temperature even though the temperature is variable over the cell wall. The reference junctions of the thermocouples are in good thermal contact with the metal block, the temperature of which is uniform and virtually constant. Assuming the arrangement of the thermocouples meet the two above conditions, Tian (1923) has shown that the area under the emf-time curve is proportional to the heat released. As pointed out by Calvet (1956) the calorimeter directly records the calorific power W (i.e., the rate of heat production, $dQ\ dt^{-1}$, where Q represents the heat output and t the time) produced at each instant by the process being investigated. This curve, after minor corrections, represents W as a function of time and is called the thermokinetic curve. The thermokinetic curve not only gives the total quantity of heat Q, produced in any time interval, but also provides a knowledge of the kinetics of the process. Thus the major part of the heat generated is conducted to the external jacket, and it is the heat flux which is measured. Of course in the ultimate sense temperature change is being measured, since if no temperature change occurred, no heat would flow. However, the temperature changes which do occur are very small and this is one of the reasons the calorimeter is well suited to kinetic studies. One of the earlier forms of a Calvet calorimetric element contained 144 iron-constantan thermocouples. These were symetrically arranged in twelve layers, each layer containing twelve thermocouples. The size, number and length of these thermocouples were based on theoretical considerations so as to obtain maximum sensitivity in the steady state, i.e., for slow reactions. A portion of these thermocouples were wired in such a way that by a switching arrangement they could be used for Peltier compensation. Thus in the case of slow reactions the instrument could be operated essentially in a null fashion. Since Peltier compensation can be precisely regulated and measured, heats of slow reactions could be measured with an accuracy approaching a few tenths of 1 %. In addition, Peltier compensation can be used to appreciably reduce the time required for thermal equilibration.

A present day version of a Calvet calorimetric element may contain approximately 500 chromel-constantan thermocouples. By means of a switching arrangement the number of thermocouples in the detecting circuit can be varied so as to have three sensitivities. Peltier compensation can also be used.

A complete calorimetric assembly of such a Calvet unit is represented in Fig. 7, Chapter I. The metal block containing the calorimetric units is housed in an air thermostat. The thermostat consists of a series of concentric cylinders of good thermal conductivity separated by materials of poor thermal conductivity. The underlying idea is that the temperature of the cylinders at any given instant can be considered uniform. Thus, if the temperature of one of the outer cylinders is controlled to a few hundredths of a degree, temperature fluctuations are successively damped as they travel inward and this serves to maintain the temperature of the central block constant to about 10^{-5}°C for long periods of time. One further point concerning the thermostat assembly deserves comment. This is the so-called equipartition cones by means of which the metal block containing the calorimetric units is suspended between the lids of the central cylinder. As has been pointed out (Calvet and Prat, 1956) as long as the twin calorimetric elements are identical in construction, and symmetrically disposed with respect to the central axis of the block, it is not necessary to maintain temperature uniformity within the block. In fact this condition is impossible to achieve. It is necessary, therefore, that the temperature distribution be symmetrical at corresponding points on the two calorimetric units It is to this end that the equipartition cones serve. They act as collimators to transform lateral thermal perturbations not confined to the symmetry plane of the units into vertical perturbations equally divided between them.

The calorimeter was originally developed for the study of slow reactions, though as will be shown there is no reason it cannot be readily used for the study of fast reactions. Calvet has defined the two uses of the calorimeter by the terms integrator and ballastic. The integrator is the instrument of maximum sensitivity for the study of slow phenomena, whereas the term ballastic refers to the instrument for the study of more rapid reactions.

It is less sensitive than the integrator in the case of slow reactions, and in general will have more thermocouples, and in some instances, shorter ones. An integrator can be transformed into a reasonably good ballastic apparatus by the simple expedient of reducing the amount of material used in the investigation. The Calvet calorimeter as generally employed uses 6 to 10 ml of solution. It is said that the calorimeter can be used to study reactions of a month or more duration (Calvet and Prat, 1963).

Many variations and applications of the calorimeter have been described by Calvet and his associates (Calvet and Prat, 1956, 1963). One development concerns the use of semiconductors as the sensing elements rather than the usual chromel-constantan thermocouples (Calvet and Guillaud, 1965). The semiconductors employed had a factor of merit approximately nineteen times greater than chromel-constantan thermocouples. However, the use of semiconductors reduced the time constant and consequently the sensitivity of the calorimeter for the study of slow reactions.

III. Theoretical Considerations

The theory of the Calvet calorimeter has been discussed in great detail elsewhere (Calvet and Prat, 1956, 1963). Therefore, only a few of the more pertinent aspects will be mentioned here. The fundamental equation of the apparatus, known as Tian's equation, is

$$W = p g^{-1} \Delta + u g^{-1} d\Delta \, dt^{-1} \tag{1}$$

where W is the calorific power released inside the cell at the moment t, p the calorific power exchanged for a difference of 1°C between the chambers, Δ the galvanometer deflection, u the heat capacity of the cell contents and g is a constant for a given calorimeter and galvanometer. During the course of calibration, by slow electrical heating, for example, once the steady state has been reached Eq. (1) assumes the form

$$W = p g^{-1} \Delta \tag{2}$$

and thus $p g^{-1}$ (the principal constant of the apparatus) can be determined. The integral of Eq. (1), or put another way, the area under the emf-time curve, regardless of its shape or duration, will be proportional to the heat liberated. However, it is from Eq. (1) that the thermokinetics of the process under investigation are derived. This equation can be put into the form

$$W = p g^{-1} (\Delta + \tau \, d\Delta \, dt^{-1}) \tag{3}$$

where $\tau = u p^{-1}$ is the time constant of the calorimeter.

The direct record of the galvanometer deflections gives Δ as a function of time whereas the thermokinetic curve relates W to time. In order to transform one curve into the other a correction term is added to each point Δ. This correction term is the product of the time constant τ and the slope of the curve at that point. In the case of very slow thermal phenomena, for which the correction term $\tau \, d\Delta \, dt^{-1}$ is very small relative to Δ, the directly recorded curve is a satisfactory representation of the thermokinetics. The suitability of the calorimeter for giving direct recordings of the thermokinetics depends on the smallness of the time constant, τ. A somewhat more convenient quantity than τ is the constant, $t_{1/2}$, which is the time required for the thermal decay curve to fall to half of its value in the steady state condition.

Tian's simple equation is not directly applicable to the study of rapid processes. However, it has been shown (Calvet and Camia, 1958) that the directly recorded curves of these rapid processes can be transformed so that they become suitable for treatment by Tian's equation. Although analytical solutions to some of the equations developed for the transformations are difficult, a graphical solution is simple and always possible. Thus, by this means it has been feasible to follow variations in W lasting for a period of 10 seconds. With more sensitive recording equipment it may be possible to analyze impulses of only 1 second.

The equations developed for calculating the conditions of optimum sensitivity of the Calvet calorimeter are based in large part on those originally derived by Whipp (1934). However, for a heat conduction type calorimeter it is more illuminating to treat the thermocouple as a means of measuring heat rather than temperature. For this purpose the following series of equations are convenient to use when the electromotive force of thermocouples is measured with a galvanometer (Stott, 1956). Thus, the heat flowing down the thermocouple wires for a temperature difference $\Delta\theta$ between the hot and cold junctions is:

$$Q = (n/l)(k_1 a + k_2 Za)\Delta\theta \tag{4}$$

where Q is the heat flow (cal/s), n is the number of couples, a is the area of the copper wire, Za is the area of the constantan wire, l is the length of each couple wire, k_1 is the thermal conductivity of the copper, and k_2 is the thermal conductivity of the constantan (the copper-constantan thermocouple has been chosen for illustration). The electrical resistance of the couples will be:

$$R = nl\left(\frac{r_1}{a} + \frac{r_2}{Za}\right) \tag{5}$$

where r_1 is the electrical resistivity of the copper wire and r_2 is the electrical resistivity of the constantan wire. The voltage developed by the thermocouples is given by:

$$V = nc\,\Delta\theta \tag{6}$$

where c is the thermal emf per deg C of the thermocouple.

For galvanometers of the same type an approximate relationship between the galvanometer resistance and the sensitivity is:

$$D = G\sqrt{(R_G)}\,I \tag{7}$$

where G is the galvanometer constant, R_G is the galvanometer resistance, and I is the current passing through the galvanometer. The following relation also holds:

$$V = (R + R_G)\,I \tag{8}$$

Combining these equations the sensitivity, S, defined as the deflection produced by unit heat flow of heat, is:

$$S = D/Q = nCG\sqrt{R_G}\left\{H\left[R_G + \frac{n^2}{H}(k_1 + Zk_2)\left(r_1 + \frac{r_2}{Z}\right)\right]\right\}^{-1} \tag{9}$$

where H is the heat transfer coefficient of the thermocouple wires defined by Eq. (4), that is

$$H = Q/\Delta\theta \tag{10}$$

For a given value of H and R_G it can be shown that the sensitivity of the calorimeter is a maximum when:

$$n^2 = R_G H\left[(k_1 + Zk_2)\left(r_1 + \frac{r_2}{Z}\right)\right]^{-1} \tag{11}$$

and hence that $R_G = R$ for maximum sensitivity.

By means of these equations it is possible to design a calorimeter of maximum sensitivity if the desired heat-transfer coefficient of the thermocouple wires is known. In choosing the heat transfer coefficient it must be borne in mind that the larger H is, the less will be the sensitivity as may be seen from Eq. (9). On the other hand if a small amount of heat is quickly released in the calorimeter, the larger H is, the larger will be the transient deflection of the galvanometer.

One further point affecting the design of the calorimeter is the ratio of the cross sectional area of the couple wires to their length, the term a/l in Eq. (4). The value of a/l follows from the values of n and R decided by Eq. (11) and inserted in Eq. (10). As the ratio of a/l is fixed, any increase in l must be accompanied by a similar increase in a. In connection with the length of the wires, it has been assumed that all the heat which enters the couple at the hot junction flows out the cold junction. This is not strictly true, as the surface of the wire always loses some heat to the surroundings at a lower temperature by radiation and convection. The longer the wire, the greater will be the loss.

However, considering the small temperature differences encountered in microcalorimetric work, radiation and convention, as factors influencing the choice of the length of the wires, can be considered relatively unimportant.

It has been pointed out previously that a large heat transfer coefficient results in a diminution of the sensitivity. This is one means whereby the calorimeter can be rendered more suitable to function as a ballastic apparatus. As seen from Eq. (4) a large heat transfer coefficient can be obtained by increasing the number of couples, using shorter couples or by a combination of both. It may be mentioned that decreasing the length of the couples presents certain complications in regard to kinetic analysis (Calvet, 1962).

IV. Modifications

Several lines of modifications of the Calvet type calorimeter have been followed in the author's laboratory. Chief among these have been:

1. A method was sought whereby the thermopile could be easily constructed and mounted. The fabrication and mounting of the thermopile in the conventional manner represents the chief constructional difficulty and presents a task of formidable magnitude to the average biochemical laboratory.
2. Reduction in the over-all size of the calorimeter and in the amounts of material required for its operation.
3. Development of a method to initiate the reaction which would insure homogenization of the reactants while developing a minimum mechanical heat. This represents one of, if not the most difficult problems in microcalorimetry.

The fabrication of thermopiles by soldering or brazing wires together can be very tedious and difficult when a large number of thermocouples is required. However, another method, the technique of electroplating (Wilson and Epps, 1919) has been extensively used and has various advantages, particularly in the ease it affords for the construction of thermopiles with large numbers of couples. Having adopted this mode, the author and his associates have described a calorimeter of the Calvet type utilizing electroplated thermocouples as the sensing elements (Evans and Carney, 1965). One of the chief features of this particular calorimeter is ease of fabrication, and though having proved useful (Brown *et al.*, 1964, 1965a, b) it has the disadvantage, among other considerations, of requiring a relatively long period for thermal equilibration (4 hours). Refined versions have been developed which conform more nearly to the criteria of suitability of a calorimeter for biochemical analysis. Although one of these has been previously described (Evans *et al.*, 1968), some of the more pertinent aspects may prove informative here.

Details of the cell holder-thermocouple assembly are shown in Fig. 1. The cell holder was machined from copper with a flange at each end. Each flange had ten holes drilled thirty-six degrees apart. The thermocouple elements were made by winding 400 turns of enameled constantan wire, 0.005 inch diameter, onto acetate tubing which had an o.d. of $\frac{1}{4}$ inch and wall thickness of 0.01 inch.

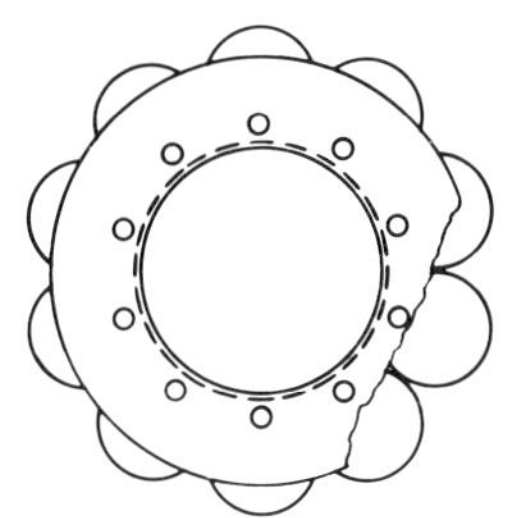

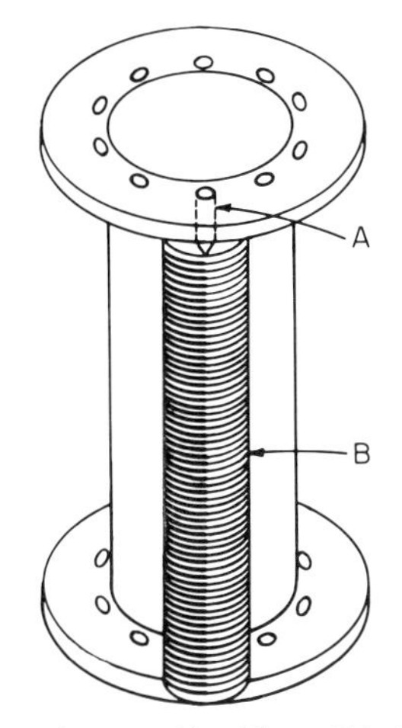

FIG. 1. Cell-holder, thermocouple assembly. The cell holder is a machined copper tube with a flange at each end; the plated thermopile B is retained by a pin A through the flange.

This tubing, approximately 2 inches in length, was held on a mandrel in the lathe while the enamel was removed from the outer periphery of the wire by means of a fine grain emery-cloth. After removal of the enamel the tube was clamped in a V-bottom jig (brass construction) so as to make good electrical contact with the wire. With the tubing held in this manner a copper coating, from acid cupric sulfate solution, of about 0.001 inch thickness was formed on one-half of the circumference of the wire coils on the tubing. Thus, each tube with the plated wire constituted a battery of four hundred thermocouples. Twenty such batteries of thermocouples were constructed, ten for each of the two cell holders.

Prior to placing the tubes containing the thermocouples onto the cell holders, both the tubes and cell holder were coated with an epoxy varnish for electrical insulation. Each individual battery of thermocouples was secured to the cell holder by means of two pins pushed through the holes in the flanges of the cell holder. Aside from firmly securing the tubes, this operation greatly facilitated the subsequent wiring arrangement. Once the ten thermocouple assemblies were fastened to the cell holder in this manner they were wired in series through a double pole, 6 position, 6 deck switch with gold contacts. This resulted in a thermopile of 4000 copper-constantan thermocouples for each

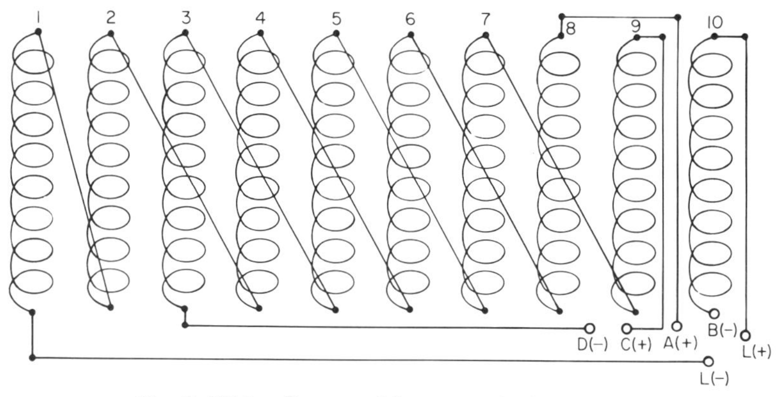

FIG. 2. Wiring diagram of thermocouple elements.

calorimetric unit. The two complete thermopiles were also wired in series, but opposition, through this same switch thereby effecting the differential arrangement. Further, by means of this switch, four of the individual thermocouple assemblies, in series (1600 thermocouples), could be isolated from the detecting circuit of each calorimetric unit for Peltier compensation. A schematic wiring diagram of the thermocouple elements contained in one calorimetric unit is shown in Fig. 2.

Once the cell holder-thermocouple assemblies were completed they were enclosed in aluminum jackets which had been machined into 120 deg sectors. These sectors were held together by two retaining rings machined to fit at each end of the jacket. The inside of the jacket was coated with epoxy varnish for electrical insulation, and so as to insure good thermal contact, the ID was machined to such size so as to result in a mild compression of the thermocouples. The exterior of the jacket was machined with a three degree taper while a matching cavity was machined into an aluminum block. This taper arrangement effectively locks the calorimetric unit into the block for good thermal contact. The two calorimetric units, can be seen in place, with the block (main heat sink) in Fig. 3.

FIG. 3. Heat sink, with two calorimetric elements, mounted in the 4-element, aluminum thermostat assembly.

The thermostat assembly consisted of four concentric aluminum cans equipped with lids. The air spaces between the two outermost cans were filled with polyurethane insulating material. A heating and cooling coil was wound on the outer surface of the can second from the outside while the sensing element of an electronic proportional temperature controller was placed on the inside. Two hollow aluminum shafts, with a diameter of approximately $1\frac{1}{2}$ inches, were fitted diametrically opposite at the equator of the outermost can of the thermostat. The entire assembly is suspended by means of these shafts which pass through pillar block bearings for rotation, whereas all lead wires to the outside pass through the interiors of the shafts. The signals from the thermopiles are fed to a DC amplifier and then to a DC recorder and a digital readout system which integrates the emf-time curve, with the DC recorder serving merely to monitor the curve. The entire calorimetric assembly is shown in Fig. 4. An idea of the size of the calorimeters can be gained from the six inch scale shown in the photographs.

FIG. 4. Calorimeter assembly.

The results of some electrical calibrations of the calorimeter with both fast and slow heating rates are shown in Table I. Here the total amounts of heat

TABLE I

ELECTRICAL CALIBRATION OF CALORIMETER

Run No.	Calibration factor $\times 10^6$ (cal/integrator count)	Deviation $\times 10^6$
1	4.4503	0.0035
2	4.4507	0.0039
3	4.4398	0.0070
4	4.4400	0.0068
5	4.4533	0.0065

liberated ranged from approximately nine up to five hundred millicalories, and the heating times from one second to in excess of 4 hours. In regard to electrical calibrations, another rather decided advantage of having means of effecting Peltier compensation is that the calorimeter can also be calibrated by this means (Calvet and Duquesne, 1964). The particular advantage arises when the calorimeter is to be operated at temperatures considerably different from the environmental. Rather serious errors, difficult to avoid, can result from conduction by the leads of the calibrating heater contained in the calorimetric cell.

Before discussing chemical calibrations of this instrument it seems pertinent to mention again the difficult problem of initiating the reaction, in particular where equal amounts of reactants are involved. In order to insure the same thermal state of the reactants prior to initiation and homogenization after, partition type cells, shown in Fig. 5, are used in conjunction with rotation of the calorimeter.

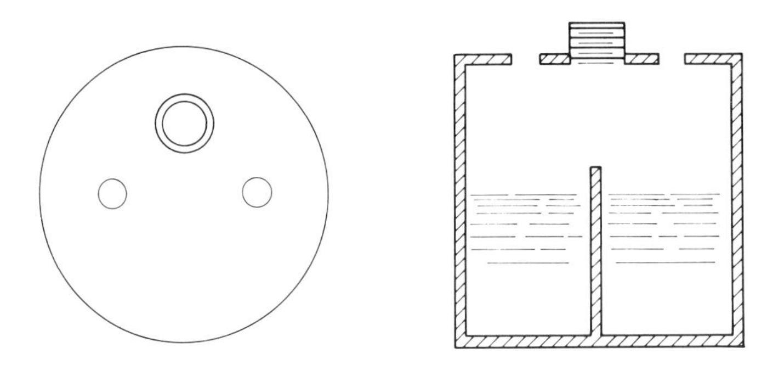

FIG. 5. Partition-type cells.

These cells were fabricated from tantalum because of its mechanical strength, chemical inertness and relatively good thermal conductivity. Several variations of the volumes of solutions are possible with these cells. Generally they are used from 1 ml down to 0.25 ml on either side of the partition, though they function equally as well with a volume excess on one side of the partition. The constant, $t_{1/2}$, varies from approximately 100 seconds down to 55 seconds, depending on the total volumes of solutions employed. Thermal equilibration requires approximately one hour after placing the cells in the calorimeter. Recent experiments show that by means of Peltier compensation this time can be reduced to about thirty minutes, and indications are that it may be possible to minimize this further. Chemical calibrations of the calorimeter were carried out using the heat of neutralization of HCl with NaOH.

Initiation and stirring of the reaction is effected by rotation of the calorimeter through a reversible motor and speed reducer at 15 rpm. The rotation is automatically controlled by a series of cams and microswitches. One cycle consists of 180 degree clockwise rotation followed by 360 counter clockwise and then 180 degree clockwise and stop. Generally four to five cycles are required to effectively homogenize the solutions when equal volumes are employed in equivalent amounts. Sturtevant (1937) has pointed out that this form of

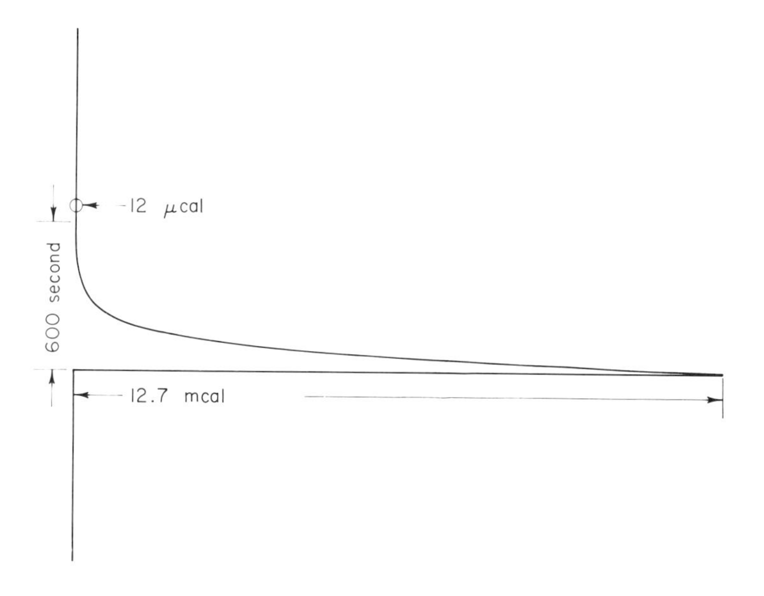

FIG. 6. Heat liberated in an instantaneous reaction.

stirring causes a disturbance, presumably due to movement of the thermocouples in the earth's magnetic field. This disturbance has been substantially reduced by means of magnetic shielding wrapped around the two innermost cans of the thermostat. A series of blank experiments show that the heat generated by mechanical and other effects is of the order of 13 μcal. An idea of the magnitude of this quantity can be gained from Fig. 6, which is an actual recording of the total heat liberated in an instantaneous reaction.

V. Applications

The one single factor which would appear to render the calorimetric technique of almost unlimited potential in the study of the usually complex reactions of biochemistry, or for that matter, indeed, in the study of any reaction, is the fact that nearly all reactions proceed with an appreciable change in heat content. Many of the physical methods on which attention has been focused in the study of macromolecules are based upon changes in such properties as density, optical rotatory power, refractive index and electrical conductivity. Most of these methods are restricted in applicability and many require the use of relatively concentrated solutions in order that an appreciable change in the observed property can be obtained. The great sensitivity of the calorimetric method makes it possible to work with dilute solutions. In all physical methods it is necessary to assume that the value of the observed property changes linearly with the extent of the reaction. This is known to be not true in some cases. In the calorimetric method direct evidence of the validity of this fundamental assumption can be obtained. Thus, if the heat capacity of the reacting solution remains constant during the reaction the heat liberated will be strictly proportional to the extent of the reaction, providing no other reaction takes place (Sturtevant, 1941). It is, of course, this last provision, coupled with the ubiquity of the evolution or absorption of heat during the course of nearly all reactions, within which the inherent weakness of calorimetry lies. However, in many instances it is possible to prove that the heat liberated is related to one specific reaction. Even though it is not related in a specific way to the chemical changes which occur, it is in a manner which provides a firm outline that must not be overstepped and can be filled in as further knowledge is acquired.

It has been implied by myself and by my colleagues, that nonthermochemical data can be obtained from calorimetric measurements. Thus, the heat evolved or absorbed in a reaction is proportional to the extent of the reaction so that kinetic studies as well as the heat liberated in a given time interval can be measured by calorimetric observations. Antigen-antibody reactions, polymerization and clotting of the fibrin monomer (Sturtevant *et al.*, 1955), membrane transport (Brown *et al.*, 1965a), and irreversible processes (Durell and Sturtevant, 1957; Mudd *et al.*, 1966) have been studied. There is another aspect of the potential of calorimetry in biochemistry and allied fields. This might be arbitrarily called the "qualitative aspect" and is perhaps best exemplified by the works of Prat and associates (Calvet and Prat, 1956, 1963; and Chapter IX this work). Numerous applications are possible since the variation of thermic output constitutes a reliable and sensitive means of checking the smallest changes in physiological functions, metabolism and other specialized activities occurring in an organism. Further the influence of chemical, physical and biotic factors can be analyzed.

REFERENCES

Atree, R. W., Cushing, R. L., Ladd, J. A., and Pieroni, J. J. (1958). *Rev. Sci. Instr.* **29**, 491.
Benjamin, L., and Benson, G. C. (1962). *Can. J. Chem.* **40**, 601.
Brown, H. D., Evans, W. J., and Altschul, A. M. (1964). *Life Sci.* **3**, 1487.

Brown, H. D., Evans, W. J., and Altschul, A. M. (1965a). *Biochim. Biophys. Acta* **94**, 302.
Brown, H. D., Neucere, N. J., Altschul, A. M., and Evans, W. J. (1965b). *Life Sci.* **4**, 1439.
Bunsen, R. (1870). *Ann. Physik* **141**, 1.
Buzzell, A., and Sturtevant, J. M. (1951). *J. Am. Chem. Soc.* **73**, 2454.
Callendar, H. L. (1910). *Proc. Phys. Soc. (London), Sect. B*, **23**, Pt. 1, 1.
Calvet, E. (1948). *Compt. Rend.* **226**, 1702.
Calvet, E. (1956). *In* "Experimental Thermochemistry" (F. D. Rossini, ed.), Vol. I, Chapt. 12. Wiley (Interscience), New York.
Calvet, E. (1962). *In* "Experimental Thermochemistry" (H. A. Skinner, ed.), Vol. II, Chapt. 17. Wiley (Interscience), New York.
Calvet, E., and Camia, F. (1958). *J. Chim. Phys.* **55**, 818.
Calvet, E., and Duquesne, R. (1964). *J. Chim. Phys.* **61**, 303.
Calvet, E., and Guillaud, C. (1965). *Compt. Rend.* **260**, 525.
Calvet, E., and Prat, H. (1956). "Microcalorimetrie." Masson, Paris.
Calvet, E., and Prat, H. (1963). "Recent Progress in Microcalorimetry" (Transl. by H. A. Skinner). Macmillan, New York.
Durrell, J., and Sturtevant, J. M. (1957). *Biochim. Biophys. Acta* **26**, 282.
Evans, W. J., and Carney, W. B. (1965). *Anal. Biochem.* **11**, 449.
Evans, W. J., McCourtney, E. J., and Carney, W. B. (1968). *Anal. Chem.* **40**, 262.
Gucker, F. T., Pickard, H. B., and Planck, R. W. (1939). *J. Am. Chem. Soc.* **61**, 459.
Hill, A. V. (1932). *Proc. Roy. Soc. (London)* **B111**, 106.
Hill, A. V. (1933a). *Proc. Roy. Soc. (London)* **B113**, 356.
Hill, A. V. (1933b). *Proc. Roy. Soc. (London)* **B113**, 345.
Kitzinger, C., and Benzinger, T. M. (1960). *Methods Biochem. Analy.* **8**, 309.
Lange, E. W., and Messner, Z. (1927). *Z. Elektrochem.* **33**, 431.
Lange, E. W., and Robinson, A. L. (1931). *Chem. Rev.* **9**, 89.
Mudd, S. H., Klee, W. A., and Ross, P. D. (1966). *Biochemistry* **5**, 1653.
Richards, T. W. (1909). *J. Am. Chem. Soc.* **31**, 1275.
Stott, J. B. (1956). *J. Sci. Instr.* **33**, 58.
Sturtevant, J. M. (1937). *J. Am. Chem. Soc.* **59**, 1528.
Sturtevant, J. M. (1941). *J. Phys. Chem.* **45**, 127.
Sturtevant, J. M. (1959). *In* "Technique of Organic Chemistry" (A. Weissberger, ed.), 3rd Ed. Wiley (Interscience), New York.
Sturtevant, J. M., Laskowski, M., Laskowski, M., Jr., Donnelly, T. H., and Scheraga, H. A. (1955). *J. Am. Chem. Soc.* **77**, 6168.
Swietoslawski, W. (1946). "Microcalorimetry." Reinhold, New York.
Tian, A. (1923). *Bull. Soc. Chim. France* **33**, 426.
Ward, A. F. H. (1930). *Proc. Cambridge Phil. Soc.* **26**, 278.
Whipp, B. (1934). *Phil. Mag. Ser.* 7, **18**, 745.
Wilson, H., and Epps, T. D. (1919). *Proc. Phys. Soc. (London)* **32**, 326.

CHAPTER XV

Combined Calorimetry and Spectrophotometry in Stopped-Flow Measurements

ROBERT L. BERGER

I. Introduction

Biochemical calorimetry has been discussed, in the main, throughout this book upon the basis of the thermodynamics of reversible systems. In general the measurements made give static or equilibrium values. We know, however, that the dynamics of biochemical reactions must often be treated by the methods of irreversible thermodynamics. The reaction of catalase with hydrogen peroxide is a well known example. Of even greater importance is the "dynamically stable" system in which a chain of reactions is carried out and the enthalpy, entropy, and Gibbs free energy then depend upon coupling reactions, activators, inhibitors, and control systems. Here the study of kinetics and the system response to change is requisite. So little is known about this area of thermochemistry that there is a most pressing need for instrumentation suited for measurements of such dynamically stable systems. Only in this way may we have even a general notion as to what kind of thermochemical data of the individual components is really required in order to understand some of the measurements made of the intact system. The classical example is the utilization of ATP by muscle. Several parts of this system have been worked out. It has been demonstrated that there is variation in heat released upon hydrolysis which is dependent upon how the system is coupled to other systems.

In recognition of this need, several calorimeters with various kinds of mixers have been reported and others are presently under development. In general they operate in the range of 1 to 2 seconds and can mix a small volume of one solution with a large volume of another. Recently, several equal, or near equal volume, mixers have been built for this time range. Of greater interest is the almost completely neglected time domain from 1 second to 1 nanosecond. The rest of the Section will be devoted to discussing such equipment, the problems in making such measurements and the possible rewards for the effort expended.

II. Instrumentation and Measurements

The present interest in this field dates from 1923 when Hartridge and Roughton (1923; Roughton, 1930) made the first measurements on the reactions of the blood gases with hemoglobin. In the Hartridge-Roughton

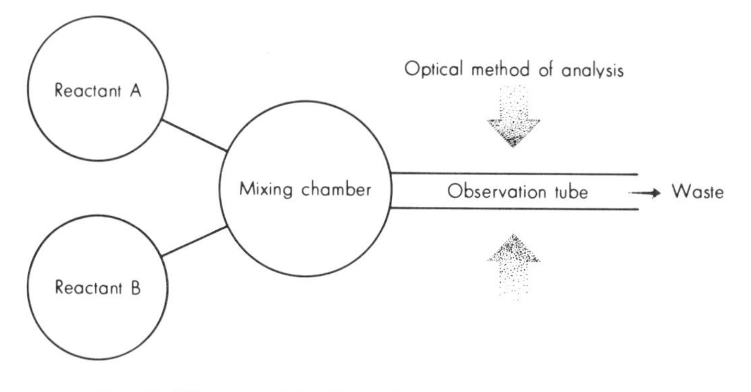

FIG. 1. Diagram of the Hartridge-Roughton flow system.

method, two solutions, as A and B of Fig. 1, are mixed quickly and flowed rapidly down an observation tube. If there is a physical change which can be followed this may be done, e.g., heat or pH, by inserting a probe into the tube and moving it to various known points. If d is the distance in centimeters between the mixing chamber and the cross-section being examined, u is the average fluid flow velocity in cm/sec (determined by collecting a measured volume over a timed interval and dividing by the cross-sectional area of the tube), then the average time the reaction has proceeded after being mixed is d divided by u. If the rate of flow is constant and turbulent, that is to say that a straight front of mixed solution is passing down the tube, then an ordinary concentration-time curve may be readily plotted from observations at 5 or more points along the tube. Fig. 2 shows such a curve for the reaction of carbon monoxide with hemoglobin. The solid points were determined from thermocouple outputs and the open points by optical observations. This method has the advantage that slow detectors may be used since a few tenths of a second to several seconds are generally required to take a measurement. On the other hand it suffers from the disadvantage that large volumes of solution are needed and the flow introduces some noise. For example, in a 3 mm diameter tube a flow rate of 420 ml per minute is required for a velocity of 100 cm per second. Thus if 2 seconds are required for a measurement, 7 ml of each reagent are required. Since one generally flows A and reads, then A plus B, and B, 21 ml are required for each point. However, this flow velocity does give a resolution of 1 msecond if mixing

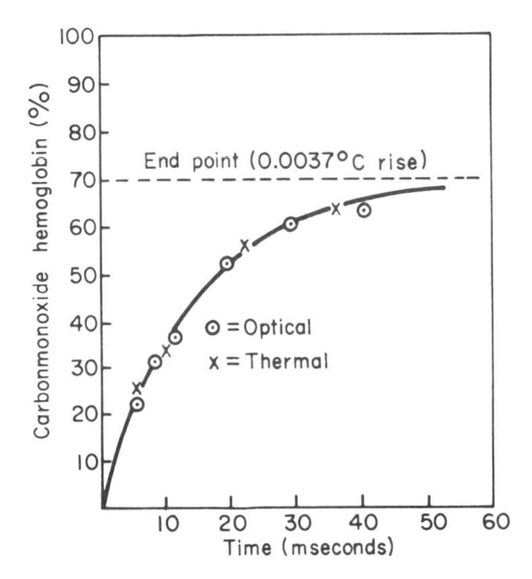

FIG. 2. Comparison of optical and thermal measurements of the rate of reacting along the 0.001M CO with 0.0014M hemoglobin. Taken from Weissberger.

occurs in this length of time. Details of such a system using a thermocouple and a Keithley nanovoltmeter will be described below. Much higher flow velocities have been achieved; it is quite easy to go to 10 meters per second, though this requires 210 ml of solutions per point. An increase in velocity does increase the turbulence but not the rate of mixing so that one must go further down the observation tube before 99% mixing occurs. It turns out that with the usual tangential mixers, described later, a time resolution of only 0.5 mseconds is about the limit. Recent improvements in mixers and flow systems have allowed us to extend this to 46 μseconds.

Alternatively, all observations can be made at a fixed point along the observation tube by varying the rate of flow, either in discrete steps as in Millikan's (1936) procedure, or continuously as in the accelerated flow method of Chance (1940). This latter scheme has had many technical difficulties which have not yet been solved. However, if the flow of the reacting liquid in the observation tube is suddenly arrested (within a time interval which is short compared to the half-life of the reaction, i.e., the time for half of the physical change to occur), then the progress of the chemical reaction in the stopped fluid may be followed. This has the disadvantage of requiring very fast detectors. The details of these methods may be found in any of the several books written on the subject (Roughton and Chance, 1963; Caldin, 1964; Gutfreund, 1965a). Recently improvements have been made in the ability to flow fast and to stop fast so that sub-millisecond time resolution is now possible (Berger *et al.*, 1968).

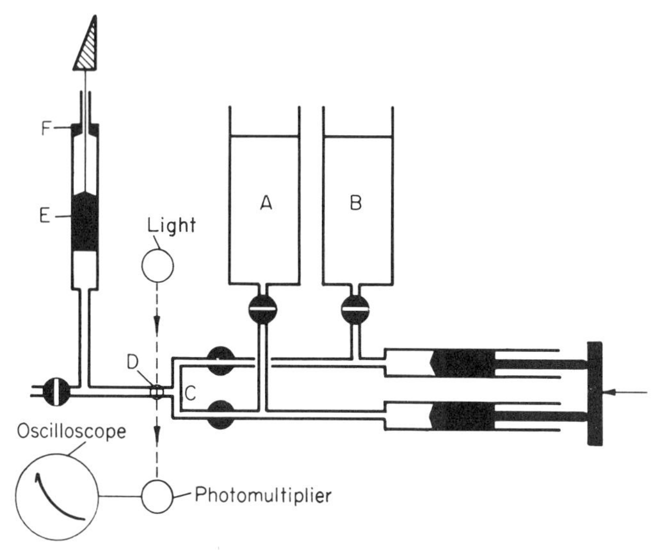

FIG. 3. Stopped flow apparatus (modified after Gibson). From Caldin (1964).

An understanding of the stopped-flow method can be gained from a discussion of Fig. 3. Solutions A and B are driven through mixer C into observation tube D, and into syringe E, which fills and hits the stop F, bringing the solutions to a halt. This produces a trigger pulse which starts the oscilloscope sweep. The time course of the reaction is displayed upon the oscilloscope screen. Suitably calibrated, the deflections of the oscilloscope indicate optical transmission changes. Since the oscilloscope sweep time is accurately known, one automatically plots transmission versus time. If I_0 is the oscilloscope deflection from dark to light, then the optical density is given by ln I/I_0. Since the optical density is proportional to the concentration, by taking the slope of a plot of ln c/c_0 versus time over the first half of its course, the rate constant may be determined in simple first order kinetics. Details of data handling and interpretations in more complex cases have been treated in several excellent texts (Caldin, 1964; Gutfreund, 1965b; Benson, 1960a).

Both continuous and stopped flow methods, have been applied to optical, thermal, and pH measurements and we shall discuss later typical continuous or stopped-flow measurements and the treatments of the data obtained. To make the present introduction complete, I would like to give a very brief discussion of the method of chemical relaxation by temperature-jump. A very full discussion of the theory and methods have been given by Eigen and DeMaeyer (1963), a simpler discussion appears in Caldin's (1964) book and most recently Czerlinski (1966) has devoted a monograph to the subject. Let us consider the chemical reaction

$$A + B \underset{k_2}{\overset{k_1}{\rightleftharpoons}} C \tag{1}$$

If the concentration of B is so large that the reaction rate is independent of changes in it, and k_1, the forward rate constant, is sufficiently great, the reaction effectively becomes

$$A \underset{k_2}{\overset{k_1}{\rightleftharpoons}} C \tag{2}$$

where $k_1 = k_1 b$ and b is the (invariant) concentration of B. Now providing the reaction has an enthalpy and the Van't Hoff relation holds, that is

$$\frac{d(\ln K)}{dT} = \frac{-\Delta H}{RT^2} \tag{3}$$

if we change the temperature by say 1 to 5 degrees so rapidly that the system cannot follow this change, we can then observe the reaction "relaxing" to its new equilibrium state. Thus we make our observations during the time just after the jump, as the system is returning to equilibrium. To obtain rate constants we must relate the measured relaxation time to these constants. Knowing the initial and final temperatures and measuring the new equilibrium constant we can obtain the enthalpy. At $t = 0$, $A = a_0$ and $C = c_0$. After the temperature jump, the concentration of $A = \bar{a}(> a_0)$ and $C = \bar{c}(< c_0)$. The concentration will change to approach $\bar{a}$ and $\bar{c}$. Let this change be $x = a - \bar{a} = c - \bar{c}$ where a and c are the concentrations at any time t, after the temperature change, T. The rate equations thus become

$$-dx/dt = k_1 a - k_2 c \tag{4}$$

at equilibrium

$$dx/dt = 0 = k_1 \bar{a} - k_2 \bar{c} \tag{5}$$

at any time t

$$a = \bar{a} - x$$

$$c = \bar{c} - x$$

$$-dx/dt = k_1(\bar{a} + x) - k_2(\bar{c} - x) \tag{6}$$

applying condition (5)

$$-dx/dt = (k_1 + k_2)\,x \tag{7}$$

Upon integration (7) becomes

$$x = x_0\,e - (k_1 + k_2)\,t \tag{8}$$

If we define $k = k_1 + k_2$, then for $kt = 1$, we see that $\ln x/x_0 = 1/e$ and the time at which this occur's is defined as the relaxation time, $\tau = 1/k$. Thus for a first order reaction

$$\tau = \frac{1}{k_1} + \frac{1}{k_2} \tag{9}$$

and for a second order reaction

$$l = k_1(\bar{a} + \bar{b}) + k_2 \tag{10}$$

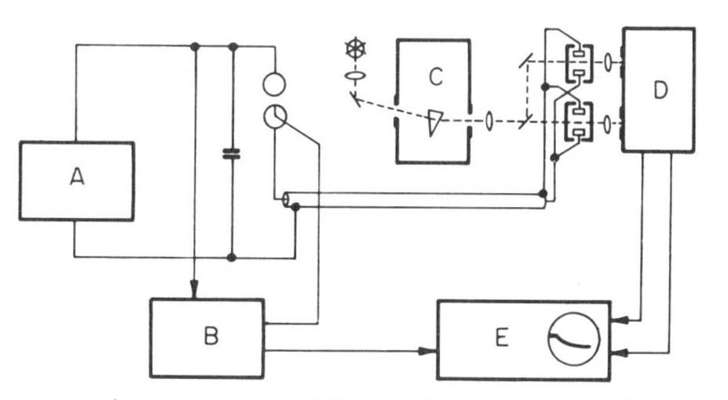

FIG. 4. Temperature-jump apparatus of Eigen, De Maeyer, and Czerlinski. A. high voltage generator; B. high voltage measurement and triggering; C. monochromator; D. photomultipliers and preamplifiers; E. differential amplifiers and oscilloscope. From Caldin (1964).

Figure 4 shows the apparatus developed by Eigen, deMaeyer and Czerlinski. The cell may be as small as 50 μl, with specially shaped electrodes in place. A voltage charges a condenser and then a trigger fires a spark gap sending the current through the cell, which will heat the liquid 4 to 10°C, if the ionic strength is about 0.1. This can be done in about 6 μseconds. Observations are then made either by fluorescence or absorption using the light source and the photomultiplier. Instrumentation using nanosecond laser pulses is under development and a combination T-jump and flow instrument is under development in the laboratories of Eigen and deMaeyer (1963) and Czerlinski (1966). These techniques have proven particularly useful for proton-transfer reactions. Hammes (1966) in a short review has tabulated some of the reactions worked out using these techniques which are of interest to biochemists.

Let us now turn our attention to the details of a continuous flow system for calorimetric measurements. The system developed at Cambridge has so far proved to be the most successful. Details may be found in the literature but a brief description here is warranted so that the reader may judge for himself the possibilities. The description given by Chipperfield (1966) is most appropriate. The apparatus is shown diagrammatically in Fig. 5. The 125 ml bottle, R and L, each contains a magnetic bar and stand on a magnetic stirrer unit. The solutions flow through the taps or through low-heat solenoid valves to the mixing chamber and then up to the 1.5 mm diameter observation tube which is

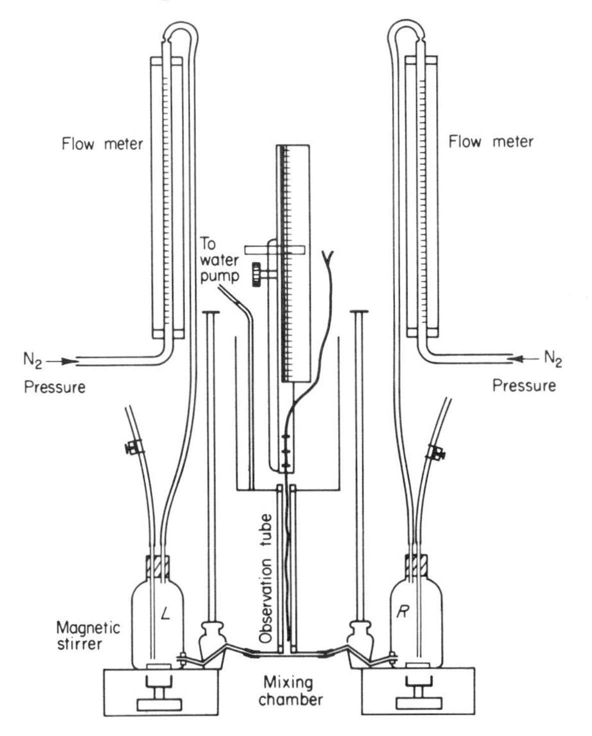

FIG. 5. A continuous flow system for calorimetric measurements.

surrounded by an air jacket. (Kel-F or stainless steel valves of this type are now commercially available from several sources including Science Products Corp., Dover, N. J.) The thermocouple can be raised and lowered in the observation tube by a rack and pinion arrangement. Solutions are driven from the bottles by compressed nitrogen. The rate of flow of gas into each bottle, measured by flow meters, is equal to the rate of flow of the solution out of the bottle. From these flow rates the mixing ratio of R to L can be calculated (to give the concentrations of reactants in the observation tube) and also the total flow rate up the observation tube (to find the time elapsed between the mixing chamber and the thermojunction). The apparatus is immersed in a 40 gallon water thermostat controlled to $\pm 0.0015°C$ at 25°C by a large toluene regulator. The voltage resulting from the difference in temperature between the thermojunction in the observation tube and a reference junction in the thermostat is amplied by a Keithley Model 148 nanovoltmeter and read on a pen recorder. If the pen recorder reads B_l when solution L is flowing alone, reads B_R when solution R is flowing alone, and reads B_M when R and L are flowing together (being mixed in the ratio x:1), then the heat due to the reaction of R and L is $B_M - (xB_R B_L)/(1 + x)$. If the bottles R and L both contained water the mixture was 0.005°C cooler than expected. This was due to R and L flowing separately at a different rate than R and L together. In all experiments the appropriate correction was made for this effect. With 125 ml of each fluid, readings could be obtained at 5 or 6 positions in the observation tube. The flow meters were calibrated by measurement of the flow of water from the bottles separately. The thermocouple was calibrated by using the apparatus to measure the heat of combination of dilute hydrochloric acid with excess of sodium hydroxide (13.5 kcal mole^{-1} at 25°C).

Heats of reaction were measured by using a wider observation tube, and measuring temperature changes at a point 8 to 10 cm from the mixing chamber. The temperature at any point in the observation tube can be used to calculate the concentration of product at this point if the heat of reaction and the concentration of reactants are known.

This apparatus with a number of simple improvements to make the measurements less tedious can be obtained commercially (Science Products Corp.) or put together from purchaseable components. A significant improvement in this apparatus can be achieved by using a Keithley Model 140 microvoltmeter and a syringe drive system controlled by magnet clutches. Thermocouples may be obtained commercially (Science Products Corp. or High Temperature Instruments Corp., Philadelphia). The latter company manufactures a multi-differential element (Model 105, 110, etc.). The response speed of the Keithley Model 140 is 10 mseconds for the small signal changes which is compatable with the thermocouple response time. This amplifier is sufficiently sensitive for the small signal changes from the thermocouple. Thus by electrically timing the syringe flow, flows of 50 to 100 mseconds can be used; this greatly reduces the fluid needed.

Chipperfield gives a number of results for the general reaction

$$CO_2 + RNH_2 \xrightarrow{k'} RNHCOO^- + H^+ \tag{11}$$

$$H^+ + RNH_2 \rightleftharpoons RNH_3^+ \tag{12}$$

where R is —CH_2COO^- for glycine. Tables 1 and 2 summarize the data. Thus for the first time a rather large volume of good thermochemical data and kinetic data were obtained on a truly fast reaction. The first observation point was 3 mseconds. By the addition of light wires (Bausch and Lomb) on a movable frame simultaneous optical measurements have been made and by replacing the thermocouple with an electrode (Ingold type 203-GA-31) pH measurements have been obtained (Rossi-Bernardi and Berger, 1968).

TABLE I

VALUES OF ΔH FOR REACTIONS (1) AND (2)[a]

Glycine		Glycyl-glycine	
Temperature (°C)	$-\Delta H$ (kcal mole^{-1})	Temperature (°C)	$-\Delta H$ (kcal mole^{-1})
10	13.1	10	13.8
18	14.1	19	14.85
25	15.4	25	15.7
30	16.2	30	16.35
35	17.1		
40	17.9		

[a] Data from Chipperfield (1966).

TABLE II

VALUES OF VELOCITY CONSTANT $k(M^{-1}\ s^{-1})$ AND ΔH FOR REACTION (1) + (2) FOR VARIOUS AMINO ACID ANIONS AT 28°C[a]

Amino acid	pK of $-NH_2$ group	$(M^{-1} s^{-1} \times 10^{-3})$	$-\Delta H$ (kcal mole^{-1})
Glycine	9.78	9.05	15.4
Glycyl-glycine	8.25	2.03	15.7
Glycine ethyl ester	7.75	2.00	15.2
α-Amino phenyl acetic acid	8.84	2.90	15.1
β-Phenyl alanine	9.16	2.98	14.8
Histidine	9.17	4.50	13.9
Valine	9.62	4.90	15.3
α-Alanine	9.87	4.30	14.4
β-Alanine	10.23	7.30	15.9
δ-Amino caproic acid	10.80	12.0	18.6

[a] Data from Chipperfield (1966).

The major disadvantage of this apparatus is the large volume of solution needed and the rather long time required to obtain a single reaction trace. In order to overcome these difficulties a stopped flow microcalorimeter was developed (Berger and Stoddart, 1965). While it is not operationally as reliable as the earlier machine, the design is one of promise. Since a considerable amount of data has been obtained using combined optical and thermocouple measurements a description of it together with some recent developments seem in order. The first model of this apparatus (Berger, 1963) was a direct descendent

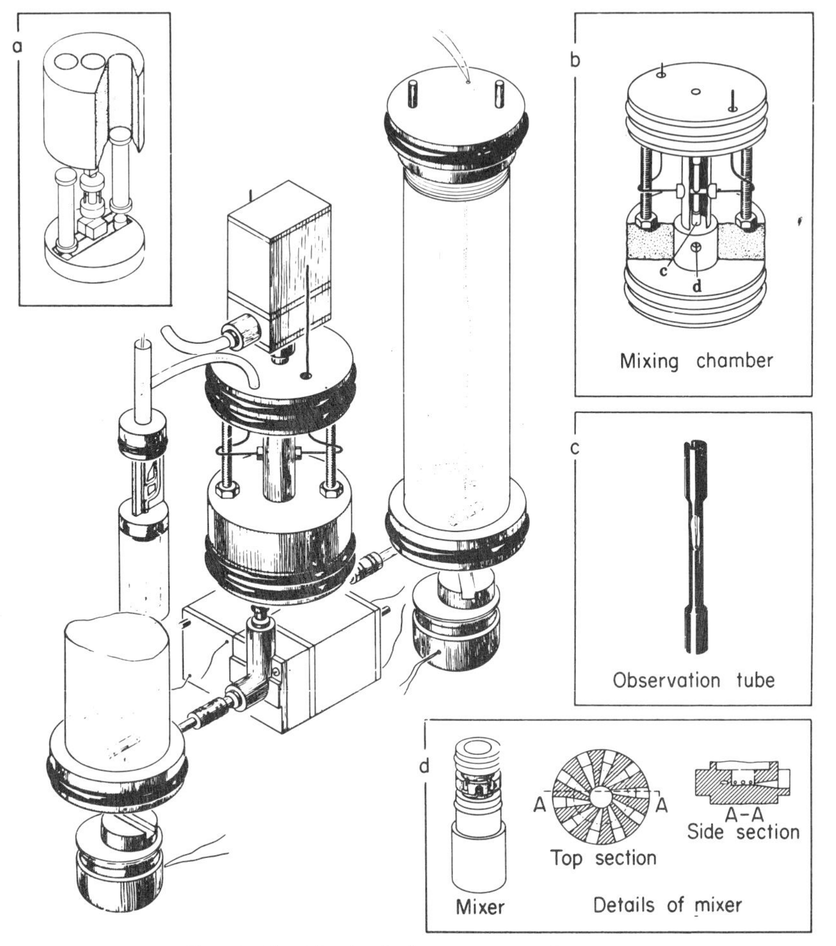

FIG. 6. Combined calorimeter-spectrophotometer.

of the Roughton-Chipperfield apparatus described above. When the speed of response of the system was increased, however, many problems not troublesome at slow detection speeds arose. In order to solve these, the apparatus shown in Fig. 6 was constructed. By immersing the system in the large aluminum block the outside bath temperature need only be kept at 0.01 °C so that ordinary commercial thermostats may be used. Though the bottles cannot be easily removed they can be cleaned by running solution through them *in situ*. The solenoid stop valve (Valcor Corp.) was found to be quite adequate for the speed of response of other parts of the system. No difference was found using a stopping syringe. Most important has been the speed of response of the thermocouple. Recent work has permitted a complete theoretical and experimental analysis and the development of a thermocouple with a 1 msecond response time when coated with Paralyene C (Balko and Berger, 1968). They must be coated since even the smallest break produces enormous streaming potentials.

Berger and Stoddart (1965) and Berger (1963), have described use of the instrument with a Hill galvanometer optical lever system. Basically it consists

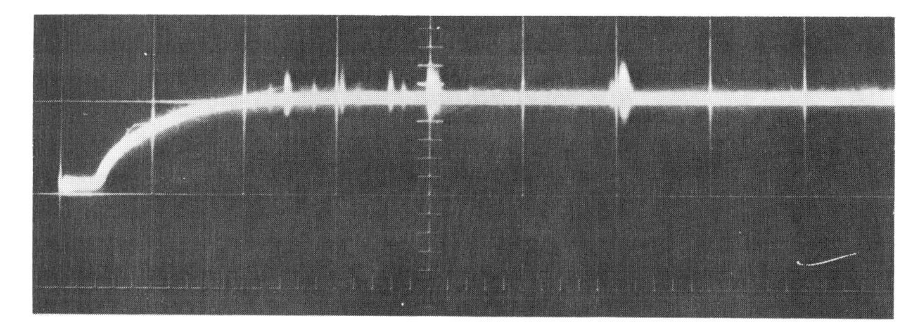

FIG. 7. Response of the stopped-flow instrument to a current pulse.

of a fast galvanometer (0.01 second response time) which has a light shining on its mirror and reflected into a differential photo-amplifier. A feedback circuit improves the response time somewhat. Figure 7 shows the response of the system to a current pulse. While this is an elaborate system to use, since great care must be taken to isolate the galvanometer from vibration, it is useable. Details of circuits, mountings, etc. are given in Berger and Stoddart (1965). An experiment is carried out by first calibrating with half-neutralized Tris (Chapter XII) in excess mixed with HCl to give the appropriate heat. At the usual concentrations, 11.36 kcal $mole^{-1}$ is produced. Corrections do need to be made for ionic strength but not temperature. Similarly, known indicator optical densities are inserted and the optical system calibrated. Either the $CoSO_4$ system recommended by the National Bureau of Standards (Mellon, 1950) may be used or, in the case of pH indicators this is mandatory, known pH solutions made up with the indicator and run into the observation tube. The solutions are then loaded and allowed to stand with occasional stirring until the bottle thermocouples indicate a temperature difference of about 0.001°C. This generally takes 1 to 2 hours. With both channels of the recorder connected (an Offner Type RS) to a sensor, calibration pulses are inserted and the pulsing sequence started. When the pulser is started, a signal is applied to all three solenoid valves for a set number of milliseconds. This time is just long enough to come up to flow velocity and then close all three valves, thus stopping the flow. This occurs in about 4 mseconds or less. About 2 ml of each reagent are required per curve. The reaction then proceeds as shown in Fig. 8. This shows the results of the reaction of sodium bicarbonate and hydrogen chloride according to the scheme:

$$H^+ + HCO_3^- \longrightarrow H_2CO_3 \xrightarrow{k_2} CO_2 + H_2O$$

This apparatus is of theoretical interest but it cannot be said to be truly operational. The most important drawback is the galvanometer system. For almost all measurements, except those at 3°C, extensive corrections are

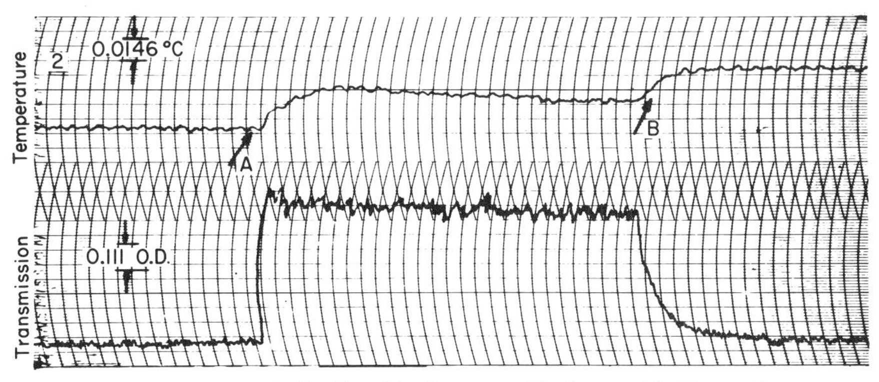

FIG. 8. Chart record of sodium bicarbonate and hydrogen chloride reaction.

required for the system response errors. A detailed discussion is given by Stoddart and Berger (1965). In order to eliminate the galvanometer we have been working on various types of sensors and amplifiers. Figure 9 shows a special transformer input amplifier employing a FET. Sensitivity is about a factor of ten less than with the galvanometer. Theoretically the peak-to-peak Johnson noise given as

$$E_{pp} > 6.45 \times 10^{-10}(RF)^{\frac{1}{2}} (300\ °K)$$

where E is in volts, R the source resistance in ohms and F the frequency bandpass is 3.2×10^{-9} or 0.0032 μV for a 4 ohm thermocouple and a 0.01 Hz to 600 Hz amplifier. Unfortunately, the peak-to-peak noise of the amplifier as measured is 0.04 μV. Thus we are an order of magnitude away from that which we should expect in theory. If we limit the bandpass to that used with the galvanometer we obtain 0.015 μV peak-to-peak which is six times that of the galvanometer. Some improvements can be made, but since in practice one often wants to measure only 2 or 3 millidegrees total temperature change, and wishes to know this to say 5%, at least 0.0001°C sensitivity is needed. At 40 μV per degree, this requires a voltage sensitivity of 0.004 μV. The rms value of the

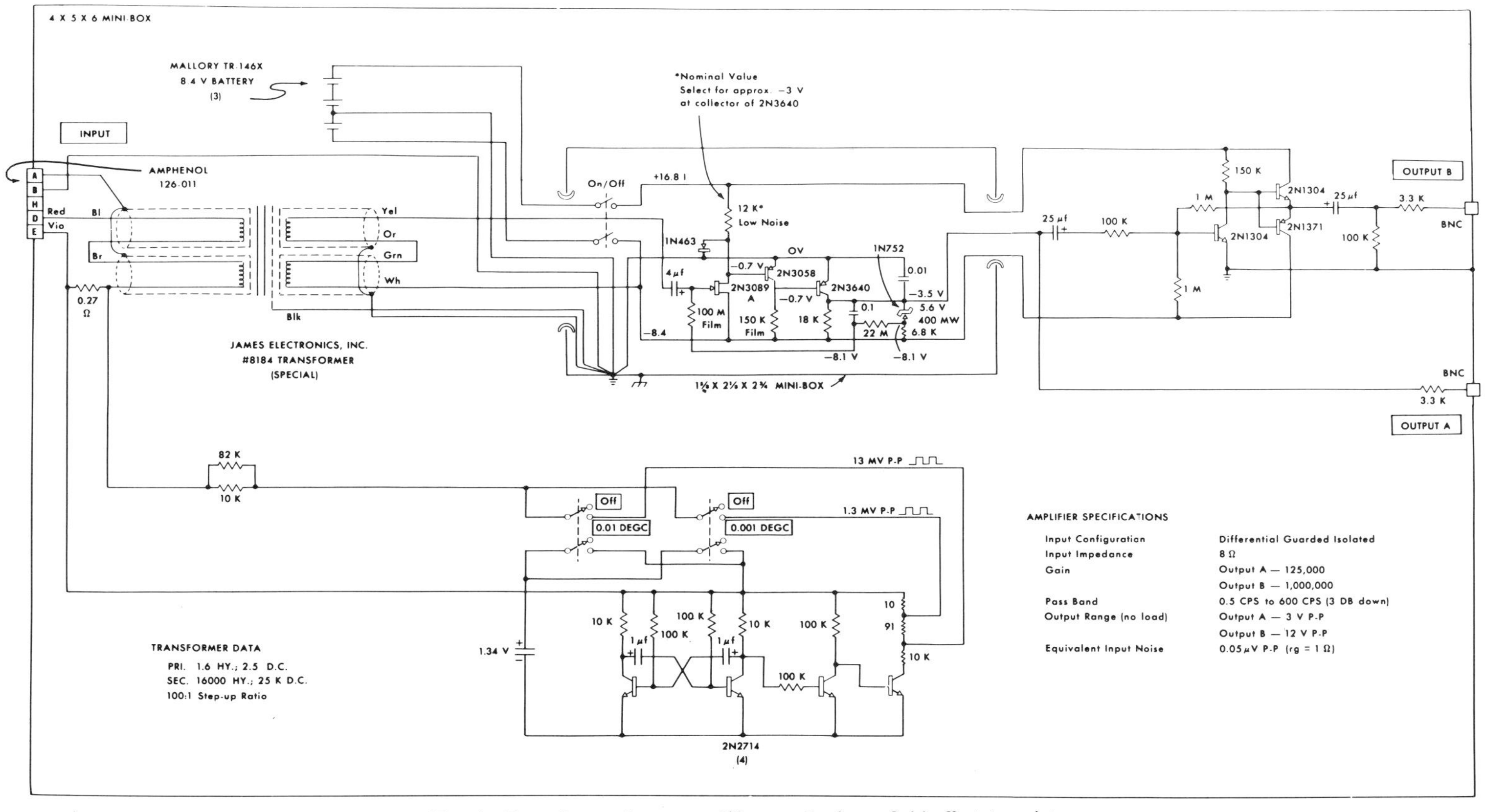

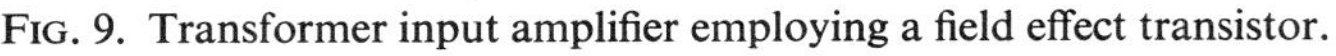
FIG. 9. Transformer input amplifier employing a field effect transistor.

voltage is about 1/5 the peak-to-peak, thus at 0.015 μV, 0.003 μV sensitivity is achieved at a bandpass of 100 Hz. Increasing the number of junctions increases the output by the square root of this number. Only physical space limitations limit the number of junctions. In other words, the galvanometer can be replaced by the amplifier provided we are willing to use ac coupling and correct the output accordingly. A correct half-life and rate constant can be obtained, but not the ΔH of the reaction. Of course, this can be done using a nanovoltmeter at the same time so that the starting and finishing temperatures are obtained.

In summary, we may say that stopped flow combined optical thermal work can be done in the temperature range of 0 to 40°C with a time resolution of about 5 mseconds and a sensitivity of 0.0001°C using 2 ml of each reagent per reaction curve. However, a number of technical problems still need to be overcome before a really reliable instrument can become available commercially. A good working instrument is available for continuous flow work using about 125 ml of each reagent per curve; 0.003 second time resolution and 0.00002°C sensitivity can be achieved.

REFERENCES

Balko, B., and Berger, R. L. (1968). *Rev. Sci. Instr.* **39**, 498.

Benson, S. W. (1960a). "The Foundation of Chemical Kinetics." McGraw-Hill, New York.

Benson, S. W. (1960b). "The Foundation of Chemical Kinetics," Pt. II, Chapt. XVIII. McGraw-Hill, New York.

Benson, S. W. (1960c). "The Foundation of Chemical Kinetics," Chapts. 4 & 5. McGraw-Hill, New York.

Berger, R. L. (1963). *In* "Temperature—Its Measurements and Control in Science Industry" (J. D. Hardy, ed.), Vol. III, Pt. 3. Reinhold, New York.

Berger, R. L., and Stoddart, L. C. (1965). *Rev. Sci. Instr.* **36**, 78.

Berger, R. L., Borchardt, W., Balko, B., and Friaut, W. (1968). *Rev. Sci. Instr.* **39**, 486.

Caldin, E. F. (1964). "Fast Reactions in Solution." Blackwell, Oxford.

Chance, B. (1940). *J. Franklin Inst.* **229**, 455, 613, 737.

Chipperfield, J. R. (1966). *Proc. Roy. Soc.* (*London*) **B164**, 401.

Czerlinski, G. (1966). "Chemical Relaxation." Dekker, New York.

Eigen, M., and DeMaeyer, L. (1963). *In* "Technique of Organic Chemistry. Vol. VII: Investigation of Rates and Mechanisms of Reactions" (S. L. Friess, E. S. Lewis, and D. Weissberger, eds.), 2nd Ed., Pt. II. Wiley (Interscience), New York.

Gutfreund, H. (1965a). "An Introduction to the Study of Enzymes." Blackwell, Oxford.

Gutfreund, H. (1965b). "An Introduction to the Study of Enzymes," Pt. I, Chapt. III, IV, V, & VI. Blackwell, Oxford.

Hammes, G. (1966). *Science* **151**, 1507.

Hartridge, H., and Roughton, F. J. W. (1923). *Proc. Roy. Soc.* (*London*) **A104**, 376.

Mellon, M. G. (1950). "Analytical Absorption Spectroscopy." Wiley, New York.

Millikan, G. D. (1936). *Proc. Roy. Soc.* (*London*) **A155**, 277.

Rossi-Bernardi, L., and Berger, R. L. (1968). *J. Biol. Chem.* **243**, 1297.

Roughton, F. J. W. (1930). *Proc. Roy. Soc. (London)* **A126**, 439.
Roughton, F. J. W., and Chance, B. (1963). *In* "Technique of Organic Chemistry. Vol. VII: Investigation of Rates and Mechanisms of Reactions" (S. L. Friess, E. S. Lewis, and D. Weissberger, eds.), 2nd Ed., Pt. II. Wiley (Interscience), New York.
Stoddart, L. C., and Berger, R. L. (1965). *Rev. Sci. Instr.* **36**, 85.

CHAPTER XVI

Multiple Calorimeters

HARRY DARROW BROWN

I. Introduction

For some biochemical applications, an important limitation of the calorimetric technique has been the relatively few experiments possible during the work day. The enzymologist in particular thinks of experimental data in terms of protocol calling for the tens or hundreds of individual experiments which are possible over few hours with conventional (sometimes automated) color chemistry. In fact, the continuous reading calorimeter obviates some need for multiple points but to this moment at least the calorimeter is not able to carry a work load, point-for-point, with the test tube or the Technicon. An additional basic limitation of the conventional single or twin calorimeter is probably of greater portent to the biochemist. This is the fact that however rapidly the equilibration of the instrument may be or how facile the set-up may be, there must be some lapse in time between the control reaction and experimental reaction. With highly labile materials, as many enzymes unfortunately are, this introduces a regrettable—or even intolerable—variable.

The object of increasing work out-put of the calorimeter and of obtaining simultaneous control and experimental reactions has been approached by the construction of multiple instruments. Calvet (1956) used 4-element instruments which were essentially double twin calorimeters. Each of the twins was completely independent electrically. Because of the great mass of the heat sink relative to the heat flow from the separate reactions, for biochemical applications, the two systems were as a practical matter separated from each other. Calvet also found it useful to oppose or in some applications to add the two sets of elements. In this way he achieved great stability in measuring phenomena producing considerable and prolonged heats. He applied this type of instrument principally to the measurement of heat of adsorption. Thus the 4-element microcalorimeter output indicated the weight of adsorbed substance. Such an application may hold value for the study of changes-of-state and of other slow physical phenomena.

II. Four-Element Instrument

Figure 1 is a photograph of the main heat sink and the scheme used in wiring the thermopiles of a 4-element calorimeter constructed in our laboratory. In this conventional use the elements are paired and one of each pair serves as a zero-blank reference, in the manner described by Evans (Chapter XIV) for the

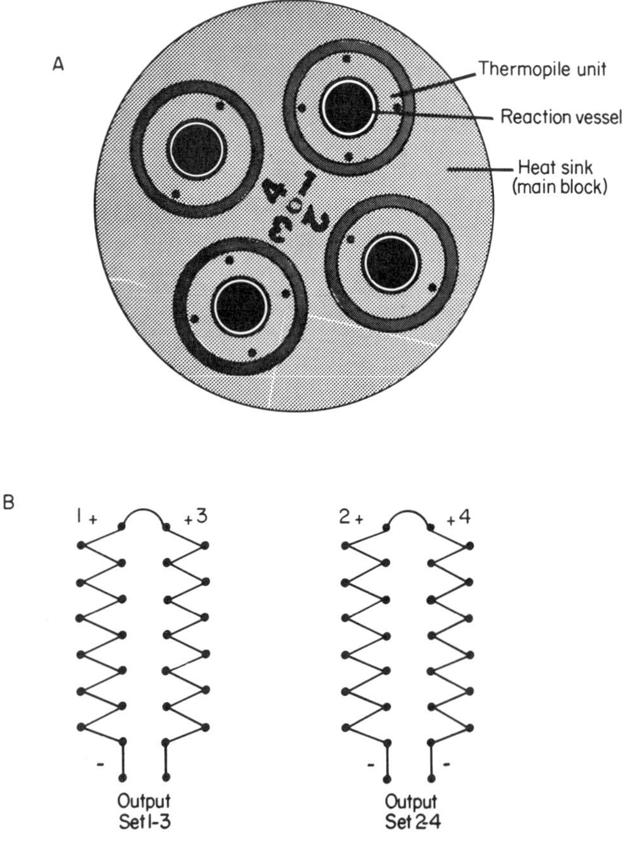

Fig. 1. A. Photograph of the block of a 4-element (classical double Tian-Calvet type) calorimeter. B. Schematic representation of series opposition arrangement of the thermopile sets.

simple twin. This entire double unit is mounted in an assembly (Fig. 2) which occupies less than $2\frac{1}{2}$ square feet of bench space and it is appropriate to many biochemical reactions. The double twin eliminates delay between the two studied reactions, but appears an uneconomic approach since two sensing elements are tied up as references and a completely double electrical system must be used.

FIG. 2. Four-element rotating calorimeter.

III. Multiplexed-Output Calorimeter

We have used a 3-element instrument and a mechanical multiplex system to approach the matter somewhat differently. In this instrument, a single element is used as reference and the emf from the two reaction sensing elements are alternately opposed to the output from the reference element. A diagram of the arrangement is given in Fig. 3 and a drawing of the instrument in Fig. 4.

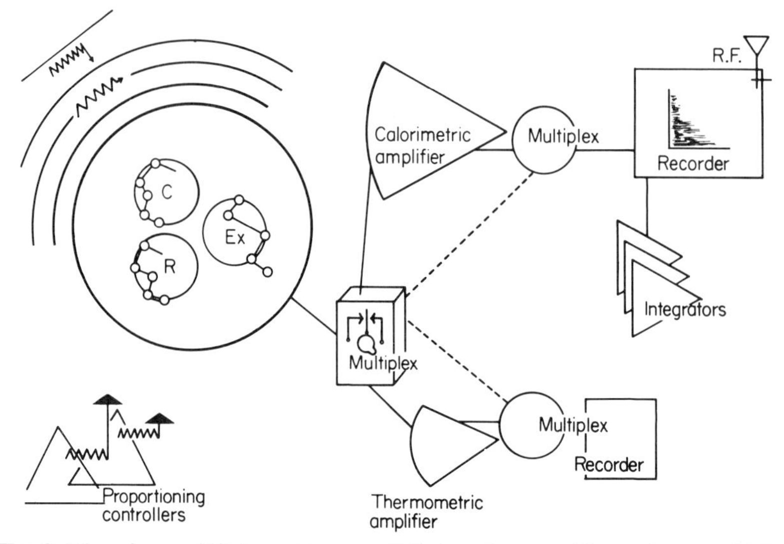

FIG. 3. The reference (R) element is sequentially in series opposition to the control (C) and experimental (Ex) thermopiles. Switches are driven by a clock-type synchronous motor and the output, through a DC chopper-stabilized amplifier, is fed to a conventional chart recorder. The individual thermopiles are also used during the cycle as thermometers.

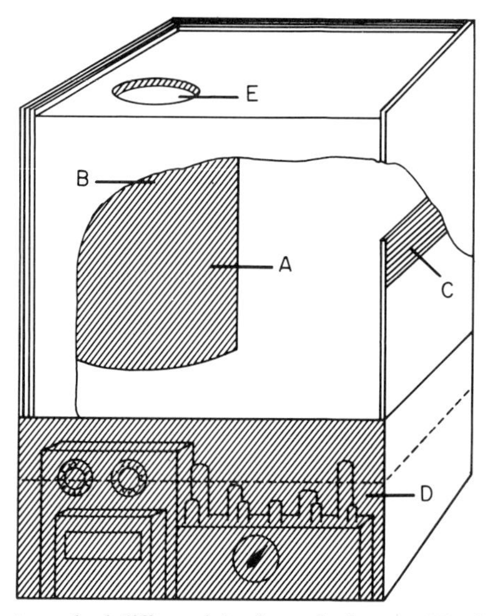

FIG. 4. Conduction-type dual-differential microcalorimeter. The instrument proper (A) has a 3-jacket thermostat conventional to Calvet calorimeters. This is turn is enclosed in an outer oven (B) isolated by copper, Mylar, steel, wood, urethane foam, and wood walls. (C) Relay switching bank (of multiplex). (D) Electronics. (E) Port for reaction cells into calorimeter.

The instruments are of the plated-coil element type (Evans and Carney, 1965) using the basic Calvet configuration. Electrically, the reference element is sequentially in series opposition to the control C and the experimental (Ex) thermopiles. The switches are driven by a clock-type synchronous motor. The output is fed through a preamplifier to a conventional chart recorder. The maximum sequence rate is determined by the full-scale response time of the recorder. We have found, however, that a 20-second cycle is sufficiently rapid. Since the two (active) elements are available to be compared with one another, this function has been included in the switch sequence. Thus the chart recording

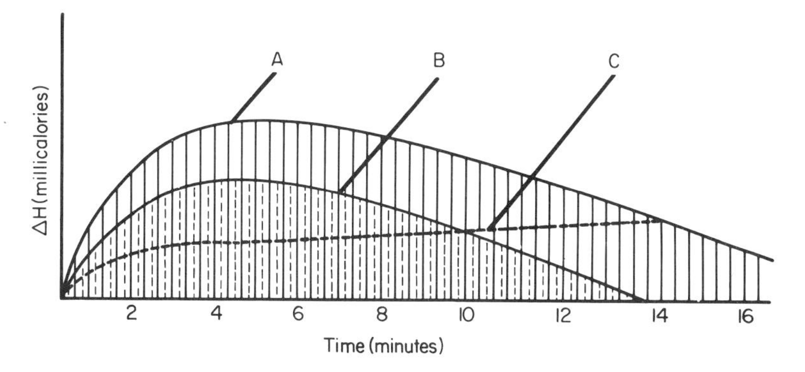

FIG. 5. Diagram of the recorder output: (A) Control reaction. (B) Experimental reaction. (C) Arithmetic difference.

contains not only the curve for the control and experimental reactions, but the curve representing the arithmetical difference between these. The appearance of the output is represented in Fig. 5. Here the solid vertical lines represent the output from the control reaction; the dashed vertical lines, the experimental reaction, and the dotted vertical lines, the electrical difference of the two active elements.

Additional advantages fall from multiplex operation. Since the switching arrangement leaves one sensing element unused in the calorimetric reading in any given moment, this thermopile is available as a thermometer. Imposed upon the use of this type of sensing element thermometrically is a very much more severe restriction of accuracy (since we must now use the block temperature as reference) than in the calorimetric reading, but the value of T obtained directly in this manner is likely to be better than an indirect approximation. The retransmitting slide wire of the chart recorder can also be used in a feed-back system to equilibrate cells rapidly using radio-frequency energy. In this use the reaction cells, into which have been built a small tuned RF circuit, are placed into the instrument at a temperature somewhat below that of the block. Thus, because the reaction cell is colder than that of the reference when the instrument is turned on, the chart recorder pen will run down scale. This movement, controlled by a limit switch, is used to initiate the transmission of an RF pulse (of maximum intensity at maximum pen travel) at the frequency of the tuned circuit in the cell being considered by the instrument at that moment. Then as the sequence switch brings the instruments to the next cell, since it too is colder than the reference, it will receive an RF pulse at its tuned frequency. As the sequence continues, the pulses will be fed as long as the recorder is driven away from the base line. When the temperature of the cell approaches that of the reference, the amount of energy radiated will be lessened by the slide wire control of the RF generator. This avoids an overshoot which would be difficult to correct. In this way the cells are rapidly brought to the block temperature.

A 4-element instrument used in multiplex makes it possible to have simultaneously information about a third reaction. Here the reaction parameters may be varied so that additional thermodynamic values can be obtained in a (single) experiment.

Perhaps larger multiple units will prove to be the type of device needed to allow us the luxury of data in the massive quantities which we, as biochemists, have come to expect.

REFERENCES

Calvet, E. (1956). *In* "Experimental Thermochemistry. Vol. I. Microcalorimetry of Slow Phenomena" (F. Rossini, ed.). Wiley (Interscience), New York.

Evans, W. J., and Carney, W. B. (1965). *Anal. Biochem.* **11**, 449.

CHAPTER XVII

Computers for Calorimetry

ROBERT L. BERGER

I. Introduction

There are two important areas of application of computer techniques to biochemical calorimetry. In the thermal detection of biochemical reactions it is necessary to correct the temperature for transient heat conduction losses in nonisothermal calorimeters. In addition, the ability to design mathematically either nonisothermal or conduction-type calorimeters is always of great importance if an optimum instrument is to be achieved. A physical simulation method will be discussed which achieves these results by the construction of a computer program based upon the direct statement of the physical laws. A second important use of the computer in biochemical calorimetry is in the correction of data for instrument response and the handling of data as it is taken. A discussion of the latter is of particular importance since the amount of data which can be obtained from a flow calorimeter is rather large. Essentially this raw data must be digested, corrected, tested, etc. by the computer before it is in a useful form. Serious questions arise as to how best to capture and store this information. Thus some discussion of hardware (i.e. black boxes) as well as of software (i.e., the actual computer FORTRAN or BASIC language programs) is necessary.

II. Application to Calorimeter Design

The calorimeter shown in Fig. 1 is a differential nonisothermal apparatus containing a ball mixer (Berger *et al.*, 1968). The analysis of this instrument has been presented elsewhere (Davids and Berger, 1968) but a brief description of the method may be useful.

The irreducible heat losses present in any calorimeter requires some correction procedure for the measured data in order to determine the total heat input to the system. This is an especially critical problem for the dynamical state, i.e., when the quantities vary with time. Both the tedious nature of hand

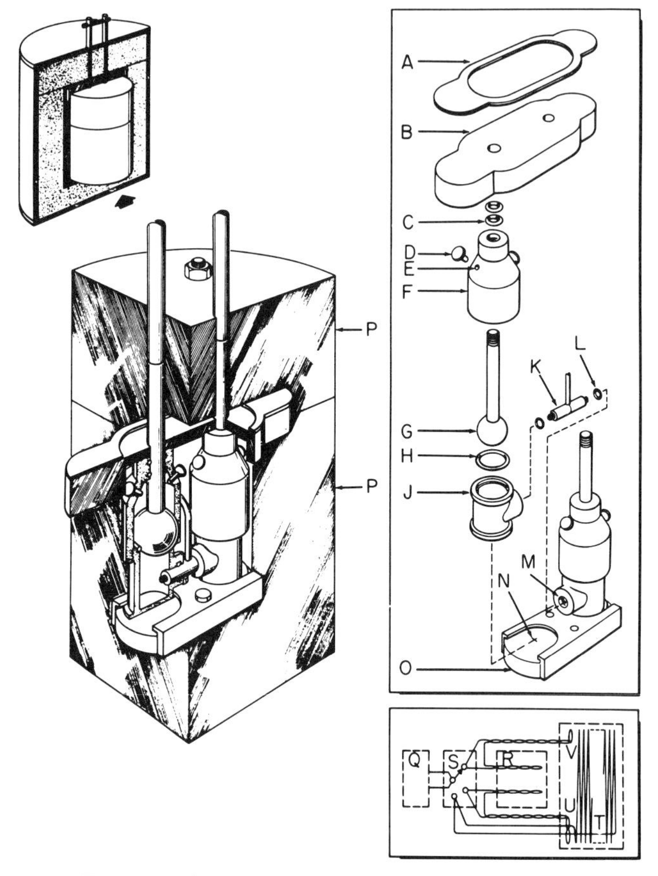

FIG. 1. Nonisothermal calorimeter using a ball mixer.

calculating such corrections as well as the ultimate aim of completely automating the measurement system made a computer program requisite. Conventionally, such corrections are made by using mathematical functions of a certain form, such as series of exponentials and curve-fitting to obtain the coefficients. The approach developed here is to use a computer model of the calorimeter to simulate the behavior of the system, and to reconstruct the ideal input that would have been measured in an adiabatic or lossless calorimeter. The advantages of this approach is that it is flexible, direct, and gives insight into the behavior of the system.

The process is cyclical in time, each cycle of computation being made for a time interval, dt, chosen small compared with the significant changes going on in the system (here $dt = 1$ second). At the beginning of the cycle, the temperature distribution and that at the probe (Upr) are known and stored in the computer. After dt has elapsed, the temperature distribution has changed due

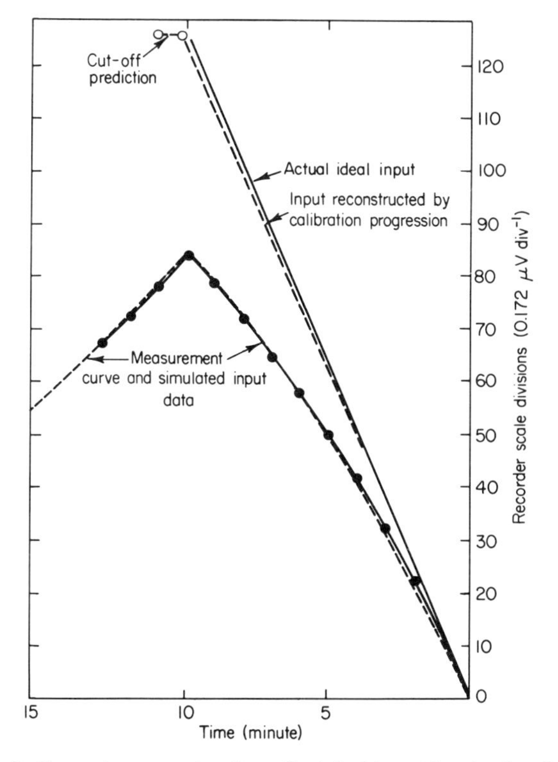

FIG. 2. Computer reconstruction of original input for simulated data.

to heat losses, and the probe temperature has changed as well. The total or actual probe temperature change du_{pr}, is thus made up of two parts: (1) that due to heat input and (2) conduction loss:

$$du_{pr_1} = du_{pr} + du_{pr_2} \tag{1}$$

Since we know the total increment from the given observed data points, and the conduction loss contribution has just been calculated, we can solve for the unknown heat input term, du_{pr}.

$$du_{pr} = du_{pr_1} - du_{pr_2} \tag{2}$$

We may also call this the ideal change which would have taken place in a lossless system. By multiplying by the known heat capacity of the fluid, we convert to heat units, thus:

$$dq = du_{pr} \times \text{cap}_{\text{fl}} \tag{3}$$

giving the heat in calories needed to take care both of the input and that lost. These heat quanta are successively cumulated and an "ideal" recorder curve is reconstructed step by step.

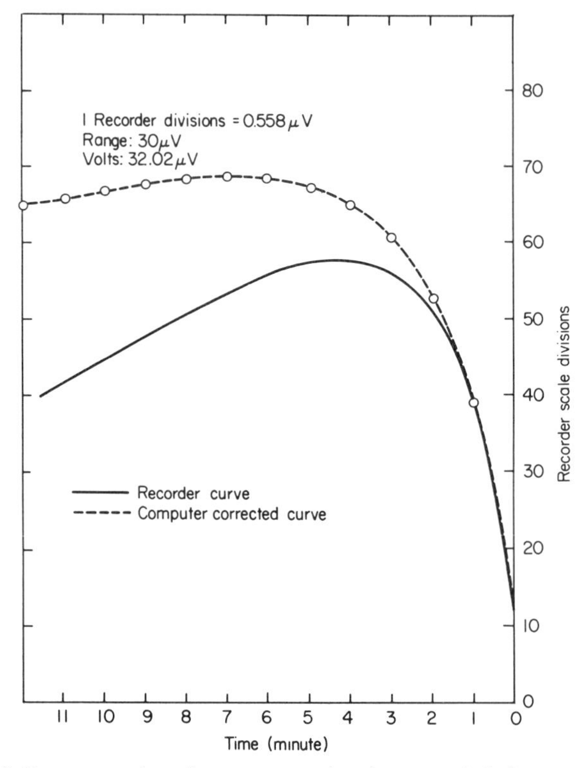

FIG. 3. Data correction of a enzyme-catalyzed transmethylation reaction.

Figure 2 shows an example of how the data is corrected for an electrical heater experiment. Figure 3 shows data correction of a transmethylation enzyme experiment.

In a similar manner a program can be worked out for the heat conduction type of calorimeter. Present methods (Calvet, 1962; Kitzinger and Benzinger, 1961) of analysis to obtain kinetics from these instruments often require considerable geometrical work. By using the methods described above a very straight-forward program has been developed for reproducing any input function (Davids and Berger, 1969).

III. Application to Data Correction and Analysis

The handling of data by the computer is rapidly gaining in popularity. This is particularly true where large amounts of data are taken or where computer calculations are to be done on the data. In most calorimetry one or both of the above is true. In addition, data often will be used in a data retrieval program.

There are essentially two ways of recording data: as an analog (recorders, magnetic tape, etc.) or as a digital signal. Most of us have grown up with recording data in an analog form and the concept of digital data requires some rethinking.

An analog signal is one which varies with time in a manner similar to the physical phenomena. Analog devices are sensors such as thermopiles. A digital signal is one which comes in discrete steps. Digital devices are sensors such as photomultipliers, that is the photocathode is digital in its output. Because of reading ease greater accuracy is possible in digital recording than in analog recording. In general any analog signal can be converted to a frequency, or pulse and counted to one part in 10^6 by conventional equipment. Other digitizing methods such as a mechanical or solid-state make-break switches, or referencing devices in which the incoming signal is compared to a fixed or stair-step reference have been developed. Each of these methods has its advantages and disadvantages, but all of them share the common virtue of being inherently more accurate than analog methods. Clearly this conversion should be made at the earliest possible point after the sensor. Unfortunately, in calorimetry an amplifier is necessary before digitization can be accomplished due to the very low voltages generated. A new amplifier has just been marketed by the Keithley Co. (Model 140), which goes a long way in helping to solve this problem. It has a dynamic range of five decades so that by using an auto-ranging digital voltmeter following it there is no danger of losing ones signal. Auto-ranging simply means that the voltmeter automatically senses that the signal will go off scale and switches to the next scale.

The present method of data handling in calorimetry involves the use of ball and disk integrators or special purpose integrators (such as that manufactured by the Infotronics Corp.), for heat conduction instruments. This gives total heat immediately but considerable work is required to obtain kinetics (Calvet, 1962; Kitzinger and Benzinger, 1961). A number of methods have been used by Sturtevant (1959) which involved measuring the error signal developed by the detectors and integrating the power supplied to keep the error signal zero. In our instrument we follow the millimicrovoltmeter with a digital voltmeter which stores its reading on punched paper tape. Data correction is done for instrument and sensor response times (Stoddart and Berger, 1965; Balko and Berger, 1968) and heat conduction losses (Berger and Davids, 1965) by processing the tape on a time-shared computer service provided by CEIR, Inc.* using

* CEIR, Inc., Bethesda, Maryland, and other cities and General Electric Corp., Bethesda, Maryland.

the computer Language BASIC† developed at Dartmouth. The service costs between $5 and $10 per hour of console time. The only drawback is the length of time it takes to submit data by way of a teletype. It takes about 1 to 1.5 seconds per data point. We have 400 data points for a 20-minute experiment, so

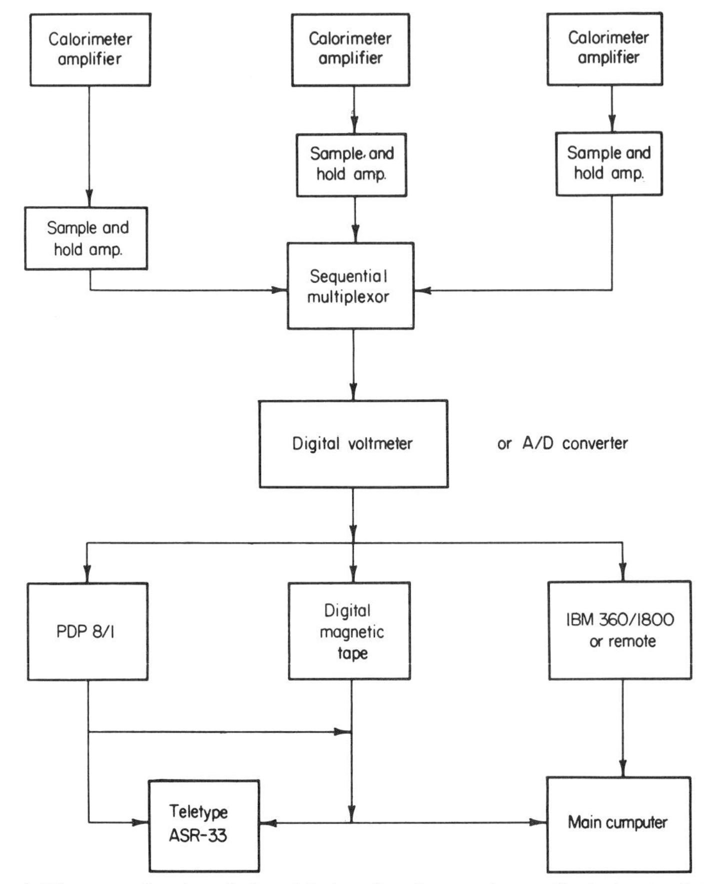

FIG. 4. Diagram of system designed to handle a large volume of calorimeter data.

7 to 8 minutes are required to submit the data. If a number of instruments were being used at once, this amount of time could become excessive. Several options are soon to be readily available at many computer centers. The most straightforward method of handling large amounts of data is to use a system such as that shown in Fig. 4. The cost of such a system is about $15,000 if the PDP8/I (Digital Equipment Co., Maynard, Mass.) or similar small computer were used. The great advantage is that the data can be processed immediately.

† Manuals are available from CEIR or General Electric.

Furthermore, many laboratory instruments can be used on such a system where a sample once a second is adequate. At this rate digital voltmeters are now available which permit 0.01 % accuracy at a 10 msecond sample rate. Input to the PDP8/I or onto tape takes less than 100 μseconds. Or, for fast calorimetry, the system can be dedicated for a few minutes and samples taken every 100 μseconds or faster. Gibson and DeSa (1968) have done this for the data handling from a stopped-flow apparatus.

REFERENCES

Balko, B., and Berger, R. L. (1968). *Rev. Sci. Instr.* **39**, 498..

Berger, R. L., and Davids, N. (1965). *Rev. Sci. Instr.* **36**, 88.

Berger, R. L., Fok Chick, Y., and Davids, N. (1968). *Rev. Sci. Instr.* **39**, 362.

Calvet, E. (1962). *In* "Experimental Thermochemistry" (C. S. Skinner, ed.), Vol. II, Chapt. 17. Wiley (Interscience), New York.

Davids, N., and Berger, R. L. (1969). Currents in Modern Biology. Special supplement on computers in biology.

Gibson, Q. H., and DeSa, J. (1968). Pittsburgh Analytical Symposium Abst. *J. anal. Chem.* **40**, 42, p. 61A.

Kitzinger, C., and Benzinger, T. H. (1961). *Methods Biochem. Analy.* **8**, 309.

Stoddart, L. C., and Berger, R. L. (1965). *Rev. Sci. Instr.* **36**, 85.

Sturtevant, J. M. (1959). *In* "Technique of Organic Chemistry" (A. Weissberger, ed.), 3rd Ed., Vol. 1, Pt. I. Wiley (Interscience), New York.

APPENDIX

Selected Values of Thermodynamic Properties*

In most cases the selections have been made from the original sources, except that the properties of most of the inorganic compounds and a few of the simple organic compounds were taken from the following compilations: Wagman *et al.* (1965), "JANAF Thermochemical Tables" (1966), "Selected Values of Properties of Chemical Compounds" (1966), Parker (1965), Harned and Owen (1958), and much additional data may be obtained from them. The following reviews were also helpful in making many of the selections: Phillips (1966), Christensen *et al.* (1967), Armstrong *et al.* (1964a, b, 1965), and Furukawa *et al.* (1966a, b, c, d, 1967). References to the published literature used in the selection are included in the tables and refer to Reference list, Chapter II.

The symbols listed in the column headed "State" have the following significance: g = ideal gas at 1 atm; l = pure liquid at 1 atm; c = pure crystalline solid at 1 atm (I, II, α, and β designate various crystalline forms); aq = hypothetical ideal solution of single species in water at unit molality; and eq, buf = An equilibrium mixture of species formed from a 1 molal solution of the species listed in a solution buffered at pH of 7. (See page 69 for calculation procedure.) The name α,β-D-glucose signifies an equilibrium mixture of α- and β-D-glucose. The names of the other sugars have a similar interpretation.

* By R. C. Wilhoit.

TABLE I

SELECTED VALUES OF THERMODYNAMIC PROPERTIES IN THE STANDARD STATES AT 25°C

Name	Formula	State	kcal mole^{-1}		cal deg^{-1} mole^{-1}		References[a]
			$\Delta Hf°$	$\Delta Gf°$	$S°$	$C_p°$	
Oxygen	O_2	g	0.0	0.0	48.996	7.016	(190, 367)
		aq	−2.89	3.9	26.5	40.	(206, 367)
Hydrogen	H_2	g	0.0	0.0	31.208	6.889	(190, 367)
		aq	−1.0	4.2	13.8		(367)
Hydrogen ion	H^+	aq	0.0	0.0	0.0	0.0	
		eq, buf	0.0	−9.55			
Hydroxyl ion	OH^-	aq	−54.975	−37.594	−2.59	−35.5	(29, 147, 152, 293, 364)
Water	H_2O	l	−68.315	−56.688	16.71	17.995	(367)
Nitrogen	N_2	g	0.0	0.0	45.77	6.961	(115, 190, 367)
		aq	−2.52	4.32	22.8	41.	(206, 367)
Ammonia	NH_3	g	−11.02	−3.96	46.03	8.52	(190, 367)
		aq	−19.19	−6.35	26.6		(367)
		eq, buf	−31.67	−9.58	−4.9		
Ammonium ion	NH_4^+	aq	−31.67	−18.97	27.1	19.1	(20, 367)
Ammonium hydroxide	NH_4OH	aq	−87.505	−63.04	43.3	−16.4	(269, 270, 292, 367)
Phosphoric acid	H_3PO_4	c	−305.7	−267.5	26.42	25.35	(96, 115, 162, 190, 367)
		aq	−307.9	−273.1	37.8		(59, 93, 95, 162, 287, 367)
		eq, buf	−309.96	−280.44	57.3		
Dihydrogen phosphate ion	$H_2PO_4^-$	aq	−310.38	−270.73	21.6		(16, 152, 236, 292, 357)
Hydrogen phosphate ion	HPO_4^{-2}	aq	−309.38	−260.91	−8.0		(18, 28, 152, 285, 367)
Phosphate ion	PO_4^{-3}	aq	−305.3	−244.02	−53.8		(60, 367)
Pyrophosphoric acid	$H_4P_2O_7$	aq	−542.2	−486.8	67.7		(367)
		eq, buf	−544.2	−502.3	113.0		
Trihydrogen pyrophosphate ion	$H_3P_2O_7^-$	aq	−544.1	−484.7	54.		(367)

Dihydrogen pyrophosphate ion	$H_2P_2O_7^{-2}$	aq	−544.6	−481.6	42.		(215, 367)
Hydrogen pyrophosphate ion	$HP_2O_7^{-3}$	aq	−544.0	−472.4	13.		(59, 215, 286, 367)
Pyrophosphate ion	$P_2O_7^{-4}$	aq	−545.	−459.6	−27.		(59, 124, 215, 367)
Magnesium ion	Mg^{+2}	aq	−110.41	−108.99	−28.2		(310)
Magnesium hydrogen phosphate	$MgHPO_4$	aq	−416.7	−373.65	14.5		(60, 62, 133, 327, 328, 353)
Carbon dioxide	CO_2	g	−94.051	−94.258	51.06	8.87	(190, 367)
		aq	−98.90	−92.26	28.1		(367)
		eq, buf	−97.49	−93.37			
Carbonic acid	H_2CO_3	aq	−167.22	−148.94	44.8		(367)
		eq, buf	−165.81	−150.06	71.3		
Bicarbonate ion	HCO_3^-	aq	−165.39	−140.26	21.8		(292, 367)
Carbonate ion	CO_3^{-2}	aq	−161.84	−126.17	−13.6		(292, 367)
Ethanol	C_2H_6O	l	−66.20	−41.63	38.49	26.76	(325)
		aq	−68.60	−43.25	35.8		(13, 29, 48, 127, 214)
2-Propanol	C_3H_8O	l	−75.97	−43.09	43.16	36.86	(300, 325)
		aq	−79.07	−44.27	36.7		(48, 87, 127, 214)
Acetaldehyde	C_2H_4O	g	−39.68	−31.78	63.15	13.4	(46, 67, 88, 216, 249, 310)
		aq	−50.75	−33.24	30.9		(30, 46, 47, 310)
Acetone	C_3H_6O	l	−58.99	−36.70	47.5	30.3	(46, 88, 201, 216, 246, 273, 274, 276, 357)
		aq	−52.99	−38.17	39.0		(24, 46, 47, 248, 381)
Glycerol	$C_3H_8O_3$	l	−160.3	−114.6	48.9	51.8	(3, 46, 122, 216, 264, 273, 276, 279, 343)
		aq	−161.7	−118.9	58.8	57.	(46, 101, 109, 203, 217, 319)
α-D-Glucose	$C_6H_{12}O_6$	c	−304.60	−217.63	50.7	52.31	(89, 177, 216, 256, 273, 275, 276)
		aq	−301.88	−218.58	63.1		(108, 157, 158, 196, 349)
β-D-Glucose	$C_6H_{12}O_6$	c	−303.07	−217.23	54.5		(177)
		aq	−302.16	−218.88	63.1		(158, 196, 349)
α,β-D-Glucose	$C_6H_{12}O_6$	aq	−302.05	−219.16	64.4	73.	(15, 32, 108, 156, 158, 349, 358)
α-D-Glucose monohydrate	$C_6H_{14}O_7$	c	−375.50	−274.75	60.3		(158, 177)

TABLE I—*continued*

Name	Formula	State	kcal mole⁻¹ $\Delta Hf°$	kcal mole⁻¹ $\Delta Gf°$	cal deg⁻¹ mole⁻¹ $S°$	cal deg⁻¹ mole⁻¹ $C_p°$	References[a]
α-Lactose	$C_{12}H_{22}O_{11}$	c	−531.0				(197, 343)
		aq	−533.55	−374.02	94.2		(39, 165, 166, 196
β-Lactose	$C_{12}H_{22}O_{11}$	c	−534.6	−374.5	92.3	98.1	(7, 63, 111)
		aq	−533.82	−374.32	94.3		(39, 166, 196)
α,β-Lactose	$C_{12}H_{22}O_{11}$	aq	−533.72	−374.60	95.5		(39, 164, 165, 166)
α-Lactose monohydrate	$C_{12}H_{24}O_{12}$	c	−606.2	−431.6	99.2	105.3	(7, 63, 164, 165, 343)
α-Maltose	$C_{12}H_{22}O_{11}$	aq	−534.96	−376.10	96.4		(39)
β-Maltose	$C_{12}H_{22}O_{11}$	aq	−534.83	−375.76	95.7		(39, 196)
α,β-Maltose	$C_{12}H_{22}O_{11}$	aq	−534.91	−376.36	97.5		(39, 86, 196, 250, 265)
β-Maltose monohydrate	$C_{12}H_{24}O_{12}$	c	−606.9	−432.5	99.8	108.5	(7, 39, 63, 216, 343)
Sucrose	$C_{12}H_{22}O_{11}$	c	−531.0	−369.1	86.1	101.7	(6, 63, 197, 216, 277, 306)
		aq	−529.60	−370.8	96.5	151.3	(15, 23, 39, 108, 137, 143, 157, 159, 307, 319, 346)
Acetic acid	$C_2H_4O_2$	l	−115.73	−93.08	38.2	29.5	(104, 131, 216, 222, 273, 276, 310)
		aq	−115.98	−96.58	49.1	37.0	(1, 50, 109, 123, 149, 219, 270, 301, 302)
		eq, buf	−116.06	−99.79	59.6		
Acetate ion	$C_2H_3O_2^-$	aq	−116.06	−90.08	27.0		(1, 46, 51, 61, 105, 151, 152, 247, 255, 295, 303, 310, 333, 366)
Butyric acid	$C_4H_8O_2$	l	−127.9	−90.6	54.1	42.9	(46, 131, 216, 222, 271, 276)
		aq	−127.3	−95.5	72.6		(46, 149)
		eq, buf	−128.0	−98.6	76.7		
Butyrate ion	$C_4H_7O_2^-$	aq	−128.0	−88.92	48.2		(46, 51, 59, 61, 70, 105, 152, 317, 370)

Myristic acid	$C_{14}H_{28}O_2$	c	−200.1				(2, 216, 352)
Palmitic acid	$C_{16}H_{32}O_2$	c, I	−212.0				(82)
		c, II	−212.9	−75.3	108.8	110.1	(2, 216, 222, 273, 276, 352, 379)
		aq		−68.8			(297)
		eq, buf		−71.5			
Palmitate ion	$C_{16}H_{31}O_2^-$	aq		−62.			
Ethyl acetate	$C_4H_8O_2$	l	−115.2	−80.7	62.8	25.2	(49, 144, 216, 277, 313)
		aq		−84.8			(53, 54, 126, 153, 191, 193, 243, 360)
Trimyristin	$C_{45}H_{86}O_6$	c, α	−503.				
		c, β	−516.	−142	298.		(58, 126, 216)
Succinic acid	$C_4H_6O_4$	c	−224.86	−178.62	42.0	36.8	(34, 167, 177, 199, 200, 272, 290, 321, 365, 377)
		aq	−218.02	−178.45	64.4	55.	(5, 34, 46, 83, 145, 242, 306)
		eq, buf	−217.19	−184.75	88.3		
Hydrogen succinate ion	$C_6H_5O_4^-$	aq	−217.23	−172.71	47.8	23.0	(12, 38, 61, 71, 194, 291, 326)
Succinate ion	$C_6H_4O_4^{-2}$	aq	−217.18	−165.02	22.2	−29.3	(12, 38, 61, 71, 90, 194, 195, 290, 326)
Fumaric acid	$C_4H_4O_4$	c	−193.75	−156.13	39.7	34.0	(34, 107, 177, 216, 233, 272, 311, 323, 377)
		aq	−185.2	−154.35	62.4		(34, 218, 306, 372, 373)
		eq, buf	−185.8	−163.57	91.3		
Hydrogen fumarate ion	$C_4H_3O_4^-$	aq	−185.1	−150.13	48.6		(12, 61, 72, 372)
Fumarate ion	$C_4H_2O_4^{-2}$	aq	−185.8	−143.85	25.2		(12, 61, 72, 372)
cis-Aconitic acid	$C_6H_6O_6$	c	−292.7				(376)
cis-Aconitate ion	$C_6H_3O_6^{-3}$	aq		−219.2			(43, 46, 207, 208)
Glycolic acid	$C_2H_4O_3$	c	−157.8				(216, 310, 337, 375)
		aq	−155.0				(84, 310)
Glycolate ion	$C_2H_3O_3^-$	aq	−154.9				(61, 152, 247, 257, 310)
L(+)-Lactic acid	$C_3H_6O_3$	c	−165.88	−125.06	34.3	30.5	(100, 175, 244, 278, 316)
		aq	−164.01	−128.77	53.0		(102, 217, 244, 316)
		eq, buf	−164.11	−133.21			

TABLE I—*continued*

Name	Formula	State	kcal mole^{-1} ΔHf°	kcal mole^{-1} ΔGf°	cal deg^{-1} mole^{-1} S°	cal deg^{-1} mole^{-1} C_p°	References[a]
L(+)-Lactate ion	$C_3H_5O_3^-$	aq	−164.11	−123.50	35.0		(14, 47, 152, 213, 237, 296, 316, 382)
DL-Lactic acid	$C_3H_6O_3$	l	−161.	−124.	45.9	50.5	(100, 278, 376)
L-Malic acid	$C_4H_6O_5$	c	−263.7				(125, 216, 377)
		aq		−213.1			(83, 218, 306)
		eq, buf		−221.1			
L-Hydrogen malate ion	$C_4H_5O_5^-$	aq		−208.4			(61)
L-Malate ion	$C_4H_4O_5^{-2}$	aq		−201.4			(34, 46, 61, 90, 207, 210, 220, 324)
DL-Malic acid	$C_4H_6O_5$	c	−264.2				(218, 377)
Citric acid	$C_6H_8O_7$	c	−369.0				(234, 340, 377)
		aq	−364.7	−297.37	78.7	76.6	(5, 73, 117, 226, 301, 304)
		eq, buf	−362.28	−307.99	122.5		
Dihydrogen citrate ion	$C_6H_7O_7^-$	aq	−363.5	−293.10	68.4	44.9	(22, 154, 232, 304, 326, 333)
Hydrogen citrate ion	$C_6H_6O_7^{-2}$	aq	−362.92	−286.61	48.6	0.2	(22, 154, 232, 304, 326)
Citrate ion	$C_6H_5O_7^{-3}$	aq	−362.12	−277.89	22.0	−60.9	(22, 43, 46, 103, 154, 195, 232, 304, 326)
Citric acid monohydrate	$C_6H_{10}O_8$	c	−439.40	−352.19	67.75	64.09	(57, 103, 240, 241)
D-Isocritric acid	$C_6H_8O_7$	aq		−295.94			(161)
		eq, buf		−306.39			
D-Dihydrogen isocitrate ion	$C_6H_7O_7^-$	aq		−291.46			(161)
D-Hydrogen isocritrate ion	$C_6H_6O_7^{-2}$	aq		−285.03			(161, 207)
D-Isocitrate ion	$C_6H_5O_7^{-3}$	aq		−276.30			(43, 46, 207, 209)
Pyruvic acid	$C_3H_4O_3$	l	−140.0				(30)
		aq	−145.2	−116.3	43.		(30, 221, 280)
		eq, buf	−142.5	−122.7			

Pyruvate ion	$C_3H_3O_3^-$	aq	−142.5	−112.9	41.		(30, 46, 64, 77, 78, 208)
Oxaloactic acid	$C_4H_4O_5$	c	−235.3				(375)
		aq		−199.0			(47, 280, 334)
		eq, buf		−209.3			
Hydrogen oxaloacetate ion	$C_4H_3O_5^-$	aq		−195.6			(281)
Oxaloacetate ion	$C_4H_2O_5^{-2}$	aq		−189.6			(46, 47, 280, 331)
α-Ketoglutaric acid	$C_5H_6O_5$	c	−245.3				(376)
α-Ketoglutarate ion	$C_5H_4O^{-2}$	aq		−189.63			(46, 47, 64, 77, 78, 208, 259, 263)
Oxalosuccinate ion	$C_5H_4O_5^{-2}$	aq		−272.2			(46, 47, 259)
Glycine	$C_2H_5O_2N$	c	−128.4	−90.27	24.74	23.71	(113, 179, 180, 181, 216, 277, 363)
		eq, buf	−125.0	−90.80	37.9		
Glycine ion	$C_2H_6O_2N^+$	aq	−126.0	−94.00	45.3	41.0	(61, 97, 152, 224, 268, 326, 348, 350)
Glycine dipolar ion	$C_2H_5O_2N^{+-}$	aq	−125.0	−90.80	37.9	8.8	(65, 66, 74, 99, 140, 142, 211, 212, 299, 329, 345, 347, 371, 383)
Glycinate ion	$C_2H_4O_2N^-$	aq	−114.4	−77.46	28.8	−13.1	(61, 97, 152, 224, 268, 326, 349, 350)
DL-Alanine	$C_3H_7O_2N$	c	−134.7	−88.9	31.6	29.1	(46, 174, 216)
L-Alanine	$C_3H_7O_2N$	c	−134.5	−88.48	30.88	29.22	(10, 115, 171, 176, 180, 181, 363)
		eq, buf	−132.6	−88.71	38.0		
L-Alanine ion	$C_3H_8O_2N^+$	aq	−133.35	−91.91/46.2	45.9	69.	(61, 152, 258, 332, 350)
L-Alanine dipolar ion	$C_3H_7O_2N^{+-}$	aq	−132.6	−88.71	38.0	33.7	(25, 46, 65, 74, 86, 135, 212, 383)
L-Alaninate ion	$C_3H_6O_2N^-$	aq	−121.8	−75.25	29.1	17.1	(61, 152, 258, 332, 350)
DL-Valine	$C_5H_{11}O_2N$	c	−147.7	−86.0	43.3		(65)
L-Valine	$C_5H_{11}O_2N$	c	−147.7	−85.8	42.72	40.35	(115, 181, 262)
		eq, buf	−147.7	−85.72	42.3		
L-Valine ion	$C_5H_{12}O_2N^+$	aq	−146.33	−88.84	57.4	−130.8	(152, 332)

TABLE I—*continued*

Name	Formula	State	kcal mole^{-1}		cal deg^{-1} mole^{-1}		References[a]
			$\Delta Hf°$	$\Delta Gf°$	$S°$	$C_p°$	
L-Valine dipolar ion	$C_5H_{11}O_2N^{+-}$	aq	−146.27	−85.72	42.3	93.	(65, 99, 238, 383)
L-Valinate ion	$C_5H_{10}O_2N^{-}$	aq	−135.62	−73.47	41.8	79.8	(152, 332)
DL-Leucine	$C_6H_{13}O_2N$	c	−155.3	−85.7	49.5	46.7	(174, 179)
L-Leucine	$C_6H_{13}O_2N$	c	−154.6	−85.2	50.10	49.8	(179, 181, 362)
		eq, buf	−153.77	−84.19	49.4		
L-Leucine ion	$C_6H_{14}O_2N^{+}$	aq	−154.16	−87.37	58.9	−157.8	(97, 152, 332)
L-Leucine dipolar ion	$C_6H_{13}O_2N^{+-}$	aq	−153.77	−84.19	49.6	121.	(65, 66, 74, 75, 211, 383)
L-Leucinate ion	$C_6H_{12}O_2N^{-}$	aq	−143.55	−70.89	39.3	107.0	(152, 332)
L-Cysteine	$C_3H_7O_2NS$	c	−127.3	−81.9	40.6	41.4	(33, 172, 173)
		eq, buf		−81.01			
L-Cysteine ion	$C_3H_8O_2NS^{+}$	aq	−83.51				(33, 128, 130)
L-Cysteine dipolar ion	$C_3H_7O_2NS^{+-}$	aq		−80.98			(33, 129, 289)
L-Hydrogen cysteinate ion	$C_3H_6O_2NS^{-}$	aq		−69.55			(33, 128, 129, 130, 380)
L-Cysteinate ion	$C_3H_5O_2NS^{-2}$	aq		−54.87			(33, 128, 380)
L-Cystine	$C_6H_{12}O_4N_2S_2$	c	−249.6	−163.9	67.06	62.60	(115, 172, 173, 182)
		eq, buf		−159.7			
L-Cystine ion	$C_6H_{14}O_4N_2S^{+2}$	aq		−163.6			(33)
L-Cystine dipolar ion	$C_6H_{13}O_4N_2S^{+2-}$	aq		−162.1			(33)
L-Cystine dipolar ion	$C_6H_{12}O_4N_2S^{+2-2}$	aq		−159.3			(33, 66, 75)
L-Cystine dipolar ion	$C_6H_{11}O_4N_2S^{+-2}$	aq					(33)
L-Cystinate ion	$C_6H_{10}O_4N_2S^{-2}$	aq		−134.4			(33)
L-Methionine	$C_5H_{11}O_2NS$	c	−181.9	−121.5	55.32	69.36	(115, 182, 363)
L-Methionine ion	$C_5H_{12}O_2NS^{+}$	aq	−178.0				(61)
L-Methionine dipolar ion	$C_5H_{11}O_2NS^{+-}$	aq	−177.9				(383)
DL-Aspartic acid	$C_4H_7O_4N$	c	−233.5	−174.3	36.9		(74, 163)
L-Aspartic acid	$C_4H_7O_4N$	c	−232.44	−174.32	40.66	37.11	(46, 115, 171, 176, 184, 262)
		eq, buf	−225.48	−176.03	69.7		

L-Aspartic acid ion	$C_4H_8O_4N^+$	aq	−228.29	−175.40	54.8		(61, 132, 322, 326, 330)
L-Aspartic acid dipolar ion	$C_4H_7O_4N^{+-}$	aq	−226.44	−171.62	51.7		(74, 163, 284, 383)
L-Aspartic acid dipolar ion	$C_4H_6O_4N^{+-2}$	aq	−225.48	−166.32	37.2		(61, 132, 205, 322, 326, 330)
L-Aspartate ion	$C_4H_5O_4N^{-2}$	aq	−216.5	−152.65	21.5		(132, 322, 326)
L-Asparagine	$C_4H_8O_3N_2$	c	−188.9	−126.9	41.7	38.4	(171, 176)
L-Asparagine dipolar ion	$C_4H_8O_3N_2^{+-}$	aq	−183.1	−125.7	57.1		(37, 66, 75, 205, 383)
L-Asparagine monohydrate	$C_4H_{10}O_4N_2$	c	−259.5	−183.4	50.10	49.7	(37, 115, 171, 176, 184, 383)
L-Glutamic acid	$C_5H_9O_4N$	c	−241.2	−174.7	44.98	41.88	(115, 171, 176, 184, 261, 262, 362)
		eq, buf	−234.7	−172.53	59.5		
L-Glutamic acid ion	$C_5H_{10}O_4N^+$	aq	−234.6	−175.58	70.2		(160, 321, 322)
L-Glutamic acid dipolar ion	$C_5H_9O_4N^{+-}$	aq	−234.7	−172.53	59.5		(46, 74, 163, 383)
L-Glutamic acid dipolar ion	$C_5H_8O_4N^{+-2}$	aq	−234.2	−166.7	41.6		(46, 160, 205, 230, 321, 322)
L-Glutamate ion	$C_5H_7O_4N^{-2}$	aq	−224.6	−153.8	30.5		(160, 230, 321, 322)
L-Glutamine	$C_5H_{10}O_3N_2$	c	−197.4	−127.2	46.63	43.93	(115, 184, 362)
		eq, buf	−192.4	−126.2	60.		
L-Glutamine dipolar ion	$C_5H_{10}O_3N_2^{+-}$	aq	−192.4	−126.2	60.		(27, 205, 227)
L-Arginine	$C_6H_{14}O_2N_4$	c	−148.6	−157.0	59.9	55.8	(174, 179)
L-Arginine dipolar ion	$C_6H_{14}O_2N_4^{+-}$	aq	−147.1				(185, 383)
Glycylglycine	$C_4H_8O_3N$	c	−178.3	−124.0	45.4	39.1	(169, 170)
		eq, buf	−175.49	−124.38		55.5	
Glycylglycine ion	$C_4H_9O_3N^+$	aq	−175.84	−128.62	68.5	69.	(202, 224, 294, 330)
Glycylglycine dipolar ion	$C_4H_8O_3N^{+-}$	aq	−175.49	−124.33	55.3	38.	(99, 282, 327)
Glycylglycinate ion	$C_4H_7O_3N^-$	aq	−164.89	−113.07	53.1		(224, 330)
Urea	CH_4ON_2	c	−79.58	−47.07	25.00	22.26	(100,122, 168, 216, 277, 314, 367)
		aq	−75.92	−48.47	42.0	19.7	(23, 36, 86, 94, 98, 109, 136, 141, 229, 319, 345, 367, 374, 378)
Creatinine	$C_4H_7ON_3$	c	−56.8	−6.8	40.0	33.2	(171, 176)
		aq		−5.53			(92)
		eq, buf		−5.53			
Creatinine ion	$C_4H_8ON_3^+$	aq		0.99			(52, 92, 134)

TABLE I—*continued*

Name	Formula	State	kcal mole^{-1}		cal deg^{-1} mole^{-1}		References[a]
			$\Delta Hf°$	$\Delta Gf°$	$S°$	$C_p°$	
Creatine	$C_4H_9O_2N_3$	c	−128.2	−63.1	45.3	41.1	(100, 171, 176, 342)
		aq		−61.95			(92, 146)
		eq, buf		−61.95			
Creatine ion	$C_4H_{10}O_2N_3^+$	aq		−65.58			(52, 80, 92)
Creatine ion	$C_4H_8O_2N_3^-$	aq		−42.5			(146)
Creatine monohydrate	$C_4H_{11}O_3N_3$	c	−199.1	−120.6	56.0	51.0	(178, 231, 342
Adenine	$C_5H_5N_5$	c	23.2	−71.8	36.1	34.2	(335, 336)
		aq		74.64			(298, 361)
		eq, buf		74.64			
Adenine ion	$C_5H_6N_5^+$	aq		68.91			(4, 59, 187, 298, 356)
Adenine ion	$C_5H_4N_5$	aq		88.11			(4, 59, 187, 228, 260, 356)
Glycerol-1-phosphoric acid	$C_3H_9O_6P$	aq		−335.7			(79, 245)
		eq, buf		−344.6			
Glycerol-1-hydrogenphosphate ion	$C_3H_8O_6P^-$	aq		−333.9			(79)
Glycerol-1-phosphate ion	$C_3H_7O_6P^{-2}$	aq		−324.8			(47, 245)
Glucose-1-phosphoric acid	$C_6H_{13}O_9P$	aq		−427.7			(69, 148, 245, 359
		eq, buf		−437.1			
Glucose-1-hydrogenphosphate ion	$C_3H_{12}O_9P^-$	aq		−426.2			(11, 62, 68, 359)
Glucose-1-phosphate ion	$C_3H_{12}O_9P^{-2}$	aq		−417.33			(68, 69, 148, 359)
Gluose-6-phosphoric acid	$C_6H_{13}O_9P$	aq		−429.6			(245, 305)
		eq, buf		−438.9			
Glucose-6-hydrogenphosphate ion	$C_6H_{12}O_9P^-$	aq		−427.6			(40, 85)
Glucose-6-phosphate ion	$C_6H_{11}O_9P^{-2}$	aq		−419.1			(40, 85, 189, 245, 305)

[a] See page 73 for Reference List to which numbers refer.

TABLE II

ENTHALPIES AND GIBBS ENERGIES OF FORMATION OF ADENOSINE PHOSPHORIC ACID SPECIES RELATIVE TO H_2ADP^{+-2} AT 25°C[a b]

Species	State	kcal mole^{-1}		Species	State	kcal mole^{-1}	
		$\Delta Hf°$	$\Delta Gf°$			$\Delta Hf°$	$\Delta Gf°$
H_2AMP	eq, buf		203.99	$MgADP^-$	aq	−103.38	−99.27
H_2AMP^{+-}	aq		209.49	H_4ATP	eq, buf	−232.97	−228.95
$HAMP^-$	aq		214.73	H_4MgATP^{+2}	eq, buf	−339.85	−344.14
MgAMP	aq		117.70	H_2ATP^{+-3}	aq	−236.33	−205.78
AMP^{-2}	aq		223.83	H_2MgATP	aq		−316.54
H_3ADP	eq, buf	3.31	−13.88	$HATP^{-3}$	aq	−232.23	−199.51
H_3MgADP^{+2}	eq, buf	−103.52	−127.47	$HMgATP^{-1}$	aq	−340.14	−313.40
H_2ADP^{+-2}	aq	0.0	0.0	ATP^{-4}	aq	−233.93	−189.03
$HADP^{-2}$	aq	4.10	5.73	$NaATP^{-3}$	aq		−254.28
HMgADP	aq	−109.66	−106.60	$MgATP^{-2}$	aq	−339.84	−305.97
ADP^{-3}	aq	2.73	15.55				

[a] Abbeviations: H_2AMP = adenosine monophosphoric acid; H_3ADP = adenosine diphosphoric acid; H_4ATP = adenosine triphosphoric acid.

[b] References: (4, 26, 27, 28, 31, 44, 45, 59, 118, 189, 191, 204, 225, 236, 339, 251, 252, 253, 254, 257, 266, 267, 285, 286, 287, 294, 305, 327, 328, 351, 354, 355, 368, 369).

TABLE III

CHANGES IN ENTHALPY AND GIBBS ENERGY FOR SOME REACTION IN SOLUTION

Reaction	ΔH°	ΔG°	References
	kcal mole^{-1}		
$C_4H_{12}O_3N^+(aq) \rightarrow C_4H_{11}O_3N(aq) + H^+(aq)$ ($C_4H_{11}O_3N$ = tris(hydroxymethyl)aminomethane)	11.35	11.01	(19, 21, 28, 81, 204, 255, 294, 351)
$C_{10}H_{14}O_4N_5{}^+(aq) \rightarrow C_{10}H_{13}O_4N_5(aq) + H^+(aq)$ ($C_{10}H_{13}O_4N_5$ = adenosine)	3.7	4.80	(4, 59, 187, 225, 236, 298)
$C_{10}H_{13}O_4N_5(aq) \rightarrow C_{10}H_{12}O_4N_5{}^-(aq) + H^+(aq)$	9.7	16.8	(188, 189, 223)
$C_{21}H_{27}O_{14}N_7P_2$(eq, buf) + H_2(g) $\rightarrow$ $C_{21}H_{28}O_{14}N_7P_2{}^-$(eq, buf) + H^+(eq, buf) ($C_{21}H_{27}O_{14}N_7P_2$ = nicotinamide adenine dinucleotide)		−4.65	(42, 47, 296, 308, 309, 320)
$C_{21}H_{28}O_{17}N_7P_3$(eq, buf) + H_2(g) $\rightarrow$ $C_{21}H_{29}O_{17}N_7P_3{}^-$(eq, buf) + H^+(eq, buf) ($C_{21}H_{28}O_{17}N_7P_3$ = nicotinamide adenine dinucleotide phosphoric acid)		−4.5	(47, 150, 263)
$C_{21}H_{36}O_{16}N_7P_3S$(eq, buf) + CH_3COOH(eq, buf) $\rightarrow$ $C_{22}H_{38}O_{17}N_7P_3$(eq, buf) + H_2O(1) ($C_{21}H_{36}O_{16}N_7P_3S$ = Coenzyme A, $C_{22}H_{38}O_{17}N_7P_3S$ = acetyl-Coenzyme A)		8.5	(43, 155, 191, 192, 235, 334)

TABLE IV

PARTIAL MOLAL PROPERTIES OF REAL AQUEOUS SOLUTIONS AT 25°C

molality (m)	γ_2	$-\log a_1$	$\bar{L}_2$ (cal mole^{-1})	$\bar{L}_1$ (cal mole^{-1})	$\bar{C}_{p2}$ (cal deg^{-1} mole^{-1})	$\bar{C}_{p1}$ (cal deg^{-1} mole^{-1})
α,β-D-Glucose References: (32, 108, 358)						
0.2	1.003	0.0016				
0.5	1.009	0.0039				
1.	1.020	0.0079	170.	−1.4		17.99
2.	1.056	0.0160	310.	−5.0	77.0	17.99
4.	1.181	0.0333	510.	−16.	82.2	17.7
6.	1.306	0.0522	670.	−30.	85.6	17.5
8.	1.423	0.0728	780.	−44.	86.6	17.3
10.	1.537	0.0946	860.	−56.		
Glycerol References: (76, 91, 101, 109, 186, 203, 217, 319)						
0.2	1.006	0.0016	−24.2	0.04	56.	17.99
0.5	1.014	0.0039			54.	18.0
1.	1.030	0.0079	190.	−6.0	53.	18.2
2.	1.050	0.0160	270.	−13.	53.	18.3
4.	1.090	0.0325	430.	−14.	53.	18.4
6.	1.125	0.0494	720.	−36.	52.	18.1
8.	1.156	0.0665	1110.	−98.	52.	17.6
10.	1.180	0.0839	1240.	−120.	51.	17.8
Glycine References: (66, 99, 138, 140, 142, 299, 318, 329, 344, 347, 383)						
0.1	0.982	0.0008	−18.5	0.022	9.7	17.99
0.2	0.962	0.0016	−38.0	0.073	10.6	17.99
0.5	0.914	0.0041	−93.9	0.404	13.3	17.98
1.	0.858	0.0084	−165.	1.36	16.6	17.93
1.5	0.816	0.0127	−220.	2.57	19.4	17.87
2.0	0.786	0.0172	−262.	3.89	22.4	17.80
2.5	0.761	0.0217	−297.	5.31	24.0	17.74
3.0	0.738	0.0262	−330.	6.96	23.6	17.71
Urea References: (36, 94, 98, 108, 109, 136, 141, 281, 319, 345, 374)						
0.2	0.983	0.0016	−33.5	0.060	21.1	17.99
0.5	0.960	0.0040	−81.0	0.357	22.1	17.99
1.	0.924	0.0081	−153.0	1.325	23.1	17.97
2.	0.866	0.0167	−275.1	4.58	24.6	17.93
4.	0.787	0.0348	−458.	14.27	26.7	17.83
6.	0.720	0.0539	−587.	25.6	28.0	17.70
8.	0.671	0.0739	−679.	37.2	29.0	17.58
10.	0.634	0.0935	−754.	49.2	29.6	17.47
15.	0.574	0.1418	−889.	79.1	30.3	17.32

Author Index

Numbers in italics refer to the pages on which the complete references are listed.

C

G

H

T

Subject Index

D

Q

R